# 深度学习
## 项目开发实践
### （TensorFlow+Sklearn+PyTorch）
### （微视频版）

王振丽　刘德民◎编著

清华大学出版社
北京

## 内 容 简 介

Python 是当今使用最为广泛的开发语言之一，被认为是开发深度学习程序的最佳语言。本书通过 9 个综合实例，详细讲解了使用 Python 语言开发大型深度学习项目的过程，这些项目在现实应用中具有极强的代表性。第 1 章讲解了 AI 智能问答系统的具体实现流程；第 2 章讲解了 AI 智能推荐系统的具体实现流程；第 3 章讲解了智能 OCR 文本检测识别系统的具体实现流程；第 4 章讲解了国际足球比赛结果预测系统的具体实现流程；第 5 章讲解了智能绘图系统的具体实现流程；第 6 章讲解了利用 ChatGPT 开发微信客服机器人的具体实现流程；第 7 章讲解了移动机器人智能物体识别系统的具体实现流程；第 8 章讲解了 AI 考勤管理系统的具体实现流程；第 9 章讲解了网络舆情数据分析系统的具体实现流程。在具体讲解每个实例时，都遵循项目的进度顺序来讲解，从接到项目到具体开发，直到最后的调试和发布，内容循序渐进，并穿插讲解了这样做的原因，深入讲解了每个重点内容的具体细节，引领读者全面掌握 Python 深度学习开发技术。

本书不但适合 Python 深度学习开发的初学者学习，也适合有一定 Python 深度学习开发基础的读者学习，还可以作为有一定基础的程序员的参考书。

本书封面贴有清华大学出版社防伪标签，无标签者不得销售。
版权所有，侵权必究。举报：010-62782989，beiqinquan@tup.tsinghua.edu.cn。

图书在版编目(CIP)数据

深度学习项目开发实践：TensorFlow+Sklearn+PyTorch：微视频版 / 王振丽，刘德民编著.
北京：清华大学出版社，2024.12. -- ISBN 978-7-302-67809-0
Ⅰ. TP181
中国国家版本馆 CIP 数据核字第 20248W22T1 号

责任编辑：魏　莹
封面设计：李　坤
责任校对：李玉茹
责任印制：沈　露

出版发行：清华大学出版社
网　　址：https://www.tup.com.cn, https://www.wqxuetang.com
地　　址：北京清华大学学研大厦 A 座　　邮　编：100084
社 总 机：010-83470000　　邮　购：010-62786544
投稿与读者服务：010-62776969, c-service@tup.tsinghua.edu.cn
质量反馈：010-62772015, zhiliang@tup.tsinghua.edu.cn

印 装 者：三河市科茂嘉荣印务有限公司
经　　销：全国新华书店
开　　本：185mm×230mm　　印　张：22.75　　字　数：552 千字
版　　次：2024 年 12 月第 1 版　　印　次：2024 年 12 月第 1 次印刷
定　　价：99.00 元

产品编号：102097-01

# 前　　言

## 项目实战的重要性

在竞争日益激烈的软件开发就业市场中，拥有良好的理论知识固然重要，但实践是将理论知识转化为实际技能的关键，它不仅能够帮助我们更好地理解和记忆所学的知识，还能够培养我们解决问题和创新的能力。

在计算机科学领域，项目实战是一种将理论知识转化为实际应用能力的实践活动。虽然课堂教学和理论学习是基础，但只有通过实际项目的实践，才能真正掌握所学的知识，并将其运用到实际场景中。项目实战的重要性具体如下。

(1) 实践锻炼：通过参与项目实战，我们将面临真实的编码挑战，从中学习解决问题的能力和技巧。实践锻炼有助于个人逐渐熟悉编程语言、开发工具和常用框架，提高编码技术和编码质量。

(2) 综合能力培养：项目实战要求我们综合运用相关的知识点和技术，从需求分析、设计到实现和测试等环节，全方位地培养我们的综合能力。

(3) 团队协作经验：项目实战通常需要与团队成员合作完成，这对培养我们的团队协作和沟通能力至关重要。通过与他人合作，我们将学会协调工作，共同解决问题，并加深对团队合作的理解和体验。

(4) 独立思考能力：通过克服项目实战中遇到的困难和挫折，将培养出我们的自信和勇气，提高独立思考和解决问题的能力。

(5) 实践经验加分：在未来求职过程中，项目实战经验将成为一大亮点。用人单位更看重具有实践经验的候选人，它们更倾向于选择那些能够快速适应工作环境并提供实际解决方案的人才。

为了帮助广大读者从一名学习编程初学者快速成长为有实践经验的开发高手，我们精心编写了本书。本书将以实战项目为素材，从项目背景和规划开始讲解，一直到项目的调试运行和维护结束，完整展示大型商业项目的运作和开发流程。

## 本书特色

1) 以实践为导向

本书的核心理念是通过实际项目的完成来学习并掌握深度学习开发技术。每个项目都非常实用，涵盖了不同领域和应用场景，帮助读者将所学的知识直接应用到实际项目中。

2) 项目新颖、框架多

本书中的 9 个实战项目贴合现实主流应用领域，都是当今开发领域的热点，并且书中的项目几乎用到了所有的深度学习开发的相关框架，例如 TensorFlow.js、PyTorch、OpenCV、Scikit-image、OpenAI、ChatGPT、TensorFlow Lite、TensorFlow、Scikit-learn、Pandas、Matplotlib 等。

3) 渐进式学习

本书按照难度逐渐增加的顺序组织项目，从简单到复杂，让读者能够循序渐进地学习和提高。每个项目都有清晰的目标和步骤，引导读者逐步实现功能。

4) 综合性项目

本书包含多个综合性项目，涉及不同的编程概念和技术。通过完成这些项目，读者将能够综合运用所学的知识，养成解决问题的能力和系统设计的思维。

5) 提供解决方案和提示

每个项目都提供了详细的解决方案和提示，帮助读者理解项目的实现细节和关键技术。这些解决方案和提示旨在启发读者思考，并提供参考，但也鼓励读者根据自己的理解和创意进行探索与实现。

6) 实用的案例

本书的项目涉及多个实际应用领域，如智能问答系统、文本检测识别系统、结果预测等。这些案例不仅有助于读者理解深度学习开发的应用范围，还能够培养读者解决实际问题的能力。

7) 强调编程实践和创造力

本书鼓励读者在学习和实践过程中发挥创造力，尝试不同的方法和解决方案。通过实践和创造，读者能够深入理解编程原理，提高解决问题的能力，并养成独立开发和创新的能力。

8) 配书资源丰富

在本书附配的资源中，不仅有书中实例的源代码和 PPT 课件(读者可扫描右侧二维码获取)，还有书中案例的全程视频讲解，视频讲解读者可扫描书中二维码获取。

扫码获取源代码

扫码获取 PPT 课件

在编写本书的过程中，我们始终本着科学、严谨的态度，力求精益求精，但疏漏之处在所难免，敬请广大读者批评、指正。

最后感谢您购买本书，希望本书能成为您编程路上的领航者，祝您学习愉快！

编 者

# 目 录

## 第 1 章 AI 智能问答系统 ... 1
### 1.1 背景介绍 ... 2
#### 1.1.1 互联网的影响 ... 2
#### 1.1.2 问答系统的发展 ... 2
### 1.2 问答系统的发展趋势：AI 问答系统 ... 4
#### 1.2.1 人工智能介绍 ... 4
#### 1.2.2 机器学习 ... 5
#### 1.2.3 深度学习 ... 5
#### 1.2.4 系统介绍 ... 6
### 1.3 技术架构 ... 6
#### 1.3.1 TensorFlow.js ... 6
#### 1.3.2 SQuAD 2.0 ... 7
#### 1.3.3 BERT ... 8
#### 1.3.4 知识蒸馏 ... 8
### 1.4 具体实现 ... 9
#### 1.4.1 编写 HTML 文件 ... 9
#### 1.4.2 脚本处理 ... 11
#### 1.4.3 加载训练模型 ... 12
#### 1.4.4 查询处理 ... 12
#### 1.4.5 文章处理 ... 14
#### 1.4.6 加载处理 ... 16
#### 1.4.7 寻找答案 ... 17
#### 1.4.8 提取最佳答案 ... 18
#### 1.4.9 将答案转换回文本 ... 20
### 1.5 调试运行 ... 20

## 第 2 章 AI 智能推荐系统 ... 23
### 2.1 背景介绍 ... 24
#### 2.1.1 推荐系统能解决什么问题 ... 24
#### 2.1.2 推荐系统的应用领域 ... 25
#### 2.1.3 推荐系统和搜索引擎 ... 25
### 2.2 项目介绍 ... 26
### 2.3 数据采集和整理 ... 26
#### 2.3.1 数据整理 ... 26
#### 2.3.2 电影详情数据 ... 30
#### 2.3.3 提取电影特征 ... 37
### 2.4 情感分析和序列化操作 ... 43
### 2.5 Web 端实时推荐 ... 44
#### 2.5.1 Flask 启动页面 ... 44
#### 2.5.2 模板文件 ... 47
#### 2.5.3 后端处理 ... 49
### 2.6 调试运行 ... 56

## 第 3 章 智能 OCR 文本检测识别系统 ... 59
### 3.1 背景介绍 ... 60
### 3.2 OCR 系统简介 ... 60
#### 3.2.1 OCR 的基本原理和使用方式 ... 60
#### 3.2.2 文字识别的基本步骤 ... 61
#### 3.2.3 深度学习对 OCR 的影响 ... 63
#### 3.2.4 与 OCR 相关的深度学习技术 ... 63
### 3.3 系统介绍 ... 64
### 3.4 准备模型 ... 65
#### 3.4.1 文本检测模型 ... 65
#### 3.4.2 文本识别模型 ... 65

| 3.5 | 创建工程 | 66 |
| 3.5.1 | 工程配置 | 66 |
| 3.5.2 | 配置应用程序 | 67 |
| 3.5.3 | 导入模型 | 68 |
| 3.6 | 具体实现 | 68 |
| 3.6.1 | 页面布局 | 69 |
| 3.6.2 | 实现主 Activity | 69 |
| 3.6.3 | 图像处理操作 | 73 |
| 3.6.4 | 运行 OCR 模型 | 76 |
| 3.7 | 调试运行 | 80 |

## 第 4 章 国际足球比赛结果预测系统 81

- 4.1 欧洲足球五大联赛 ... 82
- 4.2 模块架构 ... 83
- 4.3 准备数据 ... 83
- 4.4 数据可视化分析 ... 84
  - 4.4.1 事件收集 ... 84
  - 4.4.2 射门数据可视化 ... 87
  - 4.4.3 球队和球员数据可视化 ... 89
  - 4.4.4 联赛数据可视化 ... 96
  - 4.4.5 巴塞罗那队的进球数据饼形图 ... 100
  - 4.4.6 红牌和黄牌数据可视化 ... 101
  - 4.4.7 进球数据可视化 ... 106
  - 4.4.8 梅西和 C 罗的数据可视化 ... 114
  - 4.4.9 五大联赛的球员数量可视化 ... 118
- 4.5 比赛预测 ... 119
  - 4.5.1 读取数据 ... 119
  - 4.5.2 清洗数据 ... 121
  - 4.5.3 逻辑回归算法 ... 122
  - 4.5.4 创建梯度提升模型 ... 123
  - 4.5.5 创建随机森林分类器模型 ... 124
  - 4.5.6 不平衡处理 ... 126
- 4.6 进球预测 ... 128
  - 4.6.1 预处理 ... 128
  - 4.6.2 创建循环神经网络 ... 132

## 第 5 章 智能素描绘图系统 137

- 5.1 项目介绍 ... 138
- 5.2 需求分析 ... 138
- 5.3 功能模块 ... 139
- 5.4 预处理 ... 140
  - 5.4.1 低动态范围配置 ... 140
  - 5.4.2 图像处理和调整 ... 141
  - 5.4.3 获取原始图像的笔画 ... 143
  - 5.4.4 方向检测 ... 147
  - 5.4.5 去蓝处理 ... 150
  - 5.4.6 图像合成 ... 152
  - 5.4.7 快速排序 ... 156
  - 5.4.8 侧窗滤波 ... 157
- 5.5 开始绘图 ... 160
  - 5.5.1 基于边缘绘画的绘图程序 ... 160
  - 5.5.2 绘制铅笔画 ... 170

## 第 6 章 ChatGPT 微信客服机器人 173

- 6.1 ChatGPT 概述 ... 174
  - 6.1.1 ChatGPT 的发展历程 ... 174
  - 6.1.2 GPT 系列的演变 ... 175
  - 6.1.3 ChatGPT 的主要特点 ... 175
- 6.2 系统介绍 ... 176
- 6.3 项目结构 ... 176
- 6.4 准备工作 ... 177
  - 6.4.1 注册成为 OpenAI 会员 ... 177
  - 6.4.2 获取 API key ... 178
- 6.5 系统配置 ... 181
  - 6.5.1 基本配置 ... 181
  - 6.5.2 其他配置 ... 182

6.6 通道处理 ........................................ 188
    6.6.1 通用处理逻辑 ........................ 188
    6.6.2 微信聊天通道 ........................ 194
    6.6.3 微信公众号通道 .................... 199
6.7 对话处理 ........................................ 204
    6.7.1 OpenAI 对话 .......................... 204
    6.7.2 ChatGPT 对话 ......................... 207
    6.7.3 Baidu Unit 对话 ..................... 212
6.8 语音识别 ........................................ 213
    6.8.1 OpenAI 语音识别 .................... 213
    6.8.2 谷歌语音识别 ........................ 214
    6.8.3 百度语音识别 ........................ 215
    6.8.4 Microsoft Azure 语音识别 ...... 218
6.9 调试运行 ........................................ 219

## 第 7 章 移动机器人智能物体识别系统 .......... 221

7.1 背景介绍 ........................................ 222
7.2 物体识别 ........................................ 222
    7.2.1 物体识别介绍 ........................ 223
    7.2.2 图像特征的提取方法 ............. 223
7.3 系统介绍 ........................................ 225
7.4 准备模型 ........................................ 226
    7.4.1 模型介绍 ................................ 226
    7.4.2 自定义模型 ............................ 227
7.5 基于 Android 的机器人智能检测器 ........................................ 230
    7.5.1 准备工作 ................................ 230
    7.5.2 页面布局 ................................ 232
    7.5.3 实现主 Activity ..................... 232
    7.5.4 物体识别界面 ........................ 238
    7.5.5 相机预览界面拼接 ................ 241
    7.5.6 lib_task_api 方案 .................. 249
    7.5.7 lib_interpreter 方案 ............... 251

7.6 基于 iOS 的机器人智能检测器 ........ 256
    7.6.1 系统介绍 ................................ 256
    7.6.2 视图文件 ................................ 258
    7.6.3 相机处理 ................................ 270
    7.6.4 处理 TensorFlow Lite 模型 .... 277
7.7 调试运行 ........................................ 284

## 第 8 章 AI 考勤管理系统 .......... 285

8.1 背景介绍 ........................................ 286
8.2 系统介绍 ........................................ 286
8.3 系统需求分析 ................................ 287
    8.3.1 可行性分析 ............................ 287
    8.3.2 系统操作流程分析 ................ 287
    8.3.3 系统模块设计 ........................ 287
8.4 系统配置 ........................................ 289
    8.4.1 Django 配置文件 ................... 289
    8.4.2 路径导航文件 ........................ 289
    8.4.3 设计数据模型 ........................ 290
8.5 用户登录验证 ................................ 291
    8.5.1 登录表单页面 ........................ 291
    8.5.2 登录验证 ................................ 292
8.6 添加新员工信息 ............................ 293
    8.6.1 后台主页面 ............................ 293
    8.6.2 添加员工表单页面 ................ 295
    8.6.3 添加员工信息 ........................ 296
8.7 采集员工照片信息 ........................ 296
    8.7.1 设置采集对象 ........................ 297
    8.7.2 采集照片 ................................ 298
8.8 训练照片模型 ................................ 300
    8.8.1 前台页面 ................................ 300
    8.8.2 预测处理 ................................ 300
    8.8.3 训练数据集 ............................ 301
    8.8.4 训练可视化 ............................ 302
8.9 考勤打卡 ........................................ 303

8.9.1　上班打卡签到 303
  8.9.2　下班打卡签退 305
 8.10　查看员工考勤信息 305
  8.10.1　统计最近两周的考勤信息 306
  8.10.2　查看某员工在指定时间范围内的考勤信息 308
  8.10.3　查看指定日期的考勤信息 310
 8.11　查看本人的考勤信息 311
  8.11.1　视图函数 311
  8.11.2　模板文件 313
 8.12　调试运行 315

# 第9章　网络舆情数据分析系统 317

 9.1　系统介绍 318
  9.1.1　舆情数据分析的方式和意义 318
  9.1.2　舆情热度分析 318
 9.2　架构设计 319
  9.2.1　模块分析 319
  9.2.2　系统结构 320
 9.3　微博爬虫 320
  9.3.1　系统配置 321
  9.3.2　批量账号模拟登录 321
  9.3.3　爬取信息 323
 9.4　系统后端 334
  9.4.1　系统配置 334
  9.4.2　数据结构设计 334
  9.4.3　数据处理 339
  9.4.4　微博话题分析 345
 9.5　系统前端 352
  9.5.1　API 导航 352
  9.5.2　博文详情 353

# 第1章

## AI 智能问答系统

本章将通过一个项目的实现过程,详细讲解 TensorFlow 技术在智能问答系统中的应用。本项目使用预先训练的模型,根据给定段落的内容回答问题,该模型可用于构建以自然语言回答用户问题的系统,具体流程由 TensorFlow+TensorFlow.js+SQuAD 2.0+MobileBERT 实现。

## 1.1　背景介绍

问答系统的设计目标是用简洁、准确的答案回答用户用自然语言提出的问题。在人工智能和自然语言处理领域，问答系统都有着较长的发展历史。1950年，英国数学家图灵(A. M. Turing)在论文 *Computing Machinery and Intelligence* 中形象地指出了什么是人工智能，以及机器应该达到的智能标准。20世纪70年代，随着自然语言理解技术的发展，出现了第一个实现用普通英语与计算机对话的人机接口 LUNAR，该系统是伍德(W. Woods)于1972年开发的，用来协助地质学家查找、比较和评估阿波罗11号飞船带回的月球岩石和土壤标本的化学分析数据。

### 1.1.1　互联网的影响

传统的问答系统虽然可以对用户提出的问题给出确定的答案，但是这些问答系统的数据源是基于一个固定的文档集合，尚且不能满足用户各种各样的需求。利用互联网上的资源是有效的解决之道，互联网上具有丰富的信息，是问答系统数据源的理想资源，因此，将问答系统与互联网结合起来就变得非常必要。这也促使了基于互联网的问答系统的出现和发展。

随着因特网的快速发展，网络上流通的信息日益增加，因特网已俨然成为巨大的信息交换平台，要在如此大型的数据库中找寻有用的数据着实不易，通常会借助搜索引擎的功能来达成。然而，以关键词为主的搜索引擎常会找出所有相关的信息，其中包含许多无用的数据，用户需要浪费很多时间浏览不相关的网页。

随着互联网的发展，网络已成为人们获取信息的重要平台。目前，世界上最大的搜索引擎 Google(谷歌)能够搜索的网页数量已经超过百亿个。传统的搜索引擎存在很多不足的地方，其中主要有以下两个方面。

(1) 以关键词的逻辑组合来表达检索需求，返回的相关信息太多。

(2) 以关键词为基础的索引，停留在语言的表层，而没有触及语义，因此，检索效果很难进一步提高。

以上两点使人们在互联网的海量信息中快速准确地找到自己所需要的信息变得越来越困难。

### 1.1.2　问答系统的发展

问答系统的概念虽然提出的时间并不长，但已经发展出了一些比较成熟的系统。

美国麻省理工学院人工智能实验室于 1993 年开发出来的 START,是全世界第一个基于 Internet 的问答系统。START 系统旨在为用户提供准确的信息,它能够回答数以百万计的英语问题,主要包括与地点相关的问题(如城市、国家、湖泊、天气、地图、人口统计学、政治和经济等)、与电影相关的问题(如片名、演员和导演等)、与人物相关的问题(如出生日期、传记等),以及与词典定义相关的问题等。该系统采用基于知识库和基于信息检索的混合模式。此外,系统还保留着原来的两个知识库:START KB 和 Internet Public Library。如果用户提出的问题属于这两个知识库的范畴,START 系统就直接利用知识库中的知识返回比较准确的回答。否则,START 系统首先将问题解析得到查询的关键词,然后利用搜索引擎得到相关信息,最后将后续处理得到的准确而简洁的回答返回给用户。比如,提出一个问题"Who was Bill Gates?",START 系统回答"Co-founder, Microsoft. Born William H. Gates on October 28, 1955, Seattle, Washington."。同时系统还会返回一个关于 Bill Gates 的网页链接,如果用户希望了解更详细的信息,就可以浏览该网页。

美国华盛顿大学开发的 MULDER 系统是最早实现的基于 Internet 的全自动问答系统。该系统没有知识库,而是完全利用 Internet 上的资源获取答案。对于一个问题,MULDER 系统返回的不是唯一的答案,而是一组候选回答,并利用统计的方法给每一个回答赋予一个权重,即置信度。比如,对于问题"Who was the first American in space?",MULDER 系统返回的候选答案中,Alan Shepard 具有 70%的置信度,John Glenn 具有 15%的置信度。同时,在每一个答案下面还会给出相关的网页链接和该网页的内容摘要。

AskJeeves 是美国一个知名的商业问答系统,它能够理解并回答用户用自然语言提出的问题。除了提供文本答案,AskJeeves 还会返回一系列与问题相关的文档链接,并且提供这些文档的内容摘要。此外,AskJeeves 还可以通过多媒体文件的形式,为用户提供更丰富的信息。比如,对于问题"Who was Bill Gates?",系统在回答文本的基础上还将显示一张 Bill Gates 的照片。作为一个商用系统,AskJeeves 的服务种类很多,不仅可以查找 Web 网页,还可以将图片、新闻、产品作为数据源,从而得到所需的信息。AskJeeves 系统中的问题分析部分是依赖手工完成的,为了能够正确理解用户的查询,AskJeeves 雇用了数百名专职人员构造问题模板,并为这些问题模板中常见的问题进行了缓存。尽管系统的问题模板能够细化并明确用户的需求,但由于这些模板需要人工生成和维护,因此涉及的工作量非常大。

国内复旦大学开发的原型系统(FDUQA)已经具有了初步的效果,同时哈尔滨工业大学(金山客服)和中国科学院计算技术研究所也在从事该领域的研究。

从系统的设计与实现来看,自动问答系统一般包括三个主要组成部分:问题分析、信息检索和答案抽取。目前,国际上问答系统的研究方兴未艾,许多大的科研院所和著名公司都积极参与该领域的研究,其中比较著名的 Microsoft、IBM、MIT、University of Amsterdam、National University of Singapore、University of Zurich、University of Southern

California、Columbia University 等。国内在问答系统方面的研究相对国外较为不足，主要有中国科学院计算技术研究所、复旦大学、哈尔滨工业大学、沈阳航空工业大学、香港城市大学等单位。

近年来，基于大型预训练语言模型(如 BERT、GPT-3 等)的问答系统取得了突破性进展。这些模型在大量文本数据上进行预训练，能够捕捉到语言的深层次特征，从而在问答任务中实现优异的性能。研究者们通过微调这些模型，使其适应特定的问答场景，如医疗、法律或教育等领域。此外，一些研究工作还集中在提高模型的可解释性、鲁棒性和泛化能力，以及如何更有效地结合外部知识源来增强模型的回答能力。

## 1.2　问答系统的发展趋势：AI 问答系统

AI 智能问答指的是基于人工智能技术的问答系统，它可以通过自然语言处理、机器学习等技术快速、准确地回答用户提出的问题；能够从大量的数据中获取信息，再经过分析和处理，提供精准、高效的问答服务。AI 智能问答系统在面对大量的信息和复杂的问题时表现出了很大的优势。常见的数据来源包括互联网上和企业内部的知识库、维基百科、论坛等。通过建立问题分类和自动问题回答等功能，AI 智能问答系统还能够工作在人工客服之外，成为企业的技术支持和客户服务的重要部分，提高效率并降低成本。

扫码看视频

### 1.2.1　人工智能介绍

人工智能就是我们平常所说的 AI，全称是 Artificial Intelligence。人工智能是一门新的技术科学，旨在研究、开发用于模拟、延伸和扩展人类智能的理论、方法、技术及应用系统。人工智能是一门极具挑战性的科学，从事这项工作的人必须懂得计算机知识、心理学和哲学，其涉及的领域较广，如机器学习、计算机视觉等。总的来说，人工智能研究的一个主要目标是使机器能够胜任一些通常需要人类智能才能完成的复杂工作。

现在通常将人工智能分为弱人工智能和强人工智能。我们看到的电影里的一些人工智能大部分都是强人工智能，它们能像人类一样思考如何处理问题，甚至能在一定程度上做出比人类更好的决定；它们能够适应周围的环境，解决一些程序中没有遇到过的突发事件。但是在目前的现实世界中，大部分人工智能只具备观察和感知的能力，在经过一定的训练后，能计算一些人类不能计算的事情，却没有自适应能力，也就是它不会处理突发的情况，只能处理程序中已经写好的、已经预测到的事情，这就叫作弱人工智能。

## 1.2.2 机器学习

机器学习(Machine Learning，ML)是一门多领域交叉学科，涉及概率论、统计学、逼近论、凸分析、算法复杂度理论等多门学科。机器学习专门研究计算机如何模拟或实现人类的学习行为，以获取新的知识或技能，重新组织已有的知识结构使之不断地改善自身的性能。

机器学习是一类算法的总称，这些算法企图从大量历史数据中挖掘出隐含的规律，并用于预测或者分类。更具体地说，机器学习可以看作寻找一个函数，输入是样本数据，输出是期望的结果，只是这个函数过于复杂，以至于不太方便形式化表达。需要注意的是，机器学习的目标是使学到的函数很好地适用于"新样本"，而不仅仅是在训练样本上表现很好。学到的函数适用于新样本的能力，称为泛化(generalization)能力。

机器学习有一个显著的特点，也是机器学习最基本的做法，就是使用一个算法从大量的数据中解析并得到有用的信息，再从中学习，然后对之后真实世界中会发生的事情进行预测或做出判断。机器学习需要海量的数据来进行训练，并从这些数据中提取有用的信息，然后反馈到真实世界的用户中。

可以用一个简单的例子来说明机器学习。假设在天猫或京东购物的时候，天猫和京东会向用户推送商品信息，这些推荐的商品往往是用户很感兴趣的东西，这个过程就是通过机器学习完成的。首先，京东和天猫根据用户以前的购物订单和经常浏览的商品记录得出哪些是用户感兴趣的商品，并且购买的概率大，然后将这些商品信息定向推送给用户。

## 1.2.3 深度学习

上文介绍的机器学习是一种实现人工智能的方法，深度学习是一种实现机器学习的技术。深度学习本来并不是一种独立的学习方法，其本身也会用到有监督和无监督的学习方法来训练深度神经网络。但由于近几年该领域发展迅猛，一些特有的学习手段(如残差网络)相继被提出，因此，越来越多的人将其单独看作一种学习方法。

假设我们需要识别某张照片是狗还是猫，如果用传统机器学习的方法，会首先定义一些特征，如有没有胡须、耳朵、鼻子、嘴巴的模样等，即我们要先确定相应的"面部特征"作为机器学习的特征，以此来对对象进行分类识别。而深度学习的方法更进一步，它自动找出这个分类问题所需要的重要特征。那么，深度学习是如何做到这一点的呢？继续以猫、狗识别的例子进行说明，识别步骤如下。

(1) 首先确定出有哪些边和角跟识别出猫或狗关系最大。
(2) 根据上一步找出的很多小元素(边、角等)构建层级网络，找出它们之间的各种组合。

(3) 在构建层级网络之后，就可以确定哪些组合可以识别出猫或狗。

> 注意：其实深度学习并不是一个独立的算法，在训练神经网络的时候也通常会用到有监督学习和无监督学习。但是由于一些独特的学习方法被提出，把它看成是一种独立的学习算法也没有问题。深度学习可以大致理解成包含多个隐藏层的神经网络结构，其中的"深"字指的就是隐藏层的数量。

### 1.2.4 系统介绍

AI 智能问答的实现需要基于大数据、自然语言处理、机器学习、人机交互等多个方面的技术。例如，自然语言处理可以解决不同的语言和词汇表达的差别，机器学习可以帮助问答系统学习问题的模式和特点。提供实时、准确的服务是智能问答的一个关键需求，因此，问答系统必须经过深入的测试和优化来达到最佳的性能和用户体验。在本章的实例中，将使用人工智能技术开发一个 AI 问答系统，展示深度学习技术在 AI 问答系统中的应用。

## 1.3 技术架构

在本项目中构建了一个可以用自然语言回答用户问题的系统，本项目使用的是 SQuAD 2.0 数据集，然后使用 BERT 的压缩版本模型 MobileBERT 进行处理，最后使用 TensorFlow.js 实现机器学习开发。

扫码看视频

### 1.3.1 TensorFlow.js

TensorFlow.js 是一个开源的，基于 WebGL 硬件加速技术的 JavaScript 库，用于训练和部署机器学习模型，其设计理念借鉴了目前广受欢迎的 TensorFlow 深度学习框架。Google 推出的第一个基于 TensorFlow 的前端深度学习框架是 Deeplearning.js，使用 TypeScript 语言开发，2018 年 Google 将其重命名为 TensorFlow.js，并在 TypeScript 内核的基础上增加了 JavaScript 的接口，以及 TensorFlow 模型导入等工程，组成了 TensorFlow.js 深度学习框架。

在 JavaScript 项目中，有两种安装 TensorFlow.js 的方法，一种是通过&lt;script&gt;标签引入，另一种是通过 npm 安装。

1) 通过&lt;script&gt;标签引入

通过使用如下脚本代码，可以将 TensorFlow.js 添加到 HTML 文件中。

```
<script src="https://cdn.jsdelivr.net/npm/@tensorflow/tfjs@2.0.0/dist/tf.min.js">
</script>
```

2) 通过 npm 安装

用户可以使用 npm CLI 工具或 Yarn 工具安装 TensorFlow.js，具体命令如下：

```
yarn add @tensorflow/tfjs
```

或

```
npm install @tensorflow/tfjs
```

TensorFlow.js 可以在浏览器和 Node.js 中运行，并且在两个平台中都具有许多不同的可用配置。每个平台都有一组影响应用开发方式的独特注意事项。在浏览器中，TensorFlow.js 支持移动设备及桌面设备。每种设备都有一组特定的约束(例如，可用 WebGL API)，系统会自动为用户确定和配置这些约束。

## 1.3.2 SQuAD 2.0

SQuAD 2.0(Stanford Question Answering Dataset)是由斯坦福大学发布的一款机器阅读理解数据集，旨在推动和测试机器理解自然语言的能力。SQuAD 2.0 的主要特点如下。

- 大规模数据集：SQuAD 2.0 包含超过十万个问题，这些问题由众包工作者基于维基百科的文章提出。
- 精确匹配与模糊匹配：评估模型性能的两个主要指标是精确匹配(exact match)和模糊匹配(fuzzy match)。精确匹配要求机器提供的答案与人类提供的答案完全一致；模糊匹配则允许存在一定程度的差异，只要机器的答案包含人类答案的关键信息即可。
- 可回答与不可回答的问题：SQuAD 2.0 不仅要求模型能够回答那些在文本中有明确答案的问题，还要求模型能够识别出那些在文本中没有答案的问题，并拒绝回答这些问题，这增加了对模型理解能力的考验。
- 对人类水平的超越：科大讯飞在 SQuAD 2.0 比赛中取得了突破性进展，其提交的系统模型在精确匹配和模糊匹配两项指标上首次全面超越了人类的平均水平。
- 技术挑战：SQuAD 2.0 的设计旨在挑战现有的阅读与理解模型，促使研究者开发出更先进的算法，以更准确地模拟人类的阅读与理解过程。
- 研究与应用价值：SQuAD 2.0 作为一个标准化的测试平台，促进了机器阅读与理解领域的研究，并且对于开发能够理解并应用自然语言的智能系统具有重要意义。

SQuAD 2.0 的发布是自然语言处理领域的一个重要里程碑，它不仅为研究人员提供了一个测试和改进模型的平台，也为未来智能系统的语言理解能力设定了新的标准。

### 1.3.3 BERT

Google 在论文 *BERT: Pre-training of Deep Bidirectional Transformers for Language Understanding* 中提出了 BERT 模型,BERT 模型主要利用了 Transformer 的 Encoder 结构。总的来说,BERT 具有以下特点。

- ❑ 结构:采用了 Transformer 的 Encoder 结构,但是模型结构比 Transformer 要深。Transformer Encoder 包含 6 个 Encoder block,BERT-base 模型包含 12 个 Encoder block,BERT-large 包含 24 个 Encoder block。
- ❑ 训练:训练主要分为预训练和 Fine-tuning 两个阶段。预训练阶段与 Word2Vec、ELMo 等类似,是在大型数据集上根据一些预训练任务训练得到。Fine-tuning 阶段是后续用于一些下游任务时进行微调,例如文本分类、词性标注、问答系统等,BERT 无须调整结构就可以在不同的任务上进行微调。
- ❑ 预训练任务 1:BERT 的第一个预训练任务是 Masked LM,在句子中随机遮盖一部分单词,然后同时利用上下文的信息预测遮盖的单词,这样可以更好地根据全文理解单词的意思。
- ❑ 预训练任务 2:BERT 的第二个预训练任务是 NSP(next sentence prediction),即下一句预测任务,这个任务主要是让模型能够更好地理解句子间的关系。

### 1.3.4 知识蒸馏

本章实例使用的神经网络模型是 BERT 的压缩版本 MobileBERT,和前者相比,压缩后的 MobileBERT 运行速度提高了 4 倍,模型尺寸缩小了 75%。本项目之所以采用压缩版的 MobileBERT,目的是提高速度和节省时间,如果更深入地说,是使用了知识蒸馏技术。

近年来,神经模型几乎在所有领域都取得了成功,包括极端复杂的问题。然而,这些模型的体积巨大,有数百万(甚至数十亿)个参数,因此不能部署在边缘设备上。

知识蒸馏指的是模型压缩思想,它通过一步一步地使用一个较大的已经训练好的网络(教师网络)去教导一个较小的网络(学生网络)确切地去做什么。通过尝试复制大网络在每一层的输出(不仅仅是最终的损失),小网络被训练,以学习大网络的准确行为。

深度学习在计算机视觉、语音识别、自然语言处理等众多领域取得了令人难以置信的成绩,然而,这些模型中的绝大多数在移动设备或嵌入式设备上运行的计算成本太过昂贵。显然,模型越复杂,理论搜索空间就越大。但是,如果我们假设较小的网络也能实现相同(甚至相似)的收敛,那么教师网络的收敛空间应该与学生网络的解空间重叠。

不幸的是,仅凭这一点并不能保证学生网络与教师网络收敛在同一点。学生网络的收敛点可能与教师网络有很大的不同。但是,如果引导学生网络复制教师网络(教师网络已经在更大的解空间中进行了搜索)的行为，则其预期收敛空间会与原有的教师网络收敛空间重叠。

知识蒸馏模式下的"教师—学生网络"到底如何工作呢？其基本流程如下。

(1) 训练教师网络：首先使用完整数据集分别对高度复杂的教师网络进行训练，这个步骤需要高计算性能，因此只能离线(在高性能 GPU 上)完成。

(2) 构建对应关系：在设计学生网络时，需要建立学生网络的中间输出与教师网络的对应关系。这种对应关系可以直接将教师网络中某一层的输出信息传递给学生网络，或者在传递给学生网络之前进行一些数据增强。

(3) 通过教师网络前向传播：教师网络前向传播数据以获得所有中间输出，然后对其应用数据增强(如果有的话)。

(4) 通过学生网络反向传播：现在利用教师网络的输出和学生网络中反向传播误差的对应关系，使学生网络能够学会复制教师网络的行为。

随着 NLP 模型的大小增加到数千亿个参数，创建这些模型的更紧凑表示的重要性也随之增加。知识蒸馏成功地实现了这一点，在一个例子中，教师模型性能的 96%保留在了一个只有原来七分之一大小的模型中。然而，在设计教师模型时，知识的提炼仍然被认为是事后考虑的事情，这可能会降低效率，把潜在的性能改进留给学生模型。

此外，在最初的提炼后对小型学生模型进行微调，并要求在微调时不降低它们的表现，能够完成我们希望学生模型完成的任务。因此，与只训练教师模型相比，通过知识蒸馏训练学生模型将需要更多的训练，这在推理的时候限制了学生模型的优点。

## 1.4 具体实现

本项目将使用 TensorFlow.js 设计一个网页，在网页中有一篇文章，利用 SQuAD 2.0 数据集和神经网络模型 MobileBERT 学习文章中的知识，然后在表单中提出与文章内容有关的问题，系统会自动回答这些问题。

扫码看视频

### 1.4.1 编写 HTML 文件

编写 HTML 文件 index.html，在上方文本框中显示介绍 Nikola Tesla(尼古拉·特斯拉)的一篇文章信息，在下方文本框中输入一个与文章内容相关的问题，单击 Search 按钮后会自动输出显示这个问题的答案。文件 index.html 的具体实现代码如下：

```
<!doctype html>
<html>
<head>
```

```html
    <meta http-equiv="Content-Type" content="text/html; charset=UTF-8">
    <script src="./index.js"></script>
</head>

<body>
    <div>
        <h3>Context (you can paste your own content in the text area)</h3>
        <textarea id='context' rows="30" cols="120">Nikola Tesla (/ˈtɛslə/;[2] Serbo-Croatian: [nǐkola têsla]; Serbian Cyrillic: Никола Тесла;[a] 10 July 1856 – 7 January 1943) was a Serbian-American[4][5][6] inventor, electrical engineer, mechanical engineer, and futurist who is best known for his contributions to the design of the modern alternating current (AC) electricity supply system.[7]<br/>

        Born and raised in the Austrian Empire, Tesla studied engineering and physics in the 1870s without receiving a degree, and gained practical experience in the early 1880s working in telephony and at Continental Edison in the new electric power industry. He emigrated in 1884 to the United States, where he would become a naturalized citizen. He worked for a short time at the Edison Machine Works in New York City before he struck out on his own. With the help of partners to finance and market his ideas, Tesla set up laboratories and companies in New York to develop a range of electrical and mechanical devices. His alternating current (AC) induction motor and related polyphase AC patents, licensed by Westinghouse Electric in 1888, earned him a considerable amount of money and became the cornerstone of the polyphase system which that company would eventually market.<br/>

        Attempting to develop inventions he could patent and market, Tesla conducted a range of experiments with mechanical oscillators/generators, electrical discharge tubes, and early X-ray imaging. He also built a wireless-controlled boat, one of the first ever exhibited. Tesla became well known as an inventor and would demonstrate his achievements to celebrities and wealthy patrons at his lab, and was noted for his showmanship at public lectures. Throughout the 1890s, Tesla pursued his ideas for wireless lighting and worldwide wireless electric power distribution in his high-voltage, high-frequency power experiments in New York and Colorado Springs. In 1893, he made pronouncements on the possibility of wireless communication with his devices. Tesla tried to put these ideas to practical use in his unfinished Wardenclyffe Tower project, an intercontinental wireless communication and power transmitter, but ran out of funding before he could complete it.[8]<br/>

        After Wardenclyffe, Tesla experimented with a series of inventions in the 1910s and 1920s with varying degrees of success. Having spent most of his money, Tesla lived in a series of New York hotels, leaving behind unpaid bills. He died in New York City in January 1943.[9] Tesla's work fell into relative obscurity following his death, until 1960, when the General Conference on Weights and Measures named the SI unit of magnetic flux density the tesla in his honor.[10] There has been a resurgence in popular interest in Tesla since the 1990s.[11]</textarea>
        <h3>Question</h3>
```

```
    <input type=text id="question"> <button id="search">Search</button>
    <h3>Answers</h3>
    <div id='answer'></div>
  </div>
</body>
</html>
```

## 1.4.2 脚本处理

当用户单击 Search 按钮后会调用脚本文件 index.js，此文件的功能是获取用户在文本框中输入的问题，然后调用神经网络模型回答这个问题。文件 index.js 的具体实现代码如下：

```
import * as qna from '@tensorflow-models/qna';
import '@tensorflow/tfjs-core';
import '@tensorflow/tfjs-backend-cpu';
import '@tensorflow/tfjs-backend-webgl';

let modelPromise = {};
let search;
let input;
let contextDiv;
let answerDiv;

const process = async () => {
  const model = await modelPromise;
  const answers = await model.findAnswers(input.value, contextDiv.value);
  console.log(answers);
  answerDiv.innerHTML =
      answers.map(answer => answer.text + ' (score =' + answer.score + ')')
          .join('<br>');
};

window.onload = () => {
  modelPromise = qna.load();
  input = document.getElementById('question');
  search = document.getElementById('search');
  contextDiv = document.getElementById('context');
  answerDiv = document.getElementById('answer');
  search.onclick = process;

  input.addEventListener('keyup', async (event) => {
    if (event.key === 'Enter') {
      process();
    }
  });
};
```

在上述代码中，使用 addEventListener 监听用户输入的问题，然后调用函数 model.findAnswers()回答问题。

### 1.4.3 加载训练模型

在文件 question_and_answer.ts 中加载神经网络模型 MobileBERT，具体实现流程如下。
(1) 首先设置输入参数和最大扫描长度，代码如下：

```
const MODEL_URL = 'https://tfhub.dev/tensorflow/tfjs-model/mobilebert/1';
const INPUT_SIZE = 384;
const MAX_ANSWER_LEN = 32;
const MAX_QUERY_LEN = 64;
const MAX_SEQ_LEN = 384;
const PREDICT_ANSWER_NUM = 5;
const OUTPUT_OFFSET = 1;
const NO_ANSWER_THRESHOLD = 4.3980759382247925;
```

在上述代码中，NO_ANSWER_THRESHOLD 是确定问题是否与上下文无关的阈值，该值是由训练 SQuAD 2.0 数据集的数据生成的。
(2) 创建加载模型 MobileBERT 的接口 ModelConfig，代码如下：

```
export interface ModelConfig {
  /**
   * 设置模型自定义 url 字符串，这对于那些托管在无法访问的地区/国家的模型非常有用
   */
  modelUrl: string;
  /**
   * 是不是来自 TF-Hub 的 URL
   */
  fromTFHub?: boolean;
}
```

### 1.4.4 查询处理

编写函数 process()实现检索处理，获取用户在表单中输入的问题，然后检索文章中的所有内容。为了确保问题的完整性，如果用户没有在问题最后输入问号，系统会自动添加一个问号。具体代码如下：

```
private process(
    query: string, context: string, maxQueryLen: number, maxSeqLen: number,
    docStride = 128): Feature[] {
  //始终在查询末尾添加问号
  query = query.replace(/\?/g, '');
```

```
query = query.trim();
query = query + '?';

const queryTokens = this.tokenizer.tokenize(query);
if (queryTokens.length > maxQueryLen) {
  throw new Error(
      'The length of question token exceeds the limit (${maxQueryLen}).');
}

const origTokens = this.tokenizer.processInput(context.trim());
const tokenToOrigIndex: number[] = [];
const allDocTokens: number[] = [];
for (let i = 0; i < origTokens.length; i++) {
  const token = origTokens[i].text;
  const subTokens = this.tokenizer.tokenize(token);
  for (let j = 0; j < subTokens.length; j++) {
    const subToken = subTokens[j];
    tokenToOrigIndex.push(i);
    allDocTokens.push(subToken);
  }
}
//3个选项：[CLS]、[SEP]、[SEP]
const maxContextLen = maxSeqLen - queryTokens.length - 3;

//可能有超过最大序列长度的文档。为了解决这个问题，采用了滑动窗口的方法
//在这种方法中，以"doc\u-stride"的步幅将大块的数据移动到最大长度
const docSpans: Array<{start: number, length: number}> = [];
let startOffset = 0;
while (startOffset < allDocTokens.length) {
  let length = allDocTokens.length - startOffset;
  if (length > maxContextLen) {
    length = maxContextLen;
  }
  docSpans.push({start: startOffset, length});
  if (startOffset + length === allDocTokens.length) {
    break;
  }
  startOffset += Math.min(length, docStride);
}

const features = docSpans.map(docSpan => {
  const tokens = [];
  const segmentIds = [];
  const tokenToOrigMap: {[index: number]: number} = {};
  tokens.push(CLS_INDEX);
  segmentIds.push(0);
  for (let i = 0; i < queryTokens.length; i++) {
```

```
      const queryToken = queryTokens[i];
      tokens.push(queryToken);
      segmentIds.push(0);
    }
    tokens.push(SEP_INDEX);
    segmentIds.push(0);
    for (let i = 0; i < docSpan.length; i++) {
      const splitTokenIndex = i + docSpan.start;
      const docToken = allDocTokens[splitTokenIndex];
      tokens.push(docToken);
      segmentIds.push(1);
      tokenToOrigMap[tokens.length] = tokenToOrigIndex[splitTokenIndex];
    }
    tokens.push(SEP_INDEX);
    segmentIds.push(1);
    const inputIds = tokens;
    const inputMask = inputIds.map(id => 1);
    while ((inputIds.length < maxSeqLen)) {
      inputIds.push(0);
      inputMask.push(0);
      segmentIds.push(0);
    }
    return {inputIds, inputMask, segmentIds, origTokens, tokenToOrigMap};
  });
  return features;
}
```

## 1.4.5 文章处理

(1) 编写函数 cleanText()，功能是删除文章中的无效字符和空白。代码如下：

```
private cleanText(text: string, charOriginalIndex: number[]): string {
  const stringBuilder: string[] = [];
  let originalCharIndex = 0, newCharIndex = 0;
  for (const ch of text) {
    //跳过不能使用的字符
    if (isInvalid(ch)) {
      originalCharIndex += ch.length;
      continue;
    }
    if (isWhitespace(ch)) {
      if (stringBuilder.length > 0 &&
        stringBuilder[stringBuilder.length - 1] !== ' ') {
        stringBuilder.push(' ');
        charOriginalIndex[newCharIndex] = originalCharIndex;
        originalCharIndex += ch.length;
```

```
      } else {
        originalCharIndex += ch.length;
        continue;
      }
    } else {
      stringBuilder.push(ch);
      charOriginalIndex[newCharIndex] = originalCharIndex;
      originalCharIndex += ch.length;
    }
    newCharIndex++;
  }
  return stringBuilder.join('');
}
```

(2) 编写函数 runSplitOnPunc()，功能是拆分文本中的标点符号。代码如下：

```
private runSplitOnPunc(text: string, count: number, charOriginalIndex: number[]):
Token[] {
  const tokens: Token[] = [];
  let startNewWord = true;
  for (const ch of text) {
    if (isPunctuation(ch)) {
      tokens.push({text: ch, index: charOriginalIndex[count]});
      count += ch.length;
      startNewWord = true;
    } else {
      if (startNewWord) {
        tokens.push({text: '', index: charOriginalIndex[count]});
        startNewWord = false;
      }
      tokens[tokens.length - 1].text += ch;
      count += ch.length;
    }
  }
  return tokens;
}
```

(3) 编写函数 tokenize()，功能是为指定的词汇库生成标记。本函数使用 Google 提供的全词屏蔽模型实现，这种新技术也称为全词掩码。在这种情况下，总是一次屏蔽与一个单词对应的所有标记。代码如下：

```
tokenize(text: string): number[] {
  let outputTokens: number[] = [];
  const words = this.processInput(text);
  words.forEach(word => {
    if (word.text !== CLS_TOKEN && word.text !== SEP_TOKEN) {
      word.text = `${SEPERATOR}${word.text.normalize(NFKC_TOKEN)}`;
    }
```

```
});
for (let i = 0; i < words.length; i++) {
  const chars = [];
  for (const symbol of words[i].text) {
    chars.push(symbol);
  }
  let isUnknown = false;
  let start = 0;
  const subTokens: number[] = [];
  const charsLength = chars.length;
  while (start < charsLength) {
    let end = charsLength;
    let currIndex;
    while (start < end) {
      const substr = chars.slice(start, end).join('');
      const match = this.trie.find(substr);
      if (match != null && match.end != null) {
        currIndex = match.getWord()[2];
        break;
      }
      end = end - 1;
    }//while
    if (currIndex == null) {
      isUnknown = true;
      break;
    }
    subTokens.push(currIndex);
    start = end;
  }//while
  if (isUnknown) {
    outputTokens.push(UNK_INDEX);
  } else {
    outputTokens = outputTokens.concat(subTokens);
  }
}//for
return outputTokens;
}
```

### 1.4.6 加载处理

编写函数 load()加载数据和网页信息，首先使用函数 loadGraphModel()加载模型文件，然后使用函数 execute()执行用户输入的操作。代码如下：

```
async load() {
  this.model = await tfconv.loadGraphModel(
    this.modelConfig.modelUrl, {fromTFHub: this.modelConfig.fromTFHub});
```

```
//预热后端
const batchSize = 1;
const inputIds = tf.ones([batchSize, INPUT_SIZE], 'int32');
const segmentIds = tf.ones([1, INPUT_SIZE], 'int32');
const inputMask = tf.ones([1, INPUT_SIZE], 'int32');
this.model.execute({
  input_ids: inputIds,
  segment_ids: segmentIds,
  input_mask: inputMask,
  global_step: tf.scalar(1, 'int32')
});

this.tokenizer = await loadTokenizer();
}
```

## 1.4.7 寻找答案

编写函数 model.findAnswers()，功能是根据用户在表单中输入的问题寻找对应的答案。此函数包含如下三个参数。

- question：寻找答案的问题。
- context：从这里面查找答案。
- 返回值：返回一个数组，每个选项是一种可能的答案。

函数 model.findAnswers()的具体实现代码如下：

```
async findAnswers(question: string, context: string): Promise<Answer[]> {
  if (question == null || context == null) {
    throw new Error(
      'The input to findAnswers call is null, ' +
      'please pass a string as input.');
  }

  const features =
      this.process(question, context, MAX_QUERY_LEN, MAX_SEQ_LEN);
  const inputIdArray = features.map(f => f.inputIds);
  const segmentIdArray = features.map(f => f.segmentIds);
  const inputMaskArray = features.map(f => f.inputMask);
  const globalStep = tf.scalar(1, 'int32');
  const batchSize = features.length;
  const result = tf.tidy(() => {
    const inputIds =
        tf.tensor2d(inputIdArray, [batchSize, INPUT_SIZE], 'int32');
    const segmentIds =
        tf.tensor2d(segmentIdArray, [batchSize, INPUT_SIZE], 'int32');
    const inputMask =
```

```
      tf.tensor2d(inputMaskArray, [batchSize, INPUT_SIZE], 'int32');
  return this.model.execute(
          {
            input_ids: inputIds,
            segment_ids: segmentIds,
            input_mask: inputMask,
            global_step: globalStep
          },
          ['start_logits', 'end_logits']) as [tf.Tensor2D, tf.Tensor2D];
});
const logits = await Promise.all([result[0].array(), result[1].array()]);
//处理所有中间张量
globalStep.dispose();
result[0].dispose();
result[1].dispose();

const answers = [];
for (let i = 0; i < batchSize; i++) {
  answers.push(this.getBestAnswers(
      logits[0][i], logits[1][i], features[i].origTokens,
      features[i].tokenToOrigMap, context, i));
}
return answers.reduce((flatten, array) => flatten.concat(array), [])
    .sort((logitA, logitB) => logitB.score - logitA.score)
    .slice(0, PREDICT_ANSWER_NUM);
}
```

### 1.4.8 提取最佳答案

(1) 通过以下代码从 logits 数组和输入中查找最佳的 N 个答案和 logits。其中，参数 startLogits 表示开始答案索引，参数 endLogits 表示结束答案索引，参数 origTokens 表示通道的原始标记，参数 tokenToOrigMap 表示令牌到索引的映射。

```
QuestionAndAnswerImpl.prototype.getBestAnswers = function (startLogits, endLogits,
origTokens, tokenToOrigMap, context, docIndex) {
  var _a;
  if (docIndex === void 0) { docIndex = 0; }
  //模型使用封闭区间[开始,结束]作为索引
  var startIndexes = this.getBestIndex(startLogits);
  var endIndexes = this.getBestIndex(endLogits);
  var origResults = [];
  startIndexes.forEach(function (start) {
    endIndexes.forEach(function (end) {
      if (tokenToOrigMap[start] && tokenToOrigMap[end] && end >= start) {
        var length_2 = end - start + 1;
```

```
            if (length_2 < MAX_ANSWER_LEN) {
                origResults.push({ start: start, end: end, score:
startLogits[start] + endLogits[end] });
            }
        }
    });
});
origResults.sort(function (a, b) { return b.score - a.score; });
var answers = [];
for (var i = 0; i < origResults.length; i++) {
    if (i >= PREDICT_ANSWER_NUM ||
        origResults[i].score < NO_ANSWER_THRESHOLD) {
        break;
    }
    var convertedText = '';
    var startIndex = 0;
    var endIndex = 0;
    if (origResults[i].start > 0) {
        _a = this.convertBack(origTokens, tokenToOrigMap, origResults[i].start,
origResults[i].end, context), convertedText = _a[0], startIndex = _a[1], endIndex = _a[2];
    }
    else {
        convertedText = '';
    }
    answers.push({
        text: convertedText,
        score: origResults[i].score,
        startIndex: startIndex,
        endIndex: endIndex
    });
}
return answers;
};
```

(2) 编写函数 getBestIndex()，功能是通过神经网络模型检索文章后，系统会找到多个答案，然后根据得分高低选出其中的 5 个最佳答案。代码如下：

```
getBestIndex(logits: number[]): number[] {
  const tmpList = [];
  for (let i = 0; i < MAX_SEQ_LEN; i++) {
    tmpList.push([i, i, logits[i]]);
  }
  tmpList.sort((a, b) => b[2] - a[2]);

  const indexes = [];
  for (let i = 0; i < PREDICT_ANSWER_NUM; i++) {
      indexes.push(tmpList[i][0]);
```

```
  }
  return indexes;
}
```

### 1.4.9　将答案转换回文本

接下来，使用函数 convertBack()将问题的答案转换回原始文本形式。代码如下：

```
convertBack(
    origTokens: Token[], tokenToOrigMap: {[key: string]: number},
    start: number, end: number, context: string): [string, number, number] {
  //移位索引是 logits + offset
  const shiftedStart = start + OUTPUT_OFFSET;
  const shiftedEnd = end + OUTPUT_OFFSET;
  const startIndex = tokenToOrigMap[shiftedStart];
  const endIndex = tokenToOrigMap[shiftedEnd];
  const startCharIndex = origTokens[startIndex].index;
  const endCharIndex = endIndex < origTokens.length - 1 ?
      origTokens[endIndex + 1].index - 1 :
      origTokens[endIndex].index + origTokens[endIndex].text.length;
  return [
    context.slice(startCharIndex, endCharIndex + 1).trim(), startCharIndex,
    endCharIndex
  ];
}
```

## 1.5　调试运行

　　至此，整个实例介绍完毕，接下来开始调试本项目。本项目基于 Yarn 和 npm 进行架构调试，其中，Yarn 对代码来说是一个包管理器，可以让我们使用并分享全世界开发者(例如 JavaScript)的代码。调试本项目的基本流程如下。

扫码看视频

（1）安装 Node.js，打开 Node.js 命令行界面，输入如下命令进入项目的 qna 目录：

```
cd qna
```

（2）输入如下命令，在 qna 目录中安装 npm：

```
npm install
```

（3）输入如下命令进入子目录 demo：

```
cd qna/demo
```

(4) 输入如下命令安装本项目需要的依赖项：

```
yarn
```

(5) 输入如下命令编译依赖项：

```
yarn build-deps
```

(6) 输入如下命令启动测试服务器，并监视文件的更改变化情况：

```
yarn watch
```

经过这些调试，所有的编译运行工作全部完成，在笔者计算机中的整个编译过程如下：

```
E:\123\lv\TensorFlow\daima\tfjs-models-master\qna>cd demo

E:\123\lv\TensorFlow\daima\tfjs-models-master\qna\demo>yarn
yarn install v1.22.10
[1/5] Validating package.json...
[2/5] Resolving packages...
warning Resolution field "is-svg@4.3.1" is incompatible with requested version "is-svg@^3.0.0"
success Already up-to-date.
Done in 5.09s.
E:\123\lv\TensorFlow\daima\tfjs-models-master\qna\demo>yarn build-deps
yarn run v1.22.10
$ yarn build-qna
$ cd .. && yarn && yarn build-npm
warning package-lock.json found. Your project contains lock files generated by tools other than Yarn. It is advised not to mix package managers in order to avoid resolution inconsistencies caused by unsynchronized lock files. To clear this warning, remove package-lock.json.
[1/4] Resolving packages...
success Already up-to-date.
$ yarn build && rollup -c
$ rimraf dist && tsc

src/index.ts → dist/qna.js...
created dist/qna.js in 1m 18.9s

src/index.ts → dist/qna.min.js...
created dist/qna.min.js in 1m 1.3s

src/index.ts → dist/qna.esm.js...
created dist/qna.esm.js in 45.8s
Done in 251.88s.

E:\123\lv\TensorFlow\daima\tfjs-models-master\qna\demo>yarn watch
yarn run v1.22.10
```

```
$ cross-env NODE_ENV=development parcel index.html --no-hmr --open
√ Built in 1.81s.
```

运行上述命令成功后自动打开一个网页 http://localhost:1234/，在网页上显示本项目的执行效果。执行后在表单中输入一个问题，这个问题的答案可以在表单上方的文章中找到。例如，输入"Where was Tesla born"，然后单击 search 按钮，系统会自动输出显示这个问题的答案，如图 1-1 所示。

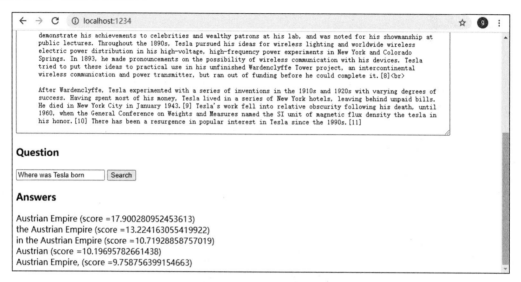

图 1-1　执行效果

# 第 2 章

## AI 智能推荐系统

推荐系统是指通过网站向用户提供商品、电影、新闻和音乐等信息的建议，帮助用户尽快找到自己感兴趣的信息。在本章的内容中，将使用 Scikit-learn、Flask 和 Pandas 开发一个电影推荐系统，同时详细介绍使用人工智能技术开发大型推荐项目的方法。

## 2.1 背景介绍

随着移动互联网的快速发展，我们步入了一个全新的信息时代。当前通过互联网提供服务的平台越来越多，提供的服务种类(如购物、视频、新闻、音乐、婚恋、社交等)也层出不穷，服务"标的物"的种类也越来越多样(亚马逊上有上百万种书)。如何让琳琅满目的"标的物"被需要它们的人发现，并满足用户的个性化需要，是企业当前面临的一大挑战。

扫码看视频

同时，随着社会的发展，受教育程度的提升，每个人都有表现自我个性的欲望。随着互联网的发展，出现了非常多的可以表达自我个性的产品，如微信朋友圈、微博、抖音、快手等，每个人的个性喜好有了极大的展示空间。另外，从进化论的角度来说，每个人都是一个差异化的个体，具有不同的性格特征，个人的生活成长环境也有极大差异，导致个人的偏好千差万别。"长尾理论"也很好地解释了为何多样化物品中的非畅销品可以满足人们多样化的需求，这些需求加起来不一定比热门商品产生的销售额小。

随着社会的进步，物质生活条件的改善，大家不必再为生存下来而担忧，所以有了越来越多的非生存需求，比如看书、看电影、购物等，而这些非生存的需求往往在很多时候是不确定的，是无意识的，自己不知道自己需要什么。生存需求对人而言显得非常强烈且明显，比如当个体处于饥饿状态时，其首要需求肯定是食物。不同于生存需求，面对非生存需求，人们实际上更愿意接受被动推荐的好的物品，比如推荐的电影符合个人口味，个体可能会很喜欢。

为了更好地为用户提供服务，在为用户提供服务的同时赚取更多的利润，越来越多的企业通过采用个性化推荐技术，辅助用户更快地发现自己喜欢的东西。企业根据用户在某产品上发生的行为记录，结合用户自身和"标的物"的信息，利用推荐技术(机器学习的一个分支)来为用户推荐可能感兴趣的物品。

### 2.1.1 推荐系统能解决什么问题

推荐系统是互联网(特别是移动互联网)快速发展的产物，随着用户规模的爆炸增长以及供应商提供的物品的种类越来越多(淘宝上有几千万种商品)，用户身边充斥着大量信息，这时推荐系统就发挥了用武之地。推荐系统本质上是在用户需求不明确的情况下，从海量的信息中为用户寻找其感兴趣信息的技术手段。推荐系统结合用户的信息(地域、年龄、性别等)、物品信息(价格、产地等)，以及用户过去对物品的行为(是否购买、是否点击、是否播放等)，利用机器学习技术构建用户兴趣模型，为用户提供精准的个性化推荐。

推荐系统很好地满足了"标的物"提供方、平台方、用户三方的需求。例如，淘宝购物的"标的物"提供方是淘宝上成千上万的店主，平台方是淘宝，用户就是在淘宝上购物的自然人或企业。通过推荐系统可以更好地将商品推荐给需要购买的用户，提升社会资源的配置效率。

## 2.1.2 推荐系统的应用领域

推荐系统广泛用于各类互联网公司，尤其是那些存在大量的供用户消费的商品的互联网平台。具体来说，推荐系统的应用领域主要有如下几类。

- 电商网站：例如淘宝、京东、亚马逊等。
- 视频：例如 Netflix、优酷、抖音、快手、电视猫等。
- 音乐：例如网易云音乐、酷狗音乐等。
- 资讯类：例如今日头条、天天快报等。
- 生活服务类：例如美团、携程、脉脉等。
- 交友类：例如陌陌、珍爱网等。

## 2.1.3 推荐系统和搜索引擎

随着搜索引擎的快速发展，它们更倾向于响应那些具有明确目的的用户查询。用户可以将对信息的需求转化为具体的关键词，提交给搜索引擎。搜索引擎随后为用户提供一系列搜索结果的列表。用户可以对这些结果进行评估和选择，这是一个主动的、有目的性的搜索过程。

但搜索引擎也面临着"马太效应"的问题。这种现象意味着，那些已经受欢迎的内容在搜索过程中会因为不断的迭代而变得更加流行。相反，那些不太受欢迎的内容可能会逐渐消失在搜索结果的深海中，难以被用户发现。

推荐系统更倾向服务于那些没有明确目的或目标模糊的用户。换言之，当用户自己都不完全清楚想要什么时，正是推荐系统大显身手之际。推荐系统通过分析用户的历史行为、兴趣偏好以及人口统计学特征等信息，并将其输入推荐算法。接着，系统运用这些算法产生一份用户可能感兴趣的项目列表。

与用户对搜索引擎的主动使用不同，用户在使用推荐系统时通常是被动的。"长尾理论"指出，人们往往只关注那些曝光率高的项目，而忽略了那些曝光率低的项目。这一理论能够很好地解释推荐系统存在的意义。实际上，研究表明，那些位于长尾、曝光率较低的项目所创造的利润，并不逊色于只销售高曝光率项目所带来的收益。

推荐系统为所有项目提供了曝光的机会，从而挖掘出长尾项目的潜在盈利能力。这不仅增加了用户发现新内容的可能性，也为企业开拓了新的收入来源。

深度学习项目开发实践(TensorFlow+Sklearn+PyTorch)(微视频版)

## 2.2 项目介绍

在本项目中,提取过去几年在全球上映的电影信息,分别使用模型进行训练,并提取用户情感数据。然后使用 Flask 开发一个 Web 网站,提供一个搜索表单供用户检索自己感兴趣的电影信息。当用户输入电影名字中的一个单词时,会自动弹出推荐的电影名字。即使用户输入的单词存在错误,也会提供推荐信息。选择某个推荐信息后,会在新页面中显示这部电影的相关信息,包括用户对这部电影的评价信息。本项目的模块结构如图 2-1 所示。

扫码看视频

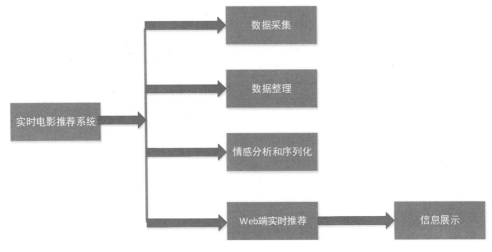

图 2-1 系统的模块结构

## 2.3 数据采集和整理

本项目使用了多个数据集文件,包含 IMDB 5000 电影数据集、2018 年电影列表、2019 年电影列表和 2020 年电影列表。在本节的内容中,将介绍整理数据并创建模型的知识。

扫码看视频

### 2.3.1 数据整理

编写文件 preprocessing 1.ipynb,基于数据集 movie_metadata.csv 整理其中的数据。文件 preprocessing 1.ipynb 的具体实现流程如下。

(1) 导入头文件和数据集文件，查看前 10 条数据。代码如下：

```
import pandas as pd
import numpy as np
data = pd.read_csv('movie_metadata.csv')
data.head(10)
```

执行后会输出数据集中的前 10 条数据，如图 2-2 所示。

| | color | director_name | num_critic_for_reviews | duration | director_facebook_likes | actor_3_facebook_likes | actor_2_name | actor_1_facebook_likes | gross |
|---|---|---|---|---|---|---|---|---|---|
| 0 | Color | James Cameron | 723.0 | 178.0 | 0.0 | 855.0 | Joel David Moore | 1000.0 | 7605058 |
| 1 | Color | Gore Verbinski | 302.0 | 169.0 | 563.0 | 1000.0 | Orlando Bloom | 40000.0 | 309404 |
| 2 | Color | Sam Mendes | 602.0 | 148.0 | 0.0 | 161.0 | Rory Kinnear | 11000.0 | 200074 |
| 3 | Color | Christopher Nolan | 813.0 | 164.0 | 22000.0 | 23000.0 | Christian Bale | 27000.0 | 4481306 |
| 4 | NaN | Doug Walker | NaN | NaN | 131.0 | NaN | Rob Walker | 131.0 | NaN |
| 5 | Color | Andrew Stanton | 462.0 | 132.0 | 475.0 | 530.0 | Samantha Morton | 640.0 | 7305867 |
| 6 | Color | Sam Raimi | 392.0 | 156.0 | 0.0 | 4000.0 | James Franco | 24000.0 | 3365303 |
| 7 | Color | Nathan Greno | 324.0 | 100.0 | 15.0 | 284.0 | Donna Murphy | 799.0 | 2008072 |
| 8 | Color | Joss Whedon | 635.0 | 141.0 | 0.0 | 19000.0 | Robert Downey Jr. | 26000.0 | 4589991 |
| 9 | Color | David Yates | 375.0 | 153.0 | 282.0 | 10000.0 | Daniel Radcliffe | 25000.0 | 3019565 |

图 2-2　输出前 10 条数据

(2) 查看数据集矩阵的长度。代码如下：

```
data.shape
```

执行后会输出：

```
(5043, 28)
```

(3) 返回数据集索引列表。代码如下：

```
data.columns
```

执行后会输出：

```
Index(['color', 'director_name', 'num_critic_for_reviews', 'duration',
       'director_facebook_likes', 'actor_3_facebook_likes', 'actor_2_name',
       'actor_1_facebook_likes', 'gross', 'genres', 'actor_1_name',
       'movie_title', 'num_voted_users', 'cast_total_facebook_likes',
       'actor_3_name', 'facenumber_in_poster', 'plot_keywords',
       'movie_imdb_link', 'num_user_for_reviews', 'language', 'country',
       'content_rating', 'budget', 'title_year', 'actor_2_facebook_likes',
       'imdb_score', 'aspect_ratio', 'movie_facebook_likes'],
      dtype='object')
```

(4) 统计近年来的电影数量。代码如下：

```
import matplotlib.pyplot as plt
data.title_year.value_counts(dropna=False).sort_index().plot(kind='barh',figsize=
(15,16))
plt.show()
```

执行效果如图 2-3 所示。由此可见，最早的电影数据是 1916 年。

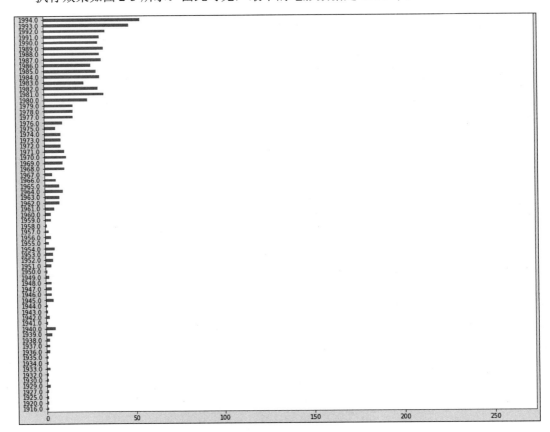

图 2-3　近年来的电影数量

(5) 查看数据集中的前 10 条数据，只提取其中的几个字段。代码如下：

```
data = data.loc[:,['director_name',
'actor_1_name','actor_2_name','actor_3_name',
'genres','movie_title']]
```

执行后会输出：

| | director_name | actor_1_name | actor_2_name | actor_3_name | genres | movie_title |
|---|---|---|---|---|---|---|
| 0 | James Cameron | CCH Pounder | Joel David Moore | Wes Studi | Action\|Adventure\|Fantasy\|Sci-Fi | Avatar |
| 1 | Gore Verbinski | Johnny Depp | Orlando Bloom | Jack Davenport | Action\|Adventure\|Fantasy | Pirates of the Caribbean: At World's End |
| 2 | Sam Mendes | Christoph Waltz | Rory Kinnear | Stephanie Sigman | Action\|Adventure\|Thriller | Spectre |
| 3 | Christopher Nolan | Tom Hardy | Christian Bale | Joseph Gordon-Levitt | Action\|Thriller | The Dark Knight Rises |
| 4 | Doug Walker | Doug Walker | Rob Walker | NaN | Documentary | Star Wars: Episode VII - The Force Awakens … |
| 5 | Andrew Stanton | Daryl Sabara | Samantha Morton | Polly Walker | Action\|Adventure\|Sci-Fi | John Carter |
| 6 | Sam Raimi | J.K. Simmons | James Franco | Kirsten Dunst | Action\|Adventure\|Romance | Spider-Man 3 |
| 7 | Nathan Greno | Brad Garrett | Donna Murphy | M.C. Gainey | Adventure\|Animation\|Comedy\|Family\|Fantasy\|Musi… | Tangled |
| 8 | Joss Whedon | Chris Hemsworth | Robert Downey Jr. | Scarlett Johansson | Action\|Adventure\|Sci-Fi | Avengers: Age of Ultron |
| 9 | David Yates | Alan Rickman | Daniel Radcliffe | Rupert Grint | Adventure\|Family\|Fantasy\|Mystery | Harry Potter and the Half-Blood Prince |

(6) 如果数据集中的某个值为空，则替换为 unknown。代码如下：

```
data['actor_1_name'] = data['actor_1_name'].replace(np.nan, 'unknown')
data['actor_2_name'] = data['actor_2_name'].replace(np.nan, 'unknown')
data['actor_3_name'] = data['actor_3_name'].replace(np.nan, 'unknown')
data['director_name'] = data['director_name'].replace(np.nan, 'unknown')
data
```

执行后会输出：

| | director_name | actor_1_name | actor_2_name | actor_3_name | genres | movie_title |
|---|---|---|---|---|---|---|
| 0 | James Cameron | CCH Pounder | Joel David Moore | Wes Studi | Action\|Adventure\|Fantasy\|Sci-Fi | Avatar |
| 1 | Gore Verbinski | Johnny Depp | Orlando Bloom | Jack Davenport | Action\|Adventure\|Fantasy | Pirates of the Caribbean: At World's End |
| 2 | Sam Mendes | Christoph Waltz | Rory Kinnear | Stephanie Sigman | Action\|Adventure\|Thriller | Spectre |
| 3 | Christopher Nolan | Tom Hardy | Christian Bale | Joseph Gordon-Levitt | Action\|Thriller | The Dark Knight Rises |
| 4 | Doug Walker | Doug Walker | Rob Walker | unknown | Documentary | Star Wars: Episode VII - The Force Awakens … |
| … | … | … | … | … | … | … |
| 5038 | Scott Smith | Eric Mabius | Daphne Zuniga | Crystal Lowe | Comedy\|Drama | Signed Sealed Delivered |
| 5039 | unknown | Natalie Zea | Valorie Curry | Sam Underwood | Crime\|Drama\|Mystery\|Thriller | The Following |
| 5040 | Benjamin Roberds | Eva Boehnke | Maxwell Moody | David Chandler | Drama\|Horror\|Thriller | A Plague So Pleasant |
| 5041 | Daniel Hsia | Alan Ruck | Daniel Henney | Eliza Coupe | Comedy\|Drama\|Romance | Shanghai Calling |
| 5042 | Jon Gunn | John August | Brian Herzlinger | Jon Gunn | Documentary | My Date with Drew |

5043 rows × 6 columns

(7) 将 genres 列中的 "|" 替换为空格。代码如下：

```
data['genres'] = data['genres'].str.replace('|', ' ')
data
```

执行后会输出：

| | director_name | actor_1_name | actor_2_name | actor_3_name | genres | movie_title |
|---|---|---|---|---|---|---|
| 0 | James Cameron | CCH Pounder | Joel David Moore | Wes Studi | Action Adventure Fantasy Sci-Fi | Avatar |
| 1 | Gore Verbinski | Johnny Depp | Orlando Bloom | Jack Davenport | Action Adventure Fantasy | Pirates of the Caribbean: At World's End |
| 2 | Sam Mendes | Christoph Waltz | Rory Kinnear | Stephanie Sigman | Action Adventure Thriller | Spectre |
| 3 | Christopher Nolan | Tom Hardy | Christian Bale | Joseph Gordon-Levitt | Action Thriller | The Dark Knight Rises |
| 4 | Doug Walker | Doug Walker | Rob Walker | unknown | Documentary | Star Wars: Episode VII - The Force Awakens |
| ... | ... | ... | ... | ... | ... | ... |
| 5038 | Scott Smith | Eric Mabius | Daphne Zuniga | Crystal Lowe | Comedy Drama | Signed Sealed Delivered |
| 5039 | unknown | Natalie Zea | Valorie Curry | Sam Underwood | Crime Drama Mystery Thriller | The Following |
| 5040 | Benjamin Roberts | Eva Boehnke | Maxwell Moody | David Chandler | Drama Horror Thriller | A Plague So Pleasant |
| 5041 | Daniel Hsia | Alan Ruck | Daniel Henney | Eliza Coupe | Comedy Drama Romance | Shanghai Calling |
| 5042 | Jon Gunn | John August | Brian Herzlinger | Jon Gunn | Documentary | My Date with Drew |

5043 rows × 6 columns

(8) 将 movie_title 列中的数据变成小写。代码如下：

```
data['movie_title'] = data['movie_title'].str.lower()
data['movie_title'][1]
```

执行后会输出：

```
"pirates of the caribbean: at world's end\xa0"
```

(9) 删除 movie_title 结尾处的 null 终止字符。代码如下：

```
data['movie_title'] = data['movie_title'].apply(lambda x : x[:-1])
data['movie_title'][1]
```

执行后会输出：

```
"pirates of the caribbean: at world's end"
```

(10) 最后保存数据。代码如下：

```
data.to_csv('data.csv',index=False)
```

## 2.3.2 电影详情数据

编写文件 preprocessing 2.ipynb，基于数据集文件 credits.csv 和 movies_metadata.csv 获取电影信息的详细数据。文件 preprocessing 2.ipynb 的具体实现流程如下。

(1) 读取数据集文件 credits.csv 中的数据。代码如下：

```
credits = pd.read_csv('credits.csv')
credits
```

执行后会输出：

| | cast | crew | id |
|---|---|---|---|
| 0 | [{'cast_id': 14, 'character': 'Woody (voice)', 'credit_id': '52fe4284c3... | [{'credit_id': '52fe4284c3a36847f8024f49', 'department': 'Directing', '... | 862 |
| 1 | [{'cast_id': 1, 'character': 'Alan Parrish', 'credit_id': '52fe44bfc3a3... | [{'credit_id': '52fe44bfc3a36847f80a7cd1', 'department': 'Production', '... | 8844 |
| 2 | [{'cast_id': 2, 'character': 'Max Goldman', 'credit_id': '52fe466a92514... | [{'credit_id': '52fe466a9251416c75077a89', 'department': 'Directing', '... | 15602 |
| 3 | [{'cast_id': 1, 'character': '"Savannah 'Vannah' Jackson"', 'credit_id': ... | [{'credit_id': '52fe44779251416c91011acb', 'department': 'Directing', '... | 31357 |
| 4 | [{'cast_id': 1, 'character': 'George Banks', 'credit_id': '52fe44959251... | [{'credit_id': '52fe44959251416c75039ed7', 'department': 'Sound', 'gend... | 11862 |
| ... | ... | ... | ... |
| 45471 | [{'cast_id': 0, 'character': '', 'credit_id': '5894a909925141427e0079a5... | [{'credit_id': '5894a97d925141426c00818c', 'department': 'Directing', '... | 439050 |
| 45472 | [{'cast_id': 1002, 'character': 'Sister Angela', 'credit_id': '52fe4af1... | [{'credit_id': '52fe4af1c3a36847f81e9b15', 'department': 'Directing', '... | 111109 |
| 45473 | [{'cast_id': 6, 'character': 'Emily Shaw', 'credit_id': '52fe4776c3a368... | [{'credit_id': '52fe4776c3a368484e0c8387', 'department': 'Directing', '... | 67758 |
| 45474 | [{'cast_id': 2, 'character': '', 'credit_id': '52fe4ea59251416c7515d7d5... | [{'credit_id': '533bccebc3a36844cf0011a7', 'department': 'Directing', '... | 227506 |
| 45475 | [] | [{'credit_id': '593e676c92514105b702e68e', 'department': 'Directing', '... | 461257 |

45476 rows × 3 columns

(2) 读取数据集文件 movies_metadata.csv 中的内容，然后根据年份统计信息。代码如下：

```
meta = pd.read_csv('movies_metadata.csv')
meta['release_date'] = pd.to_datetime(meta['release_date'], errors='coerce')
meta['year'] = meta['release_date'].dt.year


meta['year'].value_counts().sort_index()
```

执行后会输出：

```
1874.0       1
1878.0       1
1883.0       1
1887.0       1
1888.0       2
            ...
2015.0    1905
2016.0    1604
2017.0     532
2018.0       5
2020.0       1
Name: year, Length: 135, dtype: int64
```

(3) 因为在数据集中没有足够的 2018—2020 年的电影数据，所以只能获得 2017 年之前的电影信息。通过如下代码，预处理文件中 2017 年及以前年份的电影数据。

```
new_meta = meta.loc[meta.year <= 2017,['genres','id','title','year']]
new_meta
```

执行后会输出：

| | genres | id | title | year |
|---|---|---|---|---|
| 0 | [{'id': 16, 'name': 'Animation'}, {'id': 35, 'name': 'Comedy'}, {'id': ... | 862 | Toy Story | 1995.0 |
| 1 | [{'id': 12, 'name': 'Adventure'}, {'id': 14, 'name': 'Fantasy'}, {'id': ... | 8844 | Jumanji | 1995.0 |
| 2 | [{'id': 10749, 'name': 'Romance'}, {'id': 35, 'name': 'Comedy'}] | 15602 | Grumpier Old Men | 1995.0 |
| 3 | [{'id': 35, 'name': 'Comedy'}, {'id': 18, 'name': 'Drama'}, {'id': 1074... | 31357 | Waiting to Exhale | 1995.0 |
| 4 | [{'id': 35, 'name': 'Comedy'}] | 11862 | Father of the Bride Part II | 1995.0 |
| ... | ... | ... | ... | ... |
| 45460 | [{'id': 18, 'name': 'Drama'}, {'id': 28, 'name': 'Action'}, {'id': 1074... | 30840 | Robin Hood | 1991.0 |
| 45462 | [{'id': 18, 'name': 'Drama'}] | 111109 | Century of Birthing | 2011.0 |
| 45463 | [{'id': 28, 'name': 'Action'}, {'id': 18, 'name': 'Drama'}, {'id': 53, ... | 67758 | Betrayal | 2003.0 |
| 45464 | [] | 227506 | Satan Triumphant | 1917.0 |
| 45465 | [] | 461257 | Queerama | 2017.0 |

45370 rows × 4 columns

(4) 在数据中添加 cast 和 crew 两列。代码如下：

```
new_meta['id'] = new_meta['id'].astype(int)
data = pd.merge(new_meta, credits, on='id')

pd.set_option('display.max_colwidth', 75)
data
```

执行后会输出：

| | genres | id | title | year | cast | crew |
|---|---|---|---|---|---|---|
| 0 | [{'id': 16, 'name': 'Animation'}, {'id': 35, 'name': 'Comedy'}, {'id': ... | 862 | Toy Story | 1995.0 | [{'cast_id': 14, 'character': 'Woody (voice)', 'credit_id': '52fe4284c3... | [{'credit_id': '52fe4284c3a36847f8024f49', 'department': 'Directing', '... |
| 1 | [{'id': 12, 'name': 'Adventure'}, {'id': 14, 'name': 'Fantasy'}, {'id': ... | 8844 | Jumanji | 1995.0 | [{'cast_id': 1, 'character': 'Alan Parrish', 'credit_id': '52fe44bfc3a3... | [{'credit_id': '52fe44bfc3a36847f80a7cd1', 'department': 'Production', ... |
| 2 | [{'id': 10749, 'name': 'Romance'}, {'id': 35, 'name': 'Comedy'}] | 15602 | Grumpier Old Men | 1995.0 | [{'cast_id': 2, 'character': 'Max Goldman', 'credit_id': '52fe466a92514... | [{'credit_id': '52fe466a9251416c75077a89', 'department': 'Directing', '... |
| 3 | [{'id': 35, 'name': 'Comedy'}, {'id': 18, 'name': 'Drama'}, {'id': 1074... | 31357 | Waiting to Exhale | 1995.0 | [{'cast_id': 1, 'character': "Savannah 'Vannah' Jackson", 'credit_id': ... | [{'credit_id': '52fe44779251416c91011acb', 'department': 'Directing', '... |
| 4 | [{'id': 35, 'name': 'Comedy'}] | 11862 | Father of the Bride Part II | 1995.0 | [{'cast_id': 1, 'character': 'George Banks', 'credit_id': '52fe44959251... | [{'credit_id': '52fe44959251416c75039ed7', 'department': 'Sound', 'gend... |
| ... | ... | ... | ... | ... | ... | ... |
| 45440 | [{'id': 18, 'name': 'Drama'}, {'id': 28, 'name': 'Action'}, {'id': 1074... | 30840 | Robin Hood | 1991.0 | [{'cast_id': 1, 'character': 'Sir Robert Hode', 'credit_id': '52fe44439... | [{'credit_id': '52fe44439251416c9100a899', 'department': 'Directing', '... |
| 45441 | [{'id': 18, 'name': 'Drama'}] | 111109 | Century of Birthing | 2011.0 | [{'cast_id': 1002, 'character': 'Sister Angela', 'credit_id': '52fe4af1... | [{'credit_id': '52fe4af1c3a36847f81e9b15', 'department': 'Directing', '... |
| 45442 | [{'id': 28, 'name': 'Action'}, {'id': 18, 'name': 'Drama'}, {'id': 53, ... | 67758 | Betrayal | 2003.0 | [{'cast_id': 6, 'character': 'Emily Shaw', 'credit_id': '52fe4776c3a368... | [{'credit_id': '52fe4776c3a368484e0c8387', 'department': 'Directing', '... |
| 45443 | [] | 227506 | Satan Triumphant | 1917.0 | [{'cast_id': 2, 'character': '', 'credit_id': '52fe4ea59251416c7515d7d5... | [{'credit_id': '533bccebc3a36844cf0011a7', 'department': 'Directing', '... |
| 45444 | [] | 461257 | Queerama | 2017.0 | [] | [{'credit_id': '593e676c92514105b702e68e', 'department': 'Directing', '... |

(5) 计算表达式节点或包含 Python 文本或容器显示的字符串,通过函数 make_genresList() 统计电影的类型,代码如下:

```
import ast
data['genres'] = data['genres'].map(lambda x: ast.literal_eval(x))
data['cast'] = data['cast'].map(lambda x: ast.literal_eval(x))
data['crew'] = data['crew'].map(lambda x: ast.literal_eval(x))

def make_genresList(x):
    gen = []
    st = " "
    for i in x:
        if i.get('name') == 'Science Fiction':
            scifi = 'Sci-Fi'
            gen.append(scifi)
        else:
            gen.append(i.get('name'))
    if gen == []:
        return np.NaN
    else:
        return (st.join(gen))

data['genres_list'] = data['genres'].map(lambda x: make_genresList(x))

data['genres_list']
```

执行后会输出:

```
0           Animation Comedy Family
1           Adventure Fantasy Family
2                    Romance Comedy
3              Comedy Drama Romance
4                            Comedy
                   ...
45440           Drama Action Romance
45441                          Drama
45442          Action Drama Thriller
45443                            NaN
45444                            NaN
Name: genres_list, Length: 45445, dtype: object
```

(6) 编写自定义函数 get_actor1() 和 get_actor2() 获取演员 1 和演员 2 的信息。代码如下:

```
def get_actor1(x):
    casts = []
    for i in x:
        casts.append(i.get('name'))
    if casts == []:
        return np.NaN
```

```
        else:
            return (casts[0])

data['actor_1_name'] = data['cast'].map(lambda x: get_actor1(x))

def get_actor2(x):
    casts = []
    for i in x:
        casts.append(i.get('name'))
    if casts == [] or len(casts)<=1:
        return np.NaN
    else:
        return (casts[1])data['actor_2_name'] = data['cast'].map(lambda x: get_actor2(x))

data['actor_2_name'] = data['cast'].map(lambda x: get_actor2(x))

data['actor_2_name']
```

执行后会输出:

```
0             Tim Allen
1         Jonathan Hyde
2          Jack Lemmon
3       Angela Bassett
4         Diane Keaton
              ...
45440       Uma Thurman
45441       Perry Dizon
45442      Adam Baldwin
45443  Nathalie Lissenko
45444               NaN
Name: actor_2_name, Length: 45445, dtype: object
```

(7) 编写自定义函数 get_actor3()获取演员 3 的信息。代码如下:

```
def get_actor3(x):
    casts = []
    for i in x:
        casts.append(i.get('name'))
    if casts == [] or len(casts)<=2:
        return np.NaN
    else:
        return (casts[2])

data['actor_3_name'] = data['cast'].map(lambda x: get_actor3(x))

data['actor_3_name']
```

执行后会输出:

```
0            Don Rickles
1          Kirsten Dunst
2           Ann-Margret
3         Loretta Devine
4           Martin Short
                ...
45440       David Morrissey
45441        Hazel Orencio
45442       Julie du Page
45443        Pavel Pavlov
45444                 NaN
Name: actor_3_name, Length: 45445, dtype: object
```

(8) 编写自定义函数 get_directors()获取导演信息。代码如下：

```
def get_directors(x):
    dt = []
    st = " "
    for i in x:
        if i.get('job') == 'Director':
            dt.append(i.get('name'))
    if dt == []:
        return np.NaN
    else:
        return (st.join(dt))

data['director_name'] = data['crew'].map(lambda x: get_directors(x))

data['director_name']
```

执行后会输出：

```
0           John Lasseter
1            Joe Johnston
2           Howard Deutch
3         Forest Whitaker
4           Charles Shyer
                ...
45440          John Irvin
45441           Lav Diaz
45442       Mark L. Lester
45443      Yakov Protazanov
45444        Daisy Asquith
Name: director_name, Length: 45445, dtype: object
```

(9) 分别获取数据集中 director_name、actor_1_name、actor_2_name、actor_3_name、genres_list 和 title 列的信息。代码如下：

```
movie = data.loc[:,['director_name','actor_1_name','actor_2_name',
'actor_3_name','genres_list','title']]
movie
```

执行后会输出：

| | director_name | actor_1_name | actor_2_name | actor_3_name | genres_list | title |
|---|---|---|---|---|---|---|
| 0 | John Lasseter | Tom Hanks | Tim Allen | Don Rickles | Animation Comedy Family | Toy Story |
| 1 | Joe Johnston | Robin Williams | Jonathan Hyde | Kirsten Dunst | Adventure Fantasy Family | Jumanji |
| 2 | Howard Deutch | Walter Matthau | Jack Lemmon | Ann-Margret | Romance Comedy | Grumpier Old Men |
| 3 | Forest Whitaker | Whitney Houston | Angela Bassett | Loretta Devine | Comedy Drama Romance | Waiting to Exhale |
| 4 | Charles Shyer | Steve Martin | Diane Keaton | Martin Short | Comedy | Father of the Bride Part II |
| ... | ... | ... | ... | ... | ... | ... |
| 45440 | John Irvin | Patrick Bergin | Uma Thurman | David Morrissey | Drama Action Romance | Robin Hood |
| 45441 | Lav Diaz | Angel Aquino | Perry Dizon | Hazel Orencio | Drama | Century of Birthing |
| 45442 | Mark L. Lester | Erika Eleniak | Adam Baldwin | Julie du Page | Action Drama Thriller | Betrayal |
| 45443 | Yakov Protazanov | Iwan Mosschuchin | Nathalie Lissenko | Pavel Pavlov | NaN | Satan Triumphant |
| 45444 | Daisy Asquith | NaN | NaN | NaN | NaN | Queerama |

45445 rows × 6 columns

(10) 统计数据集中的数据数目。代码如下：

```
movie.isna().sum()
```

执行后会输出：

```
director_name    835
actor_1_name     2354
actor_2_name     3683
actor_3_name     4593
genres_list      2384
title            0
dtype: int64
```

(11) 将 movie_title 列中的数据改为小写，然后打印输出定制的信息。代码如下：

```
movie = movie.rename(columns={'genres_list':'genres'})
movie = movie.rename(columns={'title':'movie_title'})

movie['movie_title'] = movie['movie_title'].str.lower()

movie['comb'] = movie['actor_1_name'] + ' ' + movie['actor_2_name'] + ' '+
movie['actor_3_name'] + ' '+ movie['director_name'] +' ' + movie['genres']

movie
```

执行后会输出：

| | director_name | actor_1_name | actor_2_name | actor_3_name | genres | movie_title | comb |
|---|---|---|---|---|---|---|---|
| 0 | John Lasseter | Tom Hanks | Tim Allen | Don Rickles | Animation Comedy Family | toy story | Tom Hanks Tim Allen Don Rickles John Lasseter Animation Comedy Family |
| 1 | Joe Johnston | Robin Williams | Jonathan Hyde | Kirsten Dunst | Adventure Fantasy Family | jumanji | Robin Williams Jonathan Hyde Kirsten Dunst Joe Johnston Adventure Fanta... |
| 2 | Howard Deutch | Walter Matthau | Jack Lemmon | Ann-Margret | Romance Comedy | grumpier old men | Walter Matthau Jack Lemmon Ann-Margret Howard Deutch Romance Comedy |
| 3 | Forest Whitaker | Whitney Houston | Angela Bassett | Loretta Devine | Comedy Drama Romance | waiting to exhale | Whitney Houston Angela Bassett Loretta Devine Forest Whitaker Comedy Dr... |
| 4 | Charles Shyer | Steve Martin | Diane Keaton | Martin Short | Comedy | father of the bride part ii | Steve Martin Diane Keaton Martin Short Charles Shyer Comedy |
| ... | ... | ... | ... | ... | ... | ... | ... |
| 45438 | Ben Rock | Monty Bane | Lucy Butler | David Grammer | Horror | the burkittsville 7 | Monty Bane Lucy Butler David Grammer Ben Rock Horror |
| 45439 | Aaron Osborne | Lisa Boyle | Kena Land | Zaneta Polard | Sci-Fi | caged heat 3000 | Lisa Boyle Kena Land Zaneta Polard Aaron Osborne Sci-Fi |
| 45440 | John Irvin | Patrick Bergin | Uma Thurman | David Morrissey | Drama Action Romance | robin hood | Patrick Bergin Uma Thurman David Morrissey John Irvin Drama Action Romance |
| 45441 | Lav Diaz | Angel Aquino | Perry Dizon | Hazel Orencio | Drama | century of birthing | Angel Aquino Perry Dizon Hazel Orencio Lav Diaz Drama |
| 45442 | Mark L. Lester | Erika Eleniak | Adam Baldwin | Julie du Page | Action Drama Thriller | betrayal | Erika Eleniak Adam Baldwin Julie du Page Mark L. Lester Action Drama Th... |

39201 rows × 7 columns

（12）使用函数 drop_duplicates()，根据 movie_title 列实现去重处理。代码如下：

```
movie.drop_duplicates(subset ="movie_title", keep = 'last', inplace = True)
movie
```

执行后会输出：

| | director_name | actor_1_name | actor_2_name | actor_3_name | genres | movie_title | comb |
|---|---|---|---|---|---|---|---|
| 0 | John Lasseter | Tom Hanks | Tim Allen | Don Rickles | Animation Comedy Family | toy story | Tom Hanks Tim Allen Don Rickles John Lasseter Animation Comedy Family |
| 1 | Joe Johnston | Robin Williams | Jonathan Hyde | Kirsten Dunst | Adventure Fantasy Family | jumanji | Robin Williams Jonathan Hyde Kirsten Dunst Joe Johnston Adventure Fanta... |
| 2 | Howard Deutch | Walter Matthau | Jack Lemmon | Ann-Margret | Romance Comedy | grumpier old men | Walter Matthau Jack Lemmon Ann-Margret Howard Deutch Romance Comedy |
| 3 | Forest Whitaker | Whitney Houston | Angela Bassett | Loretta Devine | Comedy Drama Romance | waiting to exhale | Whitney Houston Angela Bassett Loretta Devine Forest Whitaker Comedy Dr... |
| 4 | Charles Shyer | Steve Martin | Diane Keaton | Martin Short | Comedy | father of the bride part ii | Steve Martin Diane Keaton Martin Short Charles Shyer Comedy |
| ... | ... | ... | ... | ... | ... | ... | ... |
| 45438 | Ben Rock | Monty Bane | Lucy Butler | David Grammer | Horror | the burkittsville 7 | Monty Bane Lucy Butler David Grammer Ben Rock Horror |
| 45439 | Aaron Osborne | Lisa Boyle | Kena Land | Zaneta Polard | Sci-Fi | caged heat 3000 | Lisa Boyle Kena Land Zaneta Polard Aaron Osborne Sci-Fi |
| 45440 | John Irvin | Patrick Bergin | Uma Thurman | David Morrissey | Drama Action Romance | robin hood | Patrick Bergin Uma Thurman David Morrissey John Irvin Drama Action Romance |
| 45441 | Lav Diaz | Angel Aquino | Perry Dizon | Hazel Orencio | Drama | century of birthing | Angel Aquino Perry Dizon Hazel Orencio Lav Diaz Drama |
| 45442 | Mark L. Lester | Erika Eleniak | Adam Baldwin | Julie du Page | Action Drama Thriller | betrayal | Erika Eleniak Adam Baldwin Julie du Page Mark L. Lester Action Drama Th... |

36341 rows × 7 columns

## 2.3.3 提取电影特征

编写文件 preprocessing 3.ipynb，功能是基于 2018 年电影数据的特征。文件 preprocessing

3.ipynb 的具体实现流程如下。

(1) 设置要读取的数据信息的 URL，然后读取并显示数据。代码如下：

```
df1 = pd.read_html(link, header=0)[2]
df2 = pd.read_html(link, header=0)[3]
df3 = pd.read_html(link, header=0)[4]
df4 = pd.read_html(link, header=0)[5]

df = df1.append(df2.append(df3.append(df4,ignore_index=True),ignore_index=True),ignore_index=True)

df
```

执行后会输出：

| | Opening | Opening.1 | Title | Production company | Cast and crew | Ref. |
|---|---|---|---|---|---|---|
| 0 | JANUARY | 5 | Insidious: The Last Key | Universal Pictures / Blumhouse Productions / S... | Adam Robitel (director); Leigh Whannell (scree... | [2] |
| 1 | JANUARY | 5 | The Strange Ones | Vertical Entertainment | Lauren Wolkstein (director); Christopher Radcl... | [3] |
| 2 | JANUARY | 5 | Stratton | Momentum Pictures | Simon West (director); Duncan Falconer, Warren... | [4] |
| 3 | JANUARY | 10 | Sweet Country | Samuel Goldwyn Films | Warwick Thornton (director); David Tranter, St... | [5] |
| 4 | JANUARY | 12 | The Commuter | Lionsgate / StudioCanal / The Picture Company | Jaume Collet-Serra (director); Byron Willinger... | [6] |
| ... | ... | ... | ... | ... | ... | ... |
| 263 | DECEMBER | 25 | Holmes & Watson | Columbia Pictures / Gary Sanchez Productions | Etan Cohen (director/screenplay); Will Ferrell... | [162] |
| 264 | DECEMBER | 25 | Vice | Annapurna Pictures / Plan B Entertainment | Adam McKay (director/screenplay); Christian Ba... | [136] |
| 265 | DECEMBER | 25 | On the Basis of Sex | Focus Features | Mimi Leder (director); Daniel Stiepleman (scre... | [223] |
| 266 | DECEMBER | 25 | Destroyer | Annapurna Pictures | Karyn Kusama (director); Phil Hay, Matt Manfre... | [256] |
| 267 | DECEMBER | 28 | Black Mirror: Bandersnatch | Netflix | David Slade (director); Charlie Brooker (scree... | [257] |

268 rows × 6 columns

(2) 登录 themoviedb 官网，进行用户注册，然后申请一个 API 密钥，如图 2-4 所示。

图 2-4　themoviedb API 密钥

(3) 编写自定义函数 get_genre()获取 themoviedb 网站中的电影信息，在此需要用到 themoviedb API 密钥。代码如下：

```python
from tmdbv3api import TMDb
import json
import requests
tmdb = TMDb()
tmdb.api_key = 'YOUR_API_KEY'

from tmdbv3api import Movie
tmdb_movie = Movie()
def get_genre(x):
    genres = []
    result = tmdb_movie.search(x)
    movie_id = result[0].id
    response = requests.get('https://api.themoviedb.org/3/movie/{}?api_key={}'.format(movie_id,tmdb.api_key))
    data_json = response.json()
    if data_json['genres']:
        genre_str = " " 
        for i in range(0,len(data_json['genres'])):
            genres.append(data_json['genres'][i]['name'])
        return genre_str.join(genres)
    else:
        np.NaN

df['genres'] = df['Title'].map(lambda x: get_genre(str(x)))

df
```

执行后会输出：

| | Opening | Opening.1 | Title | Production company | Cast and crew | Ref. | genres |
|---|---|---|---|---|---|---|---|
| 0 | JANUARY | 5 | Insidious: The Last Key | Universal Pictures / Blumhouse Productions / S... | Adam Robitel (director); Leigh Whannell (scree... | [2] | Mystery Horror Thriller |
| 1 | JANUARY | 5 | The Strange Ones | Vertical Entertainment | Lauren Wolkstein (director); Christopher Radcl... | [3] | Thriller Drama |
| 2 | JANUARY | 5 | Stratton | Momentum Pictures | Simon West (director); Duncan Falconer, Warren... | [4] | Action Thriller |
| 3 | JANUARY | 10 | Sweet Country | Samuel Goldwyn Films | Warwick Thornton (director); David Tranter, St... | [5] | Drama History Western |
| 4 | JANUARY | 12 | The Commuter | Lionsgate / StudioCanal / The Picture Company | Jaume Collet-Serra (director); Byron Willinger... | [6] | Action Thriller |
| ... | ... | ... | ... | ... | ... | ... | ... |
| 263 | DECEMBER | 25 | Holmes & Watson | Columbia Pictures / Gary Sanchez Productions | Etan Cohen (director/screenplay); Will Ferrell... | [162] | Mystery Adventure Comedy Crime |
| 264 | DECEMBER | 25 | Vice | Annapurna Pictures / Plan B Entertainment | Adam McKay (director/screenplay); Christian Ba... | [136] | Thriller Science Fiction Action Adventure |
| 265 | DECEMBER | 25 | On the Basis of Sex | Focus Features | Mimi Leder (director); Daniel Stiepleman (scre... | [223] | Drama History |
| 266 | DECEMBER | 25 | Destroyer | Annapurna Pictures | Karyn Kusama (director); Phil Hay, Matt Manfre... | [256] | Thriller Crime Drama Action |
| 267 | DECEMBER | 28 | Black Mirror: Bandersnatch | Netflix | David Slade (director); Charlie Brooker (scree... | [257] | Science Fiction Mystery Drama Thriller TV Movie |

268 rows × 7 columns

(4) 只展示 Title、Cast and crew 和 genres 列中的内容。代码如下：

```
df_2018 = df[['Title','Cast and crew','genres']]
df_2018
```

执行后会输出：

|   | Title | Cast and crew | genres |
|---|---|---|---|
| 0 | Insidious: The Last Key | Adam Robitel (director); Leigh Whannell (scree... | Mystery Horror Thriller |
| 1 | The Strange Ones | Lauren Wolkstein (director); Christopher Radcl... | Thriller Drama |
| 2 | Stratton | Simon West (director); Duncan Falconer, Warren... | Action Thriller |
| 3 | Sweet Country | Warwick Thornton (director); David Tranter, St... | Drama History Western |
| 4 | The Commuter | Jaume Collet-Serra (director); Byron Willinger... | Action Thriller |
| ... | ... | ... | ... |
| 263 | Holmes & Watson | Etan Cohen (director/screenplay); Will Ferrell... | Mystery Adventure Comedy Crime |
| 264 | Vice | Adam McKay (director/screenplay); Christian Ba... | Thriller Science Fiction Action Adventure |
| 265 | On the Basis of Sex | Mimi Leder (director); Daniel Stiepleman (scre... | Drama History |
| 266 | Destroyer | Karyn Kusama (director); Phil Hay, Matt Manfre... | Thriller Crime Drama Action |
| 267 | Black Mirror: Bandersnatch | David Slade (director); Charlie Brooker (scree... | Science Fiction Mystery Drama Thriller TV Movie |

268 rows × 3 columns

(5) 分别编写自定义函数获取导演和演员的信息。代码如下：

```
def get_director(x):
    if " (director)" in x:
        return x.split(" (director)")[0]
    elif " (directors)" in x:
        return x.split(" (directors)")[0]
    else:
        return x.split(" (director/screenplay)")[0]

df_2018['director_name'] = df_2018['Cast and crew'].map(lambda x: get_director(x))

def get_actor1(x):
    return ((x.split("screenplay); ")[-1]).split(", ")[0])

df_2018['actor_1_name'] = df_2018['Cast and crew'].map(lambda x: get_actor1(x))

def get_actor2(x):
    if len((x.split("screenplay); ")[-1]).split(", ")) < 2:
        return np.NaN
    else:
        return ((x.split("screenplay); ")[-1]).split(", ")[1])
df_2018['actor_2_name'] = df_2018['Cast and crew'].map(lambda x: get_actor2(x))

def get_actor3(x):
    if len((x.split("screenplay); ")[-1]).split(", ")) < 3:
        return np.NaN
    else:
        return ((x.split("screenplay); ")[-1]).split(", ")[2])
```

```
df_2018['actor_3_name'] = df_2018['Cast and crew'].map(lambda x: get_actor3(x))
df_2018
```

执行后会输出：

| | Title | Cast and crew | genres | director_name | actor_1_name | actor_2_name | actor_3_name |
|---|---|---|---|---|---|---|---|
| 0 | Insidious: The Last Key | Adam Robitel (director); Leigh Whannell (scree... | Mystery Horror Thriller | Adam Robitel | Lin Shaye | Angus Sampson | Leigh Whannell |
| 1 | The Strange Ones | Lauren Wolkstein (director); Christopher Radcl... | Thriller Drama | Lauren Wolkstein | Alex Pettyfer | James Freedson-Jackson | Emily Althaus |
| 2 | Stratton | Simon West (director); Duncan Falconer, Warren... | Action Thriller | Simon West | Dominic Cooper | Austin Stowell | Gemma Chan |
| 3 | Sweet Country | Warwick Thornton (director); David Tranter, St... | Drama History Western | Warwick Thornton | Bryan Brown | Sam Neill | NaN |
| 4 | The Commuter | Jaume Collet-Serra (director); Byron Willinger... | Action Thriller | Jaume Collet-Serra | Liam Neeson | Vera Farmiga | Patrick Wilson |
| ... | ... | ... | ... | ... | ... | ... | ... |
| 263 | Holmes & Watson | Etan Cohen (director/screenplay); Will Ferrell... | Mystery Adventure Comedy Crime | Etan Cohen | Will Ferrell | John C. Reilly | Rebecca Hall |
| 264 | Vice | Adam McKay (director/screenplay); Christian Ba... | Thriller Science Fiction Action Adventure | Adam McKay | Christian Bale | Amy Adams | Steve Carell |
| 265 | On the Basis of Sex | Mimi Leder (director); Daniel Stiepleman (scre... | Drama History | Mimi Leder | Felicity Jones | Armie Hammer | Justin Theroux |
| 266 | Destroyer | Karyn Kusama (director); Phil Hay, Matt Manfre... | Thriller Crime Drama Action | Karyn Kusama | Nicole Kidman | Sebastian Stan | Toby Kebbell |
| 267 | Black Mirror: Bandersnatch | David Slade (director); Charlie Brooker (scree... | Science Fiction Mystery Drama Thriller TV Movie | David Slade | Fionn Whitehead | Will Poulter | Asim Chaudhry |

268 rows × 7 columns

(6) 将 Title 列重命名为 movie_title，然后获取指定列的电影信息。代码如下：

```
df_2018 = df_2018.rename(columns={'Title':'movie_title'})
new_df18 = df_2018.loc[:,['director_name','actor_1_name','actor_2_name',
'actor_3_name','genres','movie_title']]
new_df18
```

执行后会输出：

| | director_name | actor_1_name | actor_2_name | actor_3_name | genres | movie_title |
|---|---|---|---|---|---|---|
| 0 | Adam Robitel | Lin Shaye | Angus Sampson | Leigh Whannell | Mystery Horror Thriller | Insidious: The Last Key |
| 1 | Lauren Wolkstein | Alex Pettyfer | James Freedson-Jackson | Emily Althaus | Thriller Drama | The Strange Ones |
| 2 | Simon West | Dominic Cooper | Austin Stowell | Gemma Chan | Action Thriller | Stratton |
| 3 | Warwick Thornton | Bryan Brown | Sam Neill | NaN | Drama History Western | Sweet Country |
| 4 | Jaume Collet-Serra | Liam Neeson | Vera Farmiga | Patrick Wilson | Action Thriller | The Commuter |
| ... | ... | ... | ... | ... | ... | ... |
| 263 | Etan Cohen | Will Ferrell | John C. Reilly | Rebecca Hall | Mystery Adventure Comedy Crime | Holmes & Watson |
| 264 | Adam McKay | Christian Bale | Amy Adams | Steve Carell | Thriller Science Fiction Action Adventure | Vice |
| 265 | Mimi Leder | Felicity Jones | Armie Hammer | Justin Theroux | Drama History | On the Basis of Sex |
| 266 | Karyn Kusama | Nicole Kidman | Sebastian Stan | Toby Kebbell | Thriller Crime Drama Action | Destroyer |
| 267 | David Slade | Fionn Whitehead | Will Poulter | Asim Chaudhry | Science Fiction Mystery Drama Thriller TV Movie | Black Mirror: Bandersnatch |

268 rows × 6 columns

(7) 修改其他演员的数值为 unknown，将 movie_title 列中的数据转换为小写。代码如下：

```
new_df18['actor_2_name'] = new_df18['actor_2_name'].replace(np.nan, 'unknown')
new_df18['actor_3_name'] = new_df18['actor_3_name'].replace(np.nan, 'unknown')
new_df18['movie_title'] = new_df18['movie_title'].str.lower()
new_df18['comb'] = new_df18['actor_1_name'] + ' ' + new_df18['actor_2_name'] + ' '+
new_df18['actor_3_name'] + ' '+ new_df18['director_name'] +' ' + new_df18['genres']
```

执行后会输出：

| | director_name | actor_1_name | actor_2_name | actor_3_name | genres | movie_title | comb |
|---|---|---|---|---|---|---|---|
| 0 | Adam Robitel | Lin Shaye | Angus Sampson | Leigh Whannell | Mystery Horror Thriller | insidious: the last key | Lin Shaye Angus Sampson Leigh Whannell Adam Ro... |
| 1 | Lauren Wolkstein | Alex Pettyfer | James Freedson-Jackson | Emily Althaus | Thriller Drama | the strange ones | Alex Pettyfer James Freedson-Jackson Emily Alt... |
| 2 | Simon West | Dominic Cooper | Austin Stowell | Gemma Chan | Action Thriller | stratton | Dominic Cooper Austin Stowell Gemma Chan Simon... |
| 3 | Warwick Thornton | Bryan Brown | Sam Neill | unknown | Drama History Western | sweet country | Bryan Brown Sam Neill unknown Warwick Thornton... |
| 4 | Jaume Collet-Serra | Liam Neeson | Vera Farmiga | Patrick Wilson | Action Thriller | the commuter | Liam Neeson Vera Farmiga Patrick Wilson Jaume ... |
| ... | ... | ... | ... | ... | ... | ... | ... |
| 263 | Etan Cohen | Will Ferrell | John C. Reilly | Rebecca Hall | Mystery Adventure Comedy Crime | holmes & watson | Will Ferrell John C. Reilly Rebecca Hall Etan ... |
| 264 | Adam McKay | Christian Bale | Amy Adams | Steve Carell | Thriller Science Fiction Action Adventure | vice | Christian Bale Amy Adams Steve Carell Adam McK... |
| 265 | Mimi Leder | Felicity Jones | Armie Hammer | Justin Theroux | Drama History | on the basis of sex | Felicity Jones Armie Hammer Justin Theroux Mim... |
| 266 | Karyn Kusama | Nicole Kidman | Sebastian Stan | Toby Kebbell | Thriller Crime Drama Action | destroyer | Nicole Kidman Sebastian Stan Toby Kebbell Kary... |
| 267 | David Slade | Fionn Whitehead | Will Poulter | Asim Chaudhry | Science Fiction Mystery Drama Thriller TV Movie | black mirror: bandersnatch | Fionn Whitehead Will Poulter Asim Chaudhry Dav... |

268 rows × 7 columns

(8) 按照上述流程从维基百科中提取 2019 年电影数据信息的特征，使用函数 isna()实现去重处理，并保存处理后的数据集文件为 final_data.csv。代码如下：

```
final_df.isna().sum()

director_name    0
actor_1_name     0
actor_2_name     0
actor_3_name     0
genres           4
movie_title      0
comb             4
dtype: int64

final_df = final_df.dropna(how='any')
final_df.to_csv('final_data.csv',index=False)
```

## 2.4 情感分析和序列化操作

编写文件 sentiment.ipynb，功能是使用 pickle 模块实现数据序列化操作。通过 pickle 模块的序列化操作，我们能够将程序中运行的对象信息永久保存到文件中；通过 pickle 模块的反序列化操作，我们能够从文件中创建上一次程序保存的对象。文件 sentiment.ipynb 的具体实现流程如下。

扫码看视频

(1) 使用函数 nltk.download() 下载 stopwords，然后读取文件 reviews.txt 的内容。代码如下：

```
nltk.download("stopwords")

dataset = pd.read_csv('reviews.txt',sep = '\t', names =['Reviews','Comments'])
dataset
```

执行后会输出：

|  | Reviews | Comments |
|---|---|---|
| 0 | 1 | The Da Vinci Code book is just awesome. |
| 1 | 1 | this was the first clive cussler i've ever rea... |
| 2 | 1 | i liked the Da Vinci Code a lot. |
| 3 | 1 | i liked the Da Vinci Code a lot. |
| 4 | 1 | I liked the Da Vinci Code but it ultimatly did... |
| ... | ... | ... |
| 6913 | 0 | Brokeback Mountain was boring. |
| 6914 | 0 | So Brokeback Mountain was really depressing. |
| 6915 | 0 | As I sit here, watching the MTV Movie Awards, ... |
| 6916 | 0 | Ok brokeback mountain is such a horrible movie. |
| 6917 | 0 | Oh, and Brokeback Mountain was a terrible movie. |

6918 rows × 2 columns

(2) 使用函数 TfidfVectorizer() 将文本转换为可用作估算器输入的特征向量，然后将数据保存到文件 tranform.pkl 中，并计算准确度评分。代码如下：

```
topset = set(stopwords.words('english'))

vectorizer = TfidfVectorizer(use_idf = True,lowercase = True,
strip_accents='ascii',stop_words=stopset)

X = vectorizer.fit_transform(dataset.Comments)
y = dataset.Reviews
pickle.dump(vectorizer, open('tranform.pkl', 'wb'))

X_train, X_test, y_train, y_test = train_test_split(X, y, test_size=0.20, random_state=42)
```

```
clf = naive_bayes.MultinomialNB()
clf.fit(X_train,y_train)
accuracy_score(y_test,clf.predict(X_test))*100

clf = naive_bayes.MultinomialNB()
clf.fit(X,y)
```

执行后会分别输出准确度评分：

```
97.47109826589595
98.77167630057804
```

(3) 最后将数据保存到文件 nlp_model.pkl 中。代码如下：

```
filename = 'nlp_model.pkl'
pickle.dump(clf, open(filename, 'wb'))
```

## 2.5 Web 端实时推荐

使用 Flask 编写前端程序，然后调用前面创建的文件 nlp_model.pkl 和 tranform.pkl 中的数据，在搜索电影时利用 Ajax 技术实现实时推荐功能，并通过 themoviedb API 展示要搜索的电影的详细信息。

扫码看视频

### 2.5.1 Flask 启动页面

文件 main.py 是 Flask 的主脚本，功能是调用文件 nlp_model.pkl 和 tranform.pkl 中的数据，根据在表单中输入的数据提供实时推荐功能。文件 main.py 的主要代码如下：

```
#从磁盘上加载 NLP 模型和 TF-IDF 矢量器
filename = 'nlp_model.pkl'
clf = pickle.load(open(filename, 'rb'))
vectorizer = pickle.load(open('tranform.pkl','rb'))

#将字符串列表转换为列表
def convert_to_list(my_list):
    my_list = my_list.split('","')
    my_list[0] = my_list[0].replace('["','')
    my_list[-1] = my_list[-1].replace('"]','')
    return my_list

#将数字列表转换为列表(eg. "[1,2,3]" to [1,2,3])
def convert_to_list_num(my_list):
    my_list = my_list.split(',')
```

```python
    my_list[0] = my_list[0].replace("[","")
    my_list[-1] = my_list[-1].replace("]","")
    return my_list

def get_suggestions():
    data = pd.read_csv('main_data.csv')
    return list(data['movie_title'].str.capitalize())

app = Flask(__name__)

@app.route("/")
@app.route("/home")
def home():
    suggestions = get_suggestions()
    return render_template('home.html',suggestions=suggestions)


@app.route("/recommend",methods=["POST"])
def recommend():
    #从Ajax请求中获取数据
    title = request.form['title']
    cast_ids = request.form['cast_ids']
    cast_names = request.form['cast_names']
    cast_chars = request.form['cast_chars']
    cast_bdays = request.form['cast_bdays']
    cast_bios = request.form['cast_bios']
    cast_places = request.form['cast_places']
    cast_profiles = request.form['cast_profiles']
    imdb_id = request.form['imdb_id']
    poster = request.form['poster']
    genres = request.form['genres']
    overview = request.form['overview']
    vote_average = request.form['rating']
    vote_count = request.form['vote_count']
    rel_date = request.form['rel_date']
    release_date = request.form['release_date']
    runtime = request.form['runtime']
    status = request.form['status']
    rec_movies = request.form['rec_movies']
    rec_posters = request.form['rec_posters']
    rec_movies_org = request.form['rec_movies_org']
    rec_year = request.form['rec_year']
    rec_vote = request.form['rec_vote']

    #获取自动完成的电影推荐
    suggestions = get_suggestions()
```

```python
#为每个需要转换为列表的字符串调用convert_to_list()函数
rec_movies_org = convert_to_list(rec_movies_org)
rec_movies = convert_to_list(rec_movies)
rec_posters = convert_to_list(rec_posters)
cast_names = convert_to_list(cast_names)
cast_chars = convert_to_list(cast_chars)
cast_profiles = convert_to_list(cast_profiles)
cast_bdays = convert_to_list(cast_bdays)
cast_bios = convert_to_list(cast_bios)
cast_places = convert_to_list(cast_places)

#将字符串转换为列表(eg. "[1,2,3]" to [1,2,3])
cast_ids = convert_to_list_num(cast_ids)
rec_vote = convert_to_list_num(rec_vote)
rec_year = convert_to_list_num(rec_year)

#将字符串呈现为Python字符串
for i in range(len(cast_bios)):
    cast_bios[i] = cast_bios[i].replace(r'\n', '\n').replace(r'\"','\"')

for i in range(len(cast_chars)):
    cast_chars[i] = cast_chars[i].replace(r'\n', '\n').replace(r'\"','\"')

#将多个列表组合为一个字典,该字典可以传递到HTML文件,以便轻松处理该文件,并保留信息顺序
movie_cards = {rec_posters[i]: [rec_movies[i],rec_movies_org[i],rec_vote[i],rec_year[i]] for i in range(len(rec_posters))}

casts = {cast_names[i]:[cast_ids[i], cast_chars[i], cast_profiles[i]] for i in range(len(cast_profiles))}

cast_details = {cast_names[i]:[cast_ids[i], cast_profiles[i], cast_bdays[i], cast_places[i], cast_bios[i]] for i in range(len(cast_places))}

#从IMDB站点获取用户评论的网页
sauce = urllib.request.urlopen('https://www.imdb.com/title/{}/reviews?ref_=tt_ov_rt'.format(imdb_id)).read()
soup = bs.BeautifulSoup(sauce,'lxml')
soup_result = soup.find_all("div",{"class":"text show-more__control"})

reviews_list = []            #审查清单
reviews_status = []          #留言清单(good or bad)
for reviews in soup_result:
    if reviews.string:
        reviews_list.append(reviews.string)
        #将评审传递给模型
        movie_review_list = np.array([reviews.string])
        movie_vector = vectorizer.transform(movie_review_list)
```

```python
        pred = clf.predict(movie_vector)
        reviews_status.append('Positive' if pred else 'Negative')

    #获取当前日期
    movie_rel_date = ""
    curr_date = ""
    if(rel_date):
        today = str(date.today())
        curr_date = datetime.strptime(today,'%Y-%m-%d')
        movie_rel_date = datetime.strptime(rel_date, '%Y-%m-%d')

    #将评论和审查合并到词典中
    movie_reviews = {reviews_list[i]: reviews_status[i] for i in range(len(reviews_list))}

    #将所有数据传递到HTML文件
    return render_template('recommend.html',title=title,poster=poster,overview=
overview,vote_average=vote_average,
        vote_count=vote_count,release_date=release_date,movie_rel_date=movie_rel_date,
curr_date=curr_date,runtime=runtime,status=status,genres=genres,movie_cards=
movie_cards,reviews=movie_reviews,casts=casts,cast_details=cast_details)

if __name__ == '__main__':
    app.run(debug=True)
```

## 2.5.2 模板文件

在 Flask Web 项目中，使用模板文件实现前端功能。

(1) 本 Web 项目的主页是由模板文件 home.html 实现的，功能是提供了一个表单供用户搜索电影。主要实现代码如下：

```html
<center><h1 class="app-title">电影推荐系统</h1></center>
  <div class="form-group shadow-textarea" style="margin-top: 30px;text-align: 
center;color: white;">
      <input type="text" name="movie" class="movie form-control" 
id="autoComplete" autocomplete="off" placeholder="Enter the Movie Name" 
style="background-color: #ffffff;border-color:#ffffff;width: 60%;color: #181818" 
required="required" />
      <br>
  </div>

  <div class="form-group" style="text-align: center;">
    <button class="btn btn-primary btn-block movie-button" style=
"background-color: #e50914;text-align: center;border-color: #e50914;
width:120px;" disabled="true" >Enter</button><br><br>
  </div>
```

```html
<div id="loader" class="text-center">
</div>

<div class="fail">
    <center><h3>很抱歉您请求的电影不在我们的数据库中，请检查拼写或尝试其他电影！</h3></center>
</div>

<div class="results">
  <center>
    <h2 id="name" class="text-uppercase"></h2>
  </center>
</div>

<div class="modal fade" id="myModal" tabindex="-1" role="dialog" aria-labelledby="exampleModalLabel3" aria-hidden="true">
  <div class="modal-dialog modal-md" role="document">
    <div class="modal-content">
      <div class="modal-header" style="background-color: #e50914;color: white;">
        <h5 class="modal-title" id="exampleModalLabel3">Hey there!</h5>
        <button type="button" class="close" data-dismiss="modal" aria-label="Close">
          <span aria-hidden="true" style="color: white">&times;</span>
        </button>
      </div>
      <div class="modal-body">
          <p>如果您在搜索电影时没有获得实时推荐，请不要担心，只需键入电影名称并单击Enter按钮即可。即使输入时出现了拼写错误，系统仍然会推荐。</p>
      </div>
      <div class="modal-footer" style="text-align: center;">
        <button type="button" class="btn btn-secondary" data-dismiss="modal">知道了</button>
      </div>
    </div>
  </div>
</div>

<footer class="footer">
  <br>
  <div class="social" style="margin-bottom: 8px">
  </div>
</footer>
```

(2) 编写模板文件recommend.html，功能是当用户阻碍在表单中输入某电影名并单击回车键后，会在此页面显示这部电影的详细信息。

## 2.5.3 后端处理

在 Flask Web 项目中，除了使用主文件 main.py 实现后端处理功能外，还使用 JavaScript 技术实现了后端功能。编写文件 recommend.js，功能是调用 TheMovieDB API 实现实时推荐，并根据电影名获取这部电影的详细信息。文件 recommend.js 的具体实现流程如下。

(1) 监听用户是否在电影搜索页面的文本框中输入内容，并监听是否单击 Enter 按钮。代码如下：

```javascript
$(function() {
 //按钮将被禁用，直到在文本框中输入内容
 const source = document.getElementById('autoComplete');
 const inputHandler = function(e) {
  if(e.target.value==""){
    $('.movie-button').attr('disabled', true);
  }
  else{
    $('.movie-button').attr('disabled', false);
  }
 }
 source.addEventListener('input', inputHandler);

 $('.fa-arrow-up').click(function(){
  $('html, body').animate({scrollTop:0}, 'slow');
 });

 $('.app-title').click(function(){
  window.location.href = '/';
 })

 $('.movie-button').on('click',function(){
  var my_api_key = '你的API密钥';
  var title = $('.movie').val();
  if (title=="") {
    $('.results').css('display','none');
    $('.fail').css('display','block');
  }

  if (($('.fail').text() && ($('.footer').css('position') == 'absolute'))) {
    $('.footer').css('position', 'fixed');
  }

  else{
    load_details(my_api_key,title);
  }
```

```
});
});
```

(2) 编写函数 recommendcard()，将在单击推荐的电影选项时调用此函数。代码如下：

```
function recommendcard(e){
  $("#loader").fadeIn();
  var my_api_key = '你的API密钥';
  var title = e.getAttribute('title');
  load_details(my_api_key,title);
}
```

(3) 编写函数 load_details()，功能是从 API 获取电影的详细信息(基于电影名称)。代码如下：

```
function load_details(my_api_key,title){
  $.ajax({
    type: 'GET',
    url:'https://api.themoviedb.org/3/search/movie?api_key='+my_api_key+'&query='+title,
    async: false,
    success: function(movie){
      if(movie.results.length<1){
        $('.fail').css('display','block');
        $('.results').css('display','none');
        $("#loader").delay(500).fadeOut();
      }
      else if(movie.results.length==1) {
        $("#loader").fadeIn();
        $('.fail').css('display','none');
        $('.results').delay(1000).css('display','block');
        var movie_id = movie.results[0].id;
        var movie_title = movie.results[0].title;
        var movie_title_org = movie.results[0].original_title;
        get_movie_details(movie_id,my_api_key,movie_title,movie_title_org);
      }
      else{
        var close_match = {};
        var flag=0;
        var movie_id="";
        var movie_title="";
        var movie_title_org="";
        $("#loader").fadeIn();
        $('.fail').css('display','none');
        $('.results').delay(1000).css('display','block');
        for(var count in movie.results){
          if(title==movie.results[count].original_title){
            flag = 1;
            movie_id = movie.results[count].id;
```

```
                movie_title = movie.results[count].title;
                movie_title_org = movie.results[count].original_title;
                break;
            }
            else{
                close_match[movie.results[count].title] = similarity(title,
                    movie.results[count].title);
            }
        }
        if(flag==0){
            movie_title = Object.keys(close_match).reduce(function(a, b){ return
close_match[a] > close_match[b] ? a : b });
            var index = Object.keys(close_match).indexOf(movie_title)
            movie_id = movie.results[index].id;
            movie_title_org = movie.results[index].original_title;
        }
        get_movie_details(movie_id,my_api_key,movie_title,movie_title_org);
    }
    },
    error: function(error){
        alert('出错了 - '+error);
        $("#loader").delay(100).fadeOut();
    },
});
}
```

(4) 编写函数 similarity()，功能是使用距离参数 length 获取与用户请求的电影名称最接近的匹配项。代码如下：

```
function similarity(s1, s2) {
  var longer = s1;
  var shorter = s2;
  if (s1.length < s2.length) {
    longer = s2;
    shorter = s1;
  }
  var longerLength = longer.length;
  if (longerLength == 0) {
    return 1.0;
  }
  return (longerLength - editDistance(longer, shorter)) / parseFloat(longerLength);
}

function editDistance(s1, s2) {
  s1 = s1.toLowerCase();
  s2 = s2.toLowerCase();
```

```
var costs = new Array();
for (var i = 0; i <= s1.length; i++) {
  var lastValue = i;
  for (var j = 0; j <= s2.length; j++) {
    if (i == 0)
      costs[j] = j;
    else {
      if (j > 0) {
        var newValue = costs[j - 1];
        if (s1.charAt(i - 1) != s2.charAt(j - 1))
          newValue = Math.min(Math.min(newValue, lastValue),
            costs[j]) + 1;
        costs[j - 1] = lastValue;
        lastValue = newValue;
      }
    }
  }
  if (i > 0)
    costs[s2.length] = lastValue;
}
return costs[s2.length];
}
```

(5) 编写函数 get_movie_details()，功能是根据电影 id 获取这部电影的所有详细信息。代码如下：

```
function get_movie_details(movie_id,my_api_key,movie_title,movie_title_org) {
  $.ajax({
    type:'GET',
    url:'https://api.themoviedb.org/3/movie/'+movie_id+'?api_key='+my_api_key,
    success: function(movie_details){
      show_details(movie_details,movie_title,my_api_key,movie_id,movie_title_org);
    },
    error: function(error){
      alert("API Error! - "+error);
      $("#loader").delay(500).fadeOut();
    },
  });
}
```

(6) 编写函数 show_details()，功能是将电影的详细信息传递给 Flask，以便使用 imdb_id 显示和抓取这部电影的评论信息。代码如下：

```
function show_details(movie_details,movie_title,my_api_key,movie_id,movie_title_org){
  var imdb_id = movie_details.imdb_id;
  var poster;
  if(movie_details.poster_path){
```

```javascript
    poster = 'https://image.tmdb.org/t/p/original'+movie_details.poster_path;
}
else {
    poster = 'static/default.jpg';
}
var overview = movie_details.overview;
var genres = movie_details.genres;
var rating = movie_details.vote_average;
var vote_count = movie_details.vote_count;
var release_date = movie_details.release_date;
var runtime = parseInt(movie_details.runtime);
var status = movie_details.status;
var genre_list = []
for (var genre in genres){
    genre_list.push(genres[genre].name);
}
var my_genre = genre_list.join(", ");
if(runtime%60==0){
    runtime = Math.floor(runtime/60)+" hour(s)"
}
else {
    runtime = Math.floor(runtime/60)+" hour(s) "+(runtime%60)+" min(s)"
}

//调用get_movie_cast()函数以获取所查询电影的最佳演员阵容
movie_cast = get_movie_cast(movie_id,my_api_key);

//调用get_individual_cast()函数以获取演员阵容的详细信息
ind_cast = get_individual_cast(movie_cast,my_api_key);

// 调用get_recommendations()函数,从TMDb API 获取给定id 的电影详细信息
recommendations = get_recommendations(movie_id, my_api_key);

details = {
    'title':movie_title,
    //省略代码,包含了从movie_details 对象中提取的电影基本信息,比如,电影的演员阵容和推荐电影等
}
```

(7) 编写函数get_individual_cast(),功能是获取某个演员的详细信息。对应代码如下：

```javascript
function get_individual_cast(movie_cast,my_api_key) {
    cast_bdays = [];
    cast_bios = [];
    cast_places = [];
    for(var cast_id in movie_cast.cast_ids){
        $.ajax({
            type:'GET',
            url:'https://api.themoviedb.org/3/person/'+movie_cast.cast_ids[cast_id]+
```

```
          '?api_key='+my_api_key,async:false,
          success: function(cast_details){
            cast_bdays.push((new Date(cast_details.birthday)).toDateString().
split(' ').slice(1).join(' '));
            if(cast_details.biography){
              cast_bios.push(cast_details.biography);
            }
            else {
              cast_bios.push("Not Available");
            }
            if(cast_details.place_of_birth){
              cast_places.push(cast_details.place_of_birth);
            }
            else {
              cast_places.push("Not Available");
            }
          }
        });
      }
      return {cast_bdays:cast_bdays,cast_bios:cast_bios,cast_places:cast_places};
    }
```

(8) 编写函数 get_movie_cast()，功能是获取所请求电影演员阵容的详细信息。对应代码如下：

```
function get_movie_cast(movie_id,my_api_key){
    cast_ids= [];
    cast_names = [];
    cast_chars = [];
    cast_profiles = [];
    top_10 = [0,1,2,3,4,5,6,7,8,9];
    $.ajax({
      type:'GET',
      url:"https://api.themoviedb.org/3/movie/"+movie_id+"/credits?api_key="
      +my_api_key,async:false,
      success: function(my_movie){
        if(my_movie.cast.length>0){
          if(my_movie.cast.length>=10){
            top_cast = [0,1,2,3,4,5,6,7,8,9];
          }
          else {
            top_cast = [0,1,2,3,4];
          }
          for(var my_cast in top_cast){
            cast_ids.push(my_movie.cast[my_cast].id)
            cast_names.push(my_movie.cast[my_cast].name);
            cast_chars.push(my_movie.cast[my_cast].character);
```

```
            if(my_movie.cast[my_cast].profile_path){
              cast_profiles.push("https://image.tmdb.org/t/p/original" +my_movie.cast
[my_cast].profile_path);
            }
            else {
              cast_profiles.push("static/default.jpg");
            }
          }
        }
      },
      error: function(error){
        alert("出错了! - "+error);
        $("#loader").delay(500).fadeOut();
      }
    });

    return {cast_ids:cast_ids,cast_names:cast_names,cast_chars:cast_chars,
cast_profiles:cast_profiles};
}
```

(9) 编写函数 get_recommendations()，功能是获得实时推荐的电影信息。对应代码如下：

```
function get_recommendations(movie_id, my_api_key) {
    rec_movies = [];
    rec_posters = [];
    rec_movies_org = [];
    rec_year = [];
    rec_vote = [];

    $.ajax({
      type: 'GET',
      url:"https://api.themoviedb.org/3/movie/"+movie_id+"/recommendations?api_key="
+my_api_key,async: false,
      success: function(recommend) {
        for(var recs in recommend.results) {
          rec_movies.push(recommend.results[recs].title);
          rec_movies_org.push(recommend.results[recs].original_title);
          rec_year.push(new Date(recommend.results[recs].release_date).getFullYear());
          rec_vote.push(recommend.results[recs].vote_average);
          if(recommend.results[recs].poster_path){
            rec_posters.push("https://image.tmdb.org/t/p/original" +recommend.results
[recs].poster_path);
          }
          else {
            rec_posters.push("static/default.jpg");
          }
        }
      },
```

```
    error: function(error) {
      alert("出错了! - "+error);
      $("#loader").delay(500).fadeOut();
    }
  });
  return {rec_movies:rec_movies,rec_movies_org:rec_movies_org,
rec_posters:rec_posters,rec_year:rec_year,rec_vote:rec_vote};
}
```

## 2.6 调试运行

运行 Flask 主程序文件 main.py，然后在浏览器中输入 http://127.0.0.1:5000/，显示 Web 主页，如图 2-5 所示。

扫码看视频

图 2-5　系统主页

在文本框中输入电影名中的单词，系统会实时推荐与之相关的电影名。例如，输入"love"后的实时推荐效果如图 2-6 所示。

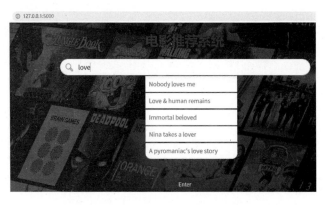

图 2-6　输入"love"后的实时推荐

如果选择实时推荐的第 3 个选项 Immortal beloved，单击 Enter 按钮后会在新页面中显示这部电影的详细信息，如图 2-7 所示。

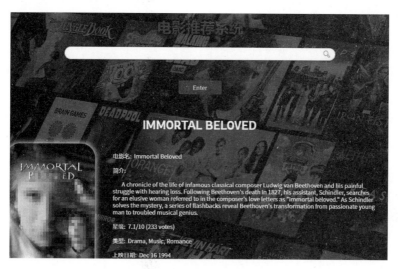

图 2-7　显示电影详细信息

# 第 3 章 智能 OCR 文本检测识别系统

OCR(optical character recognition)即光学字符识别，是指使用电子设备(如扫描仪或数码相机)检查纸上打印的字符，通过检测暗、亮的模式确定其形状，然后用字符识别方法将形状翻译成计算机文字。这个过程包括对文本资料进行扫描，并对得到的图像文件进行分析处理，以获取文字及版面信息。在本章的内容中，将详细介绍使用人工智能技术开发一个 OCR 文本检测识别系统的过程，具体过程由 OpenCV+TensorFlow Lite+TensorFlow+Android 实现。

## 3.1 背景介绍

光学文字识别的概念是在 1929 年由德国科学家 Tausheck(陶舍克)最先提出来的,后来美国科学家 Handel(汉德尔)也提出了利用技术对文字进行识别的想法。而最早对印刷体汉字识别进行研究的是 IBM 公司的 Casey(凯西)和 Nagy(纳吉),1966 年他们发表了第一篇关于汉字识别的文章,采用模板匹配法识别了 1000 个印刷体汉字。

扫码看视频

早在 20 世纪 60~70 年代,世界各国就开始有 OCR 的研究,但研究的初期,多以文字识别方法的研究为主,且识别的文字仅为 0~9 的数字。以同样拥有方块文字的日本为例,1960 年左右开始研究 OCR 的基本识别理论,初期以数字为对象,直至 1965—1970 年,开始有一些简单的产品问世,如印刷文字的邮政编码识别系统,可以识别邮件上的邮政编码,帮助邮局做区域分信作业。

20 世纪 70 年代初,日本的学者开始研究汉字识别,并做了大量的工作。中国在 OCR 技术方面的研究工作起步较晚,到 70 年代才开始研究对数字、英文字母及符号的识别,70 年代末开始进行汉字识别的研究,到 1986 年汉字识别的研究进入一个实质性的阶段,不少研究单位相继推出了中文 OCR 产品。早期的 OCR 软件,由于识别率及产品化等多方面的因素,未能达到实际要求。同时,由于硬件设备成本高,运行速度慢,也没有达到实用的程度。只有个别部门,如信息部门、新闻出版单位等使用 OCR 软件。

1986 年以后,我国的 OCR 研究有了很大进展,在汉字建模和识别方法上都有所创新,在系统研制和开发应用中都取得了丰硕的成果,不少单位相继推出了中文 OCR 产品。进入 20 世纪 90 年代以后,随着平台式扫描仪的广泛应用,以及我国信息自动化和办公自动化的普及,大大推动了 OCR 技术的进一步发展,使 OCR 的识别正确率、识别速度满足了广大用户的要求。目前,OCR 软件产品在医院、学校、企业等各大市场得到了广泛的应用。

## 3.2 OCR 系统简介

在开发 OCR 系统之前,需要先了解开发 OCR 系统需要的理论知识及流程。

### 3.2.1 OCR 的基本原理和使用方式

扫码看视频

在传统的 OCR 系统中,基本原理就是通过扫描仪将一份文稿的图像输入计算机,然后由计算机取出每个文字的图像,并将其转换成汉字的编码。其具体工作过程是,扫描仪通过电荷耦合器件 CCD 将文稿的光信号转换为电信号，再经过模拟/数字转换器转化为数字信号传输给计算机。计算机接收的是文稿的数字图像，其图像上的汉字可能是印刷汉字，也可能是手写汉字，然后对这些图像中的汉字进行识别。对于印刷体字符，首先采用光学的方式将文档资料转换成原始黑白点阵的图像文件，再通过识别软件将图像中的文字转换成文本格式，以便文字处理软件进一步加工。其中，文字识别是 OCR 的重要技术。

与其他信息数据一样，在计算机中所有扫描仪捕捉到的图文信息都是用 0、1 这两个数字来记录和进行识别的，所有信息都只是以 0、1 保存的一串串点或样本点。OCR 识别程序识别页面上的字符信息，主要通过单元模式匹配法和特征提取法两种方式来识别。

单元模式匹配法(pattern matching)是将每一个字符与保存有标准字体和字号位图的文件进行不严格的比较。如果应用程序中有一个已保存字符的大数据库，则应用程序会选取合适的字符进行正确的匹配。软件必须使用一些处理技术找出最相似的匹配，通常是通过不断试验同一个字符的不同版本来比较。有些软件可以扫描一页文本，并鉴别出定义新字体的每一个字符。有些软件则使用自己的识别技术，尽其所能鉴别页面上的字符，然后将不可识别的字符进行人工选择或直接录入。

特征提取法(feature extraction)是将每个字符分解为很多个不同的字符特征，包括斜线、水平线和曲线等。然后，又将这些特征与理解(识别)的字符进行匹配。举个简单的例子，应用程序识别到两条水平横线，它就会"认为"该字符可能是"二"。特征提取法的优点在于可以识别多种字体，例如，中文书法体就是采用特征提取法实现字符识别的。

多数 OCR 应用软件都加入了语法智能检查功能，这种功能进一步提高了识别率。它主要通过上下文检查法实现拼写和语法的纠正。在识别文字时，OCR 应用程序会进行多次的上下文衔接性检查，根据程序中已经存在的词组、固定的词序，对字符串中的用词进行检查。比较高级的应用软件会自动用其"认为"正确的词语替换错误的词语，纠正语句错误。

## 3.2.2 文字识别的基本步骤

文字识别包括以下几个步骤：图文输入、预处理、单字识别和后处理等。

### 1. 图文输入

图文输入是指通过输入设备将文档输入到计算机中，也就是实现原稿的数字化。现在用得比较普遍的设备是扫描仪。文档图像的扫描质量是 OCR 软件正确识别的前提条件。恰当地选择扫描分辨率及相关参数，是保证文字清楚、特征不丢失的关键。此外，文档应尽可能地放置端正，以保证预处理检测的倾斜角小，在进行倾斜校正后，文字图像的变形就小。这些简单的操作，会使系统的识别正确率有所提高。反之，由于扫描设置不当，文字的断笔过多，可能会分拣出半个文字的图像。文字断笔和笔画粘连会造成有些特征丢失，在将其特征与特征库比较时,会使其特征差距加大,识别错误率上升。

### 2. 预处理

扫描一幅简单的印刷文档的图像,将每一个文字图像分拣出来交给识别模块识别,这一过程称为图像预处理。预处理是指在进行文字识别之前的一些准备工作,包括图像净化处理,去掉原始图像中的显见噪声(干扰)。它的主要任务是测量文档放置的倾斜角,对文档进行版面分析,对选出的文字域进行排版确认,对横、竖排版的文字行进行切分,将每一行的文字图像分离,对标点符号进行判别等。这一阶段的工作非常重要,处理的效果将直接影响文字识别的准确率。

版面分析是对文本图像的总体分析,是将文档中的所有文字块分拣出来,区分出文本段落及排版顺序,以及图像、表格的区域,将各文字块的域界(域在图像中的始点、终点坐标)、域内的属性(横、竖排版方式)以及各文字块的连接关系作为一种数据结构,提供给识别模块自动识别。对于文本区域直接进行识别处理,对于表格区域进行专用的表格分析及识别处理,对于图像区域进行压缩或简单存储。行字切分是将大幅的图像先切割为行,再从图像行中分离出单个字符的过程。

### 3. 单字识别

单字识别是体现 OCR 文字识别的核心技术。从扫描文本中分拣出文字图像,由计算机将其图形、图像转变成文字的标准代码,是让计算机"认字"的关键,也就是所谓的识别技术。就像人脑认识文字是因为在人脑中已经保存了文字的各种特征,如文字的结构、文字的笔画等,要想让计算机来识别文字,也需要先将文字的特征等信息存储到计算机里,但要存储什么样的信息及怎样来获取这些信息是一个很复杂的过程,而且要达到非常高的识别率才能符合要求。通常采用的做法是分析文字的笔画、特征点、投影信息、点的区域分布等。

中国汉字常用的就有几千个,识别技术就是特征比较技术,通过和识别特征库的比较,找到特征最相似的字,提取该文字的标准代码,即为识别结果。比较是人们认识事物的一种基本方法,汉字识别也是通过比较找出汉字之间相同、相似、相异的地方,把握其量和质的关系,以及时间与空间的关系等。对于大字符集的汉字一般采用多级分类,多特征、全方位动态匹配求相似集,以保证分类率高、适应性强、稳定性好;细分类重点在于对相似集进行求异匹配、加权处理、结构判别、定量、定性分析,以及前后连接词关系的分析,最后进行判别。汉字识别实质上是比较科学或认知科学在人工智能方面的应用,其关键技术是识别特征库。计算机有了这样的一个特征库,才能完成认字的功能。

在图像文档的版面中,除了文字、图片,有时还会存在表格。为了使识别后的表格数字化,需要在版面分析过程中对表格域进行特殊的处理,包括对表格线的结构信息的提取,对表格内文字域的分拣,完成对表格线和对文字域的识别,并将表格线和表格内的文字进行数据化处理后可以生成不同类型的文件格式。由于文档中的表格随意性大，格式多样，有封闭式的，也有开放式的，特别是表格中的斜线，给表格分析造成了一定的困难。

### 4. 后处理

后处理是指对识别出的文字或多个识别结果采用词组方式进行上下匹配，即将单字识别的结果进行分词，与词库中的词组进行比较，以提高系统的识别率，减少误识率。

汉字字符识别是文字识别领域最为困难的一项任务，它涉及模式识别、图像处理、数字信号处理、自然语言理解、人工智能、模糊数学、信息论、计算机、中文信息处理等学科，是一门综合性技术。近几年来，印刷汉字识别系统的单字识别正确率已经超过 95%，为了进一步提高系统的总体识别率，扫描图像、图像的预处理以及识别后处理等方面的技术也都得到了深入的研究，并取得了长足的进展，有效地提高了印刷汉字识别系统的总体性能。

## 3.2.3 深度学习对 OCR 的影响

深度学习算法的应用使 OCR 识别技术得到了一次跨越式的升级，整体提升了 OCR 的识别率和识别速度。深度学习 OCR 借助神经网络可模仿人脑机制对图像、文本等数据进行分析，可以更加可靠、快速地完成海量样本的训练，得到近似专家能力的最优模型，同时可以在低质量图像、生僻字、非均匀背景、多语言混合等复杂场景中实现高效精准的识别与分类。

但是在构建复杂模型的同时，带来了高额的存储空间、计算资源消耗等问题，使其需要强大的算力支撑，因此很难落实到各个硬件平台。而如今众多的手机芯片，大多数基于 ARM 结构，相比服务端算力有限，所以移动端 OCR 算法多以牺牲一定的精度获取运行速度。深度学习的移动端部署已成为重要的发展方向。

相比于传统 OCR，在识别精度与速度上，深度学习 OCR 遥遥领先。以银行卡识别为例，深度学习 OCR 可适用于对焦不准、高噪声、低分辨率、强光影等复杂背景的图像识别，准确率提升 10%以上，同时识别速度变为原来的二分之一。

在证件分割中，深度学习网络可高效地学习到边缘情况，通过边缘检测，得到物体的边缘轮廓，然后通过边缘跟踪合并，得到证件信息，保障识别效果。

在目标检测中，可以在杂乱无序、千奇百怪的复杂场景中准确定位出主方向角度、直线、图章、文字等区域；面对缺边缺角、光斑、形变、遮挡等异常图像情况可做出提示。

## 3.2.4 与 OCR 相关的深度学习技术

### 1. LSTM+CTC

LSTM(长短期记忆网络)是一种特殊结构的循环神经网络(RNN)，它被设计来解决传统

RNN 在处理长期依赖问题时遇到的挑战。LSTM 通过引入门控机制来维持和清除信息，从而有效地捕捉时间序列中的长期依赖关系。

CTC(连接时序分类器)是一种用于序列识别问题的损失函数，它允许模型学习输入特征序列和输出标签序列之间的对齐方式，即便它们的长度可能不同。CTC 在处理诸如文本识别、语音识别等任务时尤为重要，因为它可以解决输入和输出之间的非固定对齐问题。

在进行序列识别时，CTC 能够识别并分组相邻的块，将它们视为相同的结果，即使在输出中字符可能会重复出现。此外，CTC 还可以在训练后处理结果，例如去除序列中的空隙字符和重复字符，从而解决对齐问题。

### 2. CRNN

CRNN(卷积循环神经网络)是目前比较流行的文字识别模型，它可以进行端到端的训练，无须对样本数据进行字符分割，可识别任意长度的文本序列，具有快速、高效的性能。CRNN 的基本结构如下。

1) 卷积层

卷积层用于从输入图像中提取特征序列。首先进行预处理，将所有输入图像缩放为同一高度，默认为 32，宽度可任意长；然后执行卷积操作(由类似于 VGG 的卷积、最大池化和 BN 层组成)；再从左到右提取序列特征，作为循环层的输入，每个特征向量都代表图像在一定宽度内的特征，默认为 1 像素(因为 CRNN 已将输入图像缩放为同样的高度，所以只需按一定的宽度提取特征)。

2) 循环层

循环层用于预测从卷积层获得的特征序列的标签分布，由双向 LSTM 构成循环层，预测特征序列中各特征向量的标签分布。因为 LSTM 需要时间维度，序列的 width 在模型中被视为 timesteps。Map-to-Sequence 层将误差从循环层反馈到卷积层，通过特征序列的转换，将两者连接起来。

3) 转录层

转录层通过去重、整合等操作，将从循环层获得的标签分布转换为最后的识别结果。转录层是对 LSTM 网络所预测的特征序列进行集成，并转化为最终输出结果。基于 CRNN 模型的双向 LSTM 网络层的最终连接，实现对终端的识别。

## 3.3 系统介绍

在本章讲解的 OCR 项目开发中，整个任务通常分为两个阶段。①使用文本检测模型来检测可能的文本周围的边界框。②将处理后的边界框送入文本识别模型，以确定边界框内的特定字符(在文本识别之前，还需要进行非最大

扫码看视频

抑制、透视变换等操作)。在本项目中使用两个模型实现上述两个截断的功能,这两个模型都来自 TensorFlow Hub,它们都是 FP16 量化模型。本项目的具体结构如图 3-1 所示。

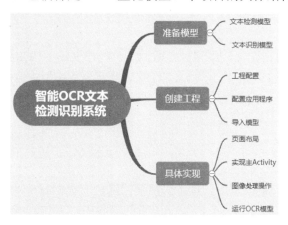

图 3-1 项目结构

## 3.4 准备模型

本项目使用的是 TensorFlow 官方提供的文本检测模型和文本识别模型,用户可以登录 TensorFlow 官方网站下载对应的模型文件。两个模型的基本信息如表 3-1 所示。

扫码看视频

表 3-1 模型信息

| 名 称 | 大小/MB |
| --- | --- |
| 文本检测模型 | 45.9 |
| 文本识别模型 | 16.8 |

### 3.4.1 文本检测模型

在本项目中,使用的文本检测模型是 TFLite 模型 east-text-detector,功能是从自然场景中进行文本检测。

### 3.4.2 文本识别模型

在本项目中,使用的文本识别模型是 TFLite 模型 Keras OCR,功能是从图像中识别文本。

Keras OCR 是一个用于检测和识别文本的模型库,以 CRAFT 作为文本检测器,以 CRNN 作为文本识别器来实现。Keras OCR 基于卷积循环神经网络技术,这是一种非常流行的文本识别模型。

Keras OCR 支持对自定义数据集进行微调,可以在 Keras OCR 的帮助下分别微调检测器和识别器。当转换为 TFLite 格式时,CTC 解码器部分从模型中删除,因为 TFLite 格式不支持它。因此,我们需要在模型的输出中显式地运行解码器以获得最终输出。

## 3.5 创建工程

在准备好 TensorFlow 模型后,接下来将使用这两个模型基于 Android 系统开发一个 OCR 文本检测识别系统。在本节的内容中,首先讲解创建一个 Android 工程的过程。

扫码看视频

### 3.5.1 工程配置

使用 Android Studio 创建一个 Android 工程,工程名为 android。工程结构如图 3-2 所示。

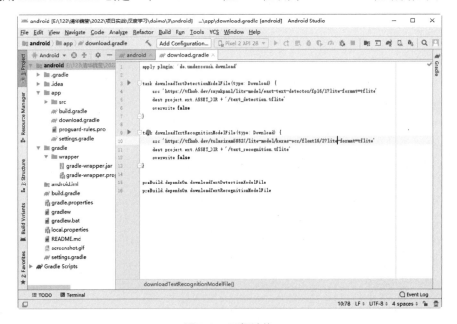

图 3-2 工程结构

## 3.5.2 配置应用程序

打开 app 模块中的文件 build.gradle，分别设置 Android 的编译版本和运行版本，在 dependencies 中设置需要引用的库文件，特别是需要引用和 TensorFlow Lite、OpenCV 相关的库。对应的代码如下：

```
apply plugin: 'com.android.application'
apply plugin: 'kotlin-android'
apply plugin: 'kotlin-android-extensions'
apply plugin: 'kotlin-kapt'

android {
    compileSdkVersion 29
    buildToolsVersion "29.0.2"
    defaultConfig {
        applicationId "org.tensorflow.lite.examples.ocr"
        minSdkVersion 21
        targetSdkVersion 29
        versionCode 1
        versionName "1.0"
        testInstrumentationRunner "androidx.test.runner.AndroidJUnitRunner"
    }
    buildTypes {
        release {
            minifyEnabled false
            proguardFiles getDefaultProguardFile('proguard-android-optimize.txt'), 'proguard-rules.pro'
        }
    }
    aaptOptions {
        noCompress "tflite"
    }
    compileOptions {
        sourceCompatibility JavaVersion.VERSION_1_8
        targetCompatibility JavaVersion.VERSION_1_8
    }
}
dependencies {
    implementation fileTree(dir: 'libs', include: ['*.jar'])

    implementation 'androidx.appcompat:appcompat:1.1.0'
    implementation 'androidx.constraintlayout:constraintlayout:1.1.3'
    implementation 'androidx.coordinatorlayout:coordinatorlayout:1.1.0'
    implementation 'androidx.legacy:legacy-support-v4:1.0.0'
    implementation 'androidx.lifecycle:lifecycle-extensions:2.2.0'
```

```
    implementation 'androidx.lifecycle:lifecycle-viewmodel:2.2.0'
    implementation 'androidx.recyclerview:recyclerview:1.1.0'
    implementation 'com.github.bumptech.glide:glide:4.11.0'
    implementation 'com.google.android.material:material:1.1.0'
    implementation 'com.google.code.gson:gson:2.8.6'
    implementation 'com.google.guava:guava:28.2-android'
    implementation 'org.jetbrains.kotlin:kotlin-stdlib:1.3.71'
    implementation 'org.jetbrains.kotlinx:kotlinx-coroutines-android:1.3.0'
    implementation 'org.jetbrains.kotlinx:kotlinx-coroutines-core:1.3.0'

    implementation 'org.tensorflow:tensorflow-lite:0.0.0-nightly-SNAPSHOT'
    implementation 'org.tensorflow:tensorflow-lite-gpu:0.0.0-nightly-SNAPSHOT'
    implementation 'org.tensorflow:tensorflow-lite-select-tf-ops:0.0.0-nightly-SNAPSHOT'
    implementation 'org.tensorflow:tensorflow-lite-support:0.0.0-nightly-SNAPSHOT'

    implementation 'com.quickbirdstudios:opencv:4.5.3.0'
}
```

### 3.5.3 导入模型

在文件 download.gradle 中设置使用的文本检测模型和文本识别模型的下载地址。具体实现代码如下：

```
task downloadTextDetectionModelFile(type: Download) {
    src 'https://tfhub.dev/sayakpaul/lite-model/east-text-detector/fp16/1?lite-format=tflite'
    dest project.ext.ASSET_DIR + '/text_detection.tflite'
    overwrite false
}

task downloadTextRecognitionModelFile(type: Download) {
    src 'https://tfhub.dev/tulasiram58827/lite-model/keras-ocr/float16/2?lite-format=tflite'
    dest project.ext.ASSET_DIR + '/text_recognition.tflite'
    overwrite false
}
```

## 3.6 具体实现

在创建 Android 工程后，接下来进入本项目的正式编码阶段。在本节的内容中，将讲解使用 Kotlin 语言开发 OCR 文本检测识别系统的过程。

扫码看视频

## 3.6.1 页面布局

(1) 本项目主界面的页面布局文件是 app/src/main/res/layout/tfe_is_activity_main.xml,功能是在 Android 屏幕上方显示预先准备的图片,在屏幕下方显示文本检测结果。

(2) 在页面布局文件 tfe_is_activity_main.xml 中,通过调用文件 app/src/main/res/layout/tfe_is_bottom_sheet_layout.xml,在主界面屏幕下方显示悬浮式配置面板,可以设置在悬浮式面板中使用 GPU 实现文本识别功能。

## 3.6.2 实现主 Activity

本项目的主 Activity 功能是由文件 app/src/main/java/org/tensorflow/lite/examples/ocr/MainActivity.kt 实现的,功能是调用前面的布局文件 tfe_is_activity_main.xml,在屏幕上方滑动显示要识别的图片,在屏幕下方显示检测结果。文件 MainActivity.kt 的具体实现流程如下。

(1) 设置系统需要的公共属性,预先准备好要使用的素材图片。对应的代码如下:

```
private const val TAG = "MainActivity"
class MainActivity : AppCompatActivity() {
  private val tfImageName = "tensorflow.jpg"
  //此处省略部分代码
  private val mutex = Mutex()
}
```

(2) 通过编写 onCreate()方法设置在程序启动时需要实例化的对象,此方法是在 Activity 创建时被系统调用的,它是 Activity 生命周期的开始。对应的代码如下:

```
override fun onCreate(savedInstanceState: Bundle?) {
  super.onCreate(savedInstanceState)
  setContentView(R.layout.tfe_is_activity_main)
  val toolbar: Toolbar = findViewById(R.id.toolbar)
  setSupportActionBar(toolbar)
  supportActionBar?.setDisplayShowTitleEnabled(false)
  tfImageView = findViewById(R.id.tf_imageview)
  androidImageView = findViewById(R.id.android_imageview)
  chromeImageView = findViewById(R.id.chrome_imageview)
  val candidateImageViews = arrayOf<ImageView>(tfImageView, androidImageView, chromeImageView)
  val assetManager = assets
  try {
    val tfInputStream: InputStream = assetManager.open(tfImageName)
    val tfBitmap = BitmapFactory.decodeStream(tfInputStream)
```

```
        tfImageView.setImageBitmap(tfBitmap)
        val androidInputStream: InputStream = assetManager.open(androidImageName)
        val androidBitmap = BitmapFactory.decodeStream(androidInputStream)
        androidImageView.setImageBitmap(androidBitmap)
        val chromeInputStream: InputStream = assetManager.open(chromeImageName)
        val chromeBitmap = BitmapFactory.decodeStream(chromeInputStream)
        chromeImageView.setImageBitmap(chromeBitmap)
    } catch (e: IOException) {
        Log.e(TAG, "Failed to open a test image")
    }

    for (iv in candidateImageViews) {
        setInputImageViewListener(iv)
    }

    resultImageView = findViewById(R.id.result_imageview)
    chipsGroup = findViewById(R.id.chips_group)
    textPromptTextView = findViewById(R.id.text_prompt)
    val useGpuSwitch: Switch = findViewById(R.id.switch_use_gpu)

    viewModel =
AndroidViewModelFactory(application).create(MLExecutionViewModel::class.java)
    viewModel.resultingBitmap.observe(
        this,
        Observer { resultImage ->
            if (resultImage != null) {
                updateUIWithResults(resultImage)
            }
            enableControls(true)
        }
    )
}
```

(3) 监听用户是否选中了悬浮面板中的 GPU 单选按钮。对应的实现代码如下：

```
mainScope.async(inferenceThread) { createModelExecutor(useGPU) }
useGpuSwitch.setOnCheckedChangeListener { _, isChecked ->
    useGPU = isChecked
    mainScope.async(inferenceThread) { createModelExecutor(useGPU) }
}
runButton = findViewById(R.id.rerun_button)
runButton.setOnClickListener {
    enableControls(false)

    mainScope.async(inferenceThread) {
        mutex.withLock {
            if (ocrModel != null) {
```

```
      viewModel.onApplyModel(baseContext, selectedImageName, ocrModel,
inferenceThread)
    } else {
      Log.d(
        TAG,
        "Skipping running OCR since the ocrModel has not been properly initialized..."
      )
    }
   }
  }
 }
 setChipsToLogView(HashMap<String, Int>())
 enableControls(true)
}
```

(4) 编写方法 setInputImageViewListener()，功能是监听用户选中的图片，如果图片被选中，则会检测图片中的文字。对应的实现代码如下：

```
@SuppressLint("ClickableViewAccessibility")
private fun setInputImageViewListener(iv: ImageView) {
 iv.setOnTouchListener(
   object : View.OnTouchListener {
     override fun onTouch(v: View, event: MotionEvent?): Boolean {
       if (v.equals(tfImageView)) {
         selectedImageName = tfImageName
         textPromptTextView.setText(getResources().getString (R.string.tfe_
using_first_image))
       } else if (v.equals(androidImageView)) {
         selectedImageName = androidImageName
         textPromptTextView.setText(getResources().getString
(R.string.tfe_using_second_image))
       } else if (v.equals(chromeImageView)) {
         selectedImageName = chromeImageName
         textPromptTextView.setText(getResources().getString (R.string.tfe_
using_third_image))
       }
       return false
     }
   }
 )
}
```

(5) 编写方法 createModelExecutor()，功能是检测并标记出图片中的文字后，调用文本识别模型进行识别。对应的实现代码如下：

```
private suspend fun createModelExecutor(useGPU: Boolean) {
 mutex.withLock {
```

```
  if (ocrModel != null) {
    ocrModel!!.close()
    ocrModel = null
  }
  try {
    ocrModel = OCRModelExecutor(this, useGPU)
  } catch (e: Exception) {
    Log.e(TAG, "Fail to create OCRModelExecutor: ${e.message}")
    val logText: TextView = findViewById(R.id.log_view)
    logText.text = e.message
  }
}
```

（6）编写方法 setChipsToLogView()，功能是设置调试日志中的信息。对应的实现代码如下：

```
private fun setChipsToLogView(itemsFound: Map<String, Int>) {
  chipsGroup.removeAllViews()

  for ((word, color) in itemsFound) {
    val chip = Chip(this)
    chip.text = word
    chip.chipBackgroundColor = getColorStateListForChip(color)
    chip.isClickable = false
    chipsGroup.addView(chip)
  }
  val labelsFoundTextView: TextView = findViewById(R.id.tfe_is_labels_found)
  if (chipsGroup.childCount == 0) {
    labelsFoundTextView.text = getString(R.string.tfe_ocr_no_text_found)
  } else {
    labelsFoundTextView.text = getString(R.string.tfe_ocr_texts_found)
  }
  chipsGroup.parent.requestLayout()
}
```

（7）编写方法 getColorStateListForChip()，功能是获取颜色值列表，识别后的文本用突出颜色显示。对应的实现代码如下：

```
private fun getColorStateListForChip(color: Int): ColorStateList {
  val states =
    arrayOf(
      intArrayOf(android.R.attr.state_enabled),   // enabled
      intArrayOf(android.R.attr.state_pressed)    // pressed
    )
  val colors = intArrayOf(color, color)
  return ColorStateList(states, colors)
}
```

### 3.6.3 图像处理操作

在本实例中需要检测图像中的文字,因而图像处理的步骤至关重要。编写文件 app/src/main/java/org/tensorflow/lite/examples/ocr/ImageUtils.kt 实现和图像处理相关的功能。具体实现流程如下:

(1) 编写方法 decodeExifOrientation(),功能是将 EXIF orientation(方向参数)枚举转换为变换矩阵。对应的代码如下:

```
abstract class ImageUtils {
  companion object {
    private fun decodeExifOrientation(orientation: Int): Matrix {
      val matrix = Matrix()

      //应用与声明的 EXIF 方向相对应的转换
      when (orientation) {
        ExifInterface.ORIENTATION_NORMAL, ExifInterface.ORIENTATION_UNDEFINED -> Unit
        ExifInterface.ORIENTATION_ROTATE_90 -> matrix.postRotate(90F)
        ExifInterface.ORIENTATION_ROTATE_180 -> matrix.postRotate(180F)
        ExifInterface.ORIENTATION_ROTATE_270 -> matrix.postRotate(270F)
        ExifInterface.ORIENTATION_FLIP_HORIZONTAL -> matrix.postScale(-1F, 1F)
        ExifInterface.ORIENTATION_FLIP_VERTICAL -> matrix.postScale(1F, -1F)
        ExifInterface.ORIENTATION_TRANSPOSE -> {
          matrix.postScale(-1F, 1F)
          matrix.postRotate(270F)
        }
        ExifInterface.ORIENTATION_TRANSVERSE -> {
          matrix.postScale(-1F, 1F)
          matrix.postRotate(90F)
        }

        //EXIF orientation 无效时出错
        else -> throw IllegalArgumentException("Invalid orientation: $orientation")
      }

      //返回生成的矩阵
      return matrix
    }
  }
}
```

(2) 编写方法 setExifOrientation(),用于设置 EXIF orientation,这样可以实现对图像的修复。EXIF orientation 参数的作用是确保无论在何种情况下拍摄的图像,观看者都可以看到以正确方向显示的照片,无须手动旋转。对应的代码如下所示。

```
fun setExifOrientation(filePath: String, value: String) {
    val exif = ExifInterface(filePath)
    exif.setAttribute(ExifInterface.TAG_ORIENTATION, value)
    exif.saveAttributes()
}
```

(3) 编写方法 computeExifOrientation()，功能是将旋转和镜像信息转换为 ExifInterface 常量之一。

(4) 编写方法 decodeBitmap()，功能是从文件中解码位图，并应用其 EXIF 中描述的转换。对应的实现代码如下：

```
fun decodeBitmap(file: File): Bitmap {
    //解码EXIF数据并检索变换矩阵
    val exif = ExifInterface(file.absolutePath)
    val transformation =
        decodeExifOrientation(
            exif.getAttributeInt(ExifInterface.TAG_ORIENTATION,
ExifInterface.ORIENTATION_ROTATE_90)
        )

    //从文件中解码位图，并使用EXIF数据进行转换
    val options = BitmapFactory.Options()
    val bitmap = BitmapFactory.decodeFile(file.absolutePath, options)
    return Bitmap.createBitmap(
        BitmapFactory.decodeFile(file.absolutePath),
        0,
        0,
        bitmap.width,
        bitmap.height,
        transformation,
        true
    )
}
```

(5) 编写方法 bitmapToTensorImageForRecognition()，功能是将位图转换为具有目标大小和标准化的识别模型的 TensorImage。各个参数的具体说明如下。

- bitmapIn：位图输入；
- width：转换后的 TensorImage 的目标宽度；
- height：转换后的 TensorImage 的目标高度；
- mean：表示要减去的均值；
- std：图像的标准差。

方法 bitmapToTensorImageForRecognition() 的具体实现代码如下：

```
fun bitmapToTensorImageForRecognition(
    bitmapIn: Bitmap,
```

```
    width: Int,
    height: Int,
    mean: Float,
    std: Float
): TensorImage {
    val imageProcessor =
        ImageProcessor.Builder()
            .add(ResizeOp(height, width, ResizeOp.ResizeMethod.BILINEAR))
            .add(TransformToGrayscaleOp())
            .add(NormalizeOp(mean, std))
            .build()
    var tensorImage = TensorImage(DataType.FLOAT32)

    tensorImage.load(bitmapIn)
    tensorImage = imageProcessor.process(tensorImage)

    return tensorImage
}
```

(6) 编写方法 bitmapToTensorImageForDetection()，功能是将位图转换为具有目标大小和标准化的检测模型的 TensorImage。方法 bitmapToTensorImageForDetection()与前面介绍的方法 bitmapToTensorImageForRecognition()功能相似，都是将位图转换为 TensorImage，并且都包含了调整大小和标准化的步骤。但是，它们在处理方式上有一些差异。

- 方法 bitmapToTensorImageForRecognition()：适用于识别模型，在转换过程中，先进行了灰度化操作，然后再进行标准化。标准化操作只接受一个标准差值，而不是数组。
- 方法 bitmapToTensorImageForDetection()：适用于检测模型，在转换过程中，没有进行灰度化操作，直接进行了标准化。标准化操作接受了均值数组和标准差值数组。

方法 bitmapToTensorImageForDetection()的具体实现代码如下：

```
fun bitmapToTensorImageForDetection(
    bitmapIn: Bitmap,
    width: Int,
    height: Int,
    means: FloatArray,
    stds: FloatArray
): TensorImage {
    val imageProcessor =
        ImageProcessor.Builder()
            .add(ResizeOp(height, width, ResizeOp.ResizeMethod.BILINEAR))
            .add(NormalizeOp(means, stds))
            .build()
    var tensorImage = TensorImage(DataType.FLOAT32)
```

```
            tensorImage.load(bitmapIn)
            tensorImage = imageProcessor.process(tensorImage)
            return tensorImage
}
```

### 3.6.4 运行 OCR 模型

编写文件 app/src/main/java/org/tensorflow/lite/examples/ocr/OCRModelExecutor.kt，功能是运行 OCR 模型，分别实现文本检测和文本识别。文件 CameraConnectionFragment.java 的具体实现流程如下。

(1) 初始化处理，验证是否支持 OpenCV。对应的代码如下：

```
init {
    try {
        if (!OpenCVLoader.initDebug()) throw Exception("Unable to load OpenCV")
        else Log.d(TAG, "OpenCV loaded")
    } catch (e: Exception) {
        val exceptionLog = "something went wrong: ${e.message}"
        Log.d(TAG, exceptionLog)
    }
}
```

(2) 创建检测解释器，检测指定范围内的图像信息。对应的代码如下：

```
detectionInterpreter = getInterpreter(context, textDetectionModel, useGPU)
//识别模型需要 Flex，因此无论用户如何选择，都会禁用 GPU 代理
recognitionInterpreter = getInterpreter(context, textRecognitionModel, false)
recognitionResult = ByteBuffer.allocateDirect(recognitionModelOutputSize * 8)
recognitionResult.order(ByteOrder.nativeOrder())
indicesMat = MatOfInt()
boundingBoxesMat = MatOfRotatedRect()
ocrResults = HashMap<String, Int>()
```

(3) 编写方法 execute()，功能是处理参数 data 指定的图像。对应的代码如下：

```
fun execute(data: Bitmap): ModelExecutionResult {
    try {
        ratioHeight = data.height.toFloat() / detectionImageHeight
        ratioWidth = data.width.toFloat() / detectionImageWidth
        ocrResults.clear()

        detectTexts(data)

        val bitmapWithBoundingBoxes = recognizeTexts(data, boundingBoxesMat, indicesMat)

        return ModelExecutionResult(bitmapWithBoundingBoxes, "OCR result", ocrResults)
    } catch (e: Exception) {
```

```
    val exceptionLog = "something went wrong: ${e.message}"
    Log.d(TAG, exceptionLog)

    val emptyBitmap = ImageUtils.createEmptyBitmap(displayImageSize, displayImageSize)
    return ModelExecutionResult(emptyBitmap, exceptionLog, HashMap<String, Int>())
  }
}
```

(4) 编写方法 detectTexts()，功能是指定参数 data 图像中的文字。

(5) 编写方法 recognizeTexts()，功能是调用模型实现文字识别功能，通过 copy() 方法返回新的 Bitmap 对象，像素格式是 ARGB_8888。在 Android 界面中显示图片时，需要的内存空间不是按图片的实际大小来计算的，而是按像素点的多少乘以每个像素点占用的空间大小来计算的。比如一个 400 像素×800 像素的图片以 ARGB_8888 格式显示则占用的内存为：(400×800×4)÷1024=1250KB。在图像中检测文字时，会使用 for 循环遍历指定区域内的每一个点，然后使用 drawLine() 方法绘制方块，将有文字的区域标记出来。最后，将有文字的位图转换为张量图像，从而实现文字识别功能。对应的实现代码如下：

```
private fun recognizeTexts(
  data: Bitmap,
  boundingBoxesMat: MatOfRotatedRect,
  indicesMat: MatOfInt
): Bitmap {
  val bitmapWithBoundingBoxes = data.copy(Bitmap.Config.ARGB_8888, true)
  val canvas = Canvas(bitmapWithBoundingBoxes)
  val paint = Paint()
  paint.style = Paint.Style.STROKE
  paint.strokeWidth = 10.toFloat()
  paint.setColor(Color.GREEN)

  for (i in indicesMat.toArray()) {
    val boundingBox = boundingBoxesMat.toArray()[i]
    val targetVertices = ArrayList<Point>()
    targetVertices.add(Point(0.toDouble(), (recognitionImageHeight - 1).toDouble()))
    targetVertices.add(Point(0.toDouble(), 0.toDouble()))
    targetVertices.add(Point((recognitionImageWidth - 1).toDouble(), 0.toDouble()))
    targetVertices.add(
      Point((recognitionImageWidth - 1).toDouble(), (recognitionImageHeight - 1).toDouble())
    )

    val srcVertices = ArrayList<Point>()

    val boundingBoxPointsMat = Mat()
    boxPoints(boundingBox, boundingBoxPointsMat)
    for (j in 0 until 4) {
      srcVertices.add(
```

```
          Point(
            boundingBoxPointsMat.get(j, 0)[0] * ratioWidth,
            boundingBoxPointsMat.get(j, 1)[0] * ratioHeight
          )
        )
        if (j != 0) {
          canvas.drawLine(
            (boundingBoxPointsMat.get(j, 0)[0] * ratioWidth).toFloat(),
            (boundingBoxPointsMat.get(j, 1)[0] * ratioHeight).toFloat(),
            (boundingBoxPointsMat.get(j - 1, 0)[0] * ratioWidth).toFloat(),
            (boundingBoxPointsMat.get(j - 1, 1)[0] * ratioHeight).toFloat(),
            paint
          )
        }
      }
      canvas.drawLine(
        (boundingBoxPointsMat.get(0, 0)[0] * ratioWidth).toFloat(),
        (boundingBoxPointsMat.get(0, 1)[0] * ratioHeight).toFloat(),
        (boundingBoxPointsMat.get(3, 0)[0] * ratioWidth).toFloat(),
        (boundingBoxPointsMat.get(3, 1)[0] * ratioHeight).toFloat(),
        paint
      )
      val srcVerticesMat =
        MatOfPoint2f(srcVertices[0], srcVertices[1], srcVertices[2], srcVertices[3])
      val targetVerticesMat =
        MatOfPoint2f(targetVertices[0], targetVertices[1], targetVertices[2],
targetVertices[3])
      val rotationMatrix = getPerspectiveTransform(srcVerticesMat, targetVerticesMat)
      val recognitionBitmapMat = Mat()
      val srcBitmapMat = Mat()
      bitmapToMat(data, srcBitmapMat)
      warpPerspective(
        srcBitmapMat,
        recognitionBitmapMat,
        rotationMatrix,
        Size(recognitionImageWidth.toDouble(), recognitionImageHeight.toDouble())
      )

      val recognitionBitmap =
        ImageUtils.createEmptyBitmap(
          recognitionImageWidth,
          recognitionImageHeight,
          0,
          Bitmap.Config.ARGB_8888
        )
      matToBitmap(recognitionBitmapMat, recognitionBitmap)

      val recognitionTensorImage =
        ImageUtils.bitmapToTensorImageForRecognition(
```

```
    recognitionBitmap,
    recognitionImageWidth,
    recognitionImageHeight,
    recognitionImageMean,
    recognitionImageStd
  )
  recognitionResult.rewind()
  recognitionInterpreter.run(recognitionTensorImage.buffer, recognitionResult)
  var recognizedText = ""
  for (k in 0 until recognitionModelOutputSize) {
    var alphabetIndex = recognitionResult.getInt(k * 8)
    if (alphabetIndex in 0..alphabets.length - 1)
      recognizedText = recognizedText + alphabets[alphabetIndex]
  }
  Log.d("Recognition result:", recognizedText)
  if (recognizedText != "") {
    ocrResults.put(recognizedText, getRandomColor())
  }
}
return bitmapWithBoundingBoxes
}
@Throws(IOException::class)
```

(6) 编写方法 loadModelFile()，功能是加载指定的模型文件。

(7) 编写方法 close()，功能是关闭识别功能，对应的代码如下：

```
override fun close() {
  detectionInterpreter.close()
  recognitionInterpreter.close()
  if (gpuDelegate != null) {
    gpuDelegate!!.close()
  }
}
```

(8) 编写方法 getRandomColor()，功能是获取 Android 的随机颜色，然后用于识别结果中的文字，以突出识别结果。对应的实现代码如下：

```
fun getRandomColor(): Int {
  val random = Random()
  return Color.argb(
    (128),
    (255 * random.nextFloat()).toInt(),
    (255 * random.nextFloat()).toInt(),
    (255 * random.nextFloat()).toInt()
  )
}

companion object {
  public const val TAG = "TfLiteOCRDemo"
```

```
    private const val textDetectionModel = "text_detection.tflite"
    private const val textRecognitionModel = "text_recognition.tflite"
    private const val numberThreads = 4
    private const val alphabets = "0123456789abcdefghijklmnopqrstuvwxyz"
    private const val displayImageSize = 257
    private const val detectionImageHeight = 320
    private const val detectionImageWidth = 320
    private val detectionImageMeans =
      floatArrayOf(103.94.toFloat(), 116.78.toFloat(), 123.68.toFloat())
    private val detectionImageStds = floatArrayOf(1.toFloat(), 1.toFloat(), 1.toFloat())
    private val detectionOutputNumRows = 80
    private val detectionOutputNumCols = 80
    private val detectionConfidenceThreshold = 0.5
    private val detectionNMSThreshold = 0.4
    private const val recognitionImageHeight = 31
    private const val recognitionImageWidth = 200
    private const val recognitionImageMean = 0.toFloat()
    private const val recognitionImageStd = 255.toFloat()
    private const val recognitionModelOutputSize = 48
}
```

## 3.7 调试运行

在 Android Studio 的顶部单击运行按钮，运行本项目，在 Android 设备中将会显示执行效果。在屏幕上方会显示要识别的图片，在下方的悬浮界面中显示识别结果。如果图像中没有文字，执行效果如图 3-3 所示；如果在图像中有文字，则显示识别结果，执行效果如图 3-4 所示。

图 3-3　没有文字时的执行效果

图 3-4　有文字时的执行效果

# 第4章

## 国际足球比赛结果预测系统

足球赛事作为一项全球性的运动盛宴，深受国内外青少年的热情追捧，并孕育出庞大的球迷社群。在本章的内容中，将详细介绍使用 Scikit-learn+NumPy+Matplotlib+Seaborn+Pandas 技术开发一个足球比赛数据可视化程序的过程，并使用机器学习技术对足球比赛结果进行预测。

## 4.1 欧洲足球五大联赛

欧洲足球五大联赛是全球最具影响力和观赏性的足球赛事。这些联赛不仅吸引了众多顶级球星加盟,也成为足球迷最喜爱的赛事。

扫码看视频

### 1. 英超联赛(English Premier League)

英超联赛常被称为"英超",是英格兰足球总会旗下的最高级职业足球联赛。其前身是"英格兰超级联赛"。英超由 20 支球队组成,由英足总负责具体运作。赛季结束时,积分榜末三位将降级至英格兰足球冠军联赛。英超以其快速的节奏、激烈的竞争和众多强队而闻名,被认为是世界上最好的联赛之一。

### 2. 西班牙足球甲级联赛(La Liga)

西班牙足球甲级联赛简称"西甲",是西班牙最高级别的足球联赛,也是欧洲乃至世界最高水平的职业足球联赛之一。目前,共有 20 支队伍参赛。皇家马德里、巴塞罗那和毕尔巴鄂竞技是三支参加了所有西甲赛季的球队。西甲还以培养出众多"金球奖"得主和顶尖足球运动员而著称。

### 3. 意大利足球甲级联赛(Serie A)

意大利足球甲级联赛简称"意甲",是意大利最高级别的足球联赛,由意甲职业联赛管理和运营。联赛采用双循环赛制,赛季结束时,积分榜末三位将降级至意大利足球乙级联赛。意乙冠军和亚军将直接升入意甲,而意乙第三名至第八名则通过附加赛争夺升级机会。

### 4. 德甲联赛(Bundesliga)

德甲联赛是德国最高水平的足球联赛。德国足协于 1962 年 7 月 28 日在多特蒙德成立德甲,联赛始于 1963-64 赛季。德甲共有 18 支球队,采用主、客场双循环赛制。赛季冠军将获得"沙拉盘"冠军奖杯;排名最后的两支球队将直接降级至德国足球乙级联赛;而排名倒数第三的球队将与德乙第三名进行附加赛,胜者将参加下赛季的德甲联赛。

### 5. 法甲联赛(Ligue 1)

法甲联赛是法国最高级别的足球联赛,由法国职业足球联盟(Ligue de Football Professionnel)管理和运营。赛季结束时,法甲最后两名球队将直接降级至法乙联赛,而法甲前两名将直接晋级;法甲第三名与第四、第五名之间的胜者将进行附加赛,以决定最后一个晋级名额。

## 4.2 模块架构

本国际足球比赛结果预测系统的基本模块架构如图 4-1 所示。

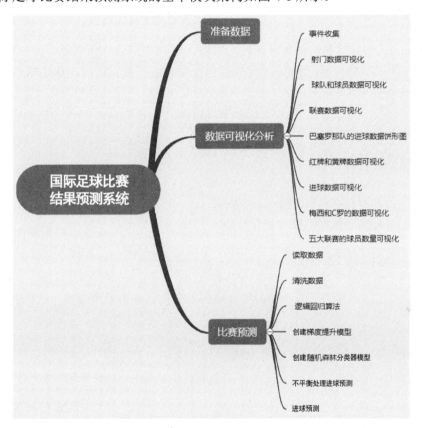

图 4-1 模块结构

## 4.3 准备数据

本项目将使用 Kaggle 网提供的足球比赛数据,在数据中主要包含 events.csv 和 ginf.csv 两个 CSV 文件,其中,ginf.csv 中保存了足球比赛的基本信息,主要包含以下字段。

❑ ID:比赛的编号。

- General：概述信息，包括所属联赛、赛季、日期和主办国。
- Teams：主、客场球队。
- Results：比赛结果，提供主、客场进球数据。
- Odds：赔率(包括主场获胜、客场获胜和平局的赔率)。

在文件 events.csv 中保存了比赛的事件信息，主要包含以下字段。

- ID：编号，包括比赛编号和事件编号。
- Teams：参加比赛的两支球队。
- Player：在比赛的球员动态，介绍球员在比赛中的重要时刻，例如射门、身体部位、机会等信息。
- Shot：射门信息，包括位置(在球场上)、结果、地点和目标等。

## 4.4 数据可视化分析

在本节的内容中，将提取 events.csv 和 ginf.csv 两个 CSV 文件中的数据，并对与足球比赛相关的数据实现可视化分析。

扫码看视频

### 4.4.1 事件收集

(1) 加载数据集，其中包含一些需要与事件数据集合并的重要数据，代码如下：

```
df_events = pd.read_csv("./events.csv")
df_ginf = pd.read_csv("./ginf.csv")
df_ginf = df_ginf[['id_odsp', 'date', 'league', 'season', 'country']]
```

在上述最后一行代码中，设置只选择需要的数据项 id_odsp 作为唯一标识符，用于合并两个数据集。

(2) 加入两个数据集，将"date(日期)"、"league(联赛)"、"赛季(season)"和"国家(country)"信息添加到主数据集，代码如下：

```
df_events = df_events.merge(df_ginf, how='left')

##用主流的联赛名称来命名联赛
leagues = {'E0': 'Premier League', 'SP1': 'La Liga',
           'I1': 'Serie A', 'F1': 'League 1', 'D1': 'Bundesliga'}

##应用映射
df_events['league'] = df_events['league'].map(leagues)
```

(3) 准备两种事件类型，都和球场中发生的比赛相关，代码如下：

```python
##事件类型1
event_type_1 = pd.Series([
    'Announcement',
    'Attempt',
    'Corner',
    'Foul',
    'Yellow card',
    'Second yellow card',
    'Red card',
    'Substitution',
    'Free kick won',
    'Offside',
    'Hand ball',
    'Penalty conceded'], index=[[item for item in range(0, 12)]])

##事件类型2
event_type2 = pd.Series(['Key Pass', 'Failed through ball', 'Sending off', 'Own goal'],
                index=[[item for item in range(12, 16)]])

## 匹配side
side = pd.Series(['Home', 'Away'], index=[[item for item in range(1, 3)]])

## 射门时球相对于球门的位置
shot_place = pd.Series([
    'Bit too high',
    'Blocked',
    'Bottom left corner',
    'Bottom right corner',
    'Centre of the goal',
    'High and wide',
    'Hits the bar',
    'Misses to the left',
    'Misses to the right',
    'Too high',
    'Top centre of the goal',
    'Top left corner',
    'Top right corner'
], index=[[item for item in range(1, 14)]])

##射门结果
shot_outcome = pd.Series(['On target', 'Off target', 'Blocked', 'Hit the bar'],
                index=[[item for item in range(1, 5)]])
##射门位置
location = pd.Series([
    'Attacking half',
    'Defensive half',
    'Centre of the box',
    'Left wing',
    'Right wing',
```

```
    'Difficult angle and long range',
    'Difficult angle on the left',
    'Difficult angle on the right',
    'Left side of the box',
    'Left side of the six yard box',
    'Right side of the box',
    'Right side of the six yard box',
    'Very close range',
    'Penalty spot',
    'Outside the box',
    'Long range',
    'More than 35 yards',
    'More than 40 yards',
    'Not recorded'
],
index=[[item for item in range(1, 20)]])

##球员射门时使用的身体部位
bodypart = pd.Series(['right foot', 'left foot', 'head'], index=[[item for item in
range(1, 4)]])

##辅助方法
assist_method = pd.Series(['None', 'Pass', 'Cross', 'Headed pass', 'Through ball'],
                index=[item for item in range(0, 5)])

##基本信息(例如常规进球、角球、任意球等)
situation = pd.Series(['Open play', 'Set piece', 'Corner', 'Free kick'],
                index=[item for item in range(1, 5)])
],

event_type_1
```

此时会输出：

```
0        Announcement
1             Attempt
2              Corner
3                Foul
4         Yellow card
5   Second yellow card
6            Red card
7        Substitution
8       Free kick won
9             Offside
10           Hand ball
11    Penalty conceded
dtype: object
```

(4) 绘制不同事件的发生次数条形图，代码如下：

```python
def plot_barplot(data, x_ticks, x_labels, y_labels, title, color='muted'):
    ##使用 whitegrid 样式绘制(也可以通过 Param 自定义)
    sns.set_style("whitegrid")    #建议的主题：darkgrid、whitegrid、dark、white 和 ticks

    ##使用自定义图形设置大小
    #plt.figure(figsize=(num, figsize)) num=10, figsize=8
    ## 绘图
    ax = sns.barplot(x = [j for j in range(0, len(data))], y=data.values, palette=color)
    ##设置从数据索引中提取数据
    ax.set_xticks([j for j in range(0, len(data))])
    ##设置图表标签
    ax.set_xticklabels(x_ticks, rotation=45)
    ax.set(xlabel = x_labels, ylabel = y_labels, title = title)
    ax.plot();
    plt.tight_layout()

##事件发生的次数
events_series = df_events['event_type'].value_counts()

##绘图
plot_barplot(events_series, event_type_1.values,"Event type", "Number of events", "Event types")
```

执行后的效果如图 4-2 所示。

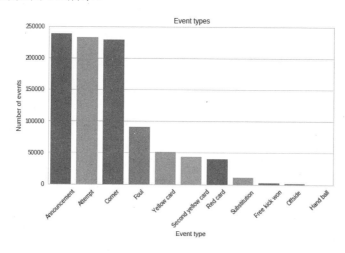

图 4-2　事件次数条形图

## 4.4.2　射门数据可视化

(1) 绘制进球的射门位置条形图，代码如下：

```
##过滤数据
df_shot_places = df_events[(df_events['event_type'] == 1) &
                   (df_events['is_goal'] == 1)]['shot_place'].value_counts()

##绘制图表
plot_barplot(df_shot_places, shot_place[[3, 4, 5, 13, 12]], 'Shot places', 'Number
    of events', 'Shot places resulting in goals')
```

执行后的效果如图4-3所示。

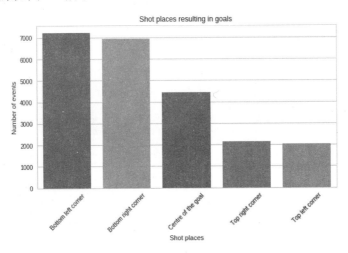

图4-3  进球的射门位置条形图

(2) 绘制没进球的射门位置条形图,代码如下:

```
df_shot_places = df_events[(df_events['event_type'] == 1) &
                   (df_events['is_goal'] == 0)]['shot_place'].value_counts()
plot_barplot(df_shot_places, shot_place, 'Shot places', 'Number of events',
    'Shot places no resulting in goals')
```

执行后的效果如图4-4所示。

(3) 按照射门位置将数据分组,并绘制出对应的条形图,代码如下:

```
##复制原始数据帧
df_shot_places_ed = df_events.copy()
df_shot_places_ed = df_events.groupby('shot_place',
as_index=False).count().sort_values('id_event', ascending=False).dropna()

##将数据帧索引映射到快照,将标签放在字典文件中
df_shot_places_ed['shot_place'] = df_shot_places_ed['shot_place'].map(shot_place)

##绘制图表
plot_barplot(df_shot_places_ed['id_event'], df_shot_places_ed['shot_place'],
```

```
            'Shot places',
            'Number of events',
            'Shot places',
            'BuGn_r')
```

执行后的效果如图 4-5 所示。

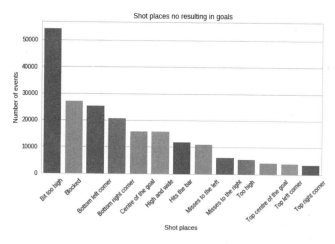

图 4-4 没进球的射门位置条形图

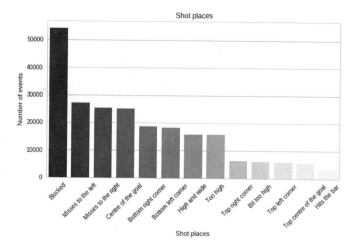

图 4-5 射门位置分组数据条形图

### 4.4.3 球队和球员数据可视化

(1) 提取西甲联赛的射门数据，绘制西甲联赛最具攻击力的球队条形图，代码如下：

```python
## 按照球队分组
grouping_by_offensive = df_events[df_events['league']=='La Liga']
[df_events['is_goal']==1].groupby('event_team')

##对值进行排序
grouping_by_offensive = grouping_by_offensive.count().sort_values(by='id_event',
ascending=False)[:10]
teams = grouping_by_offensive.index
scores = grouping_by_offensive['id_event']

##绘制图形
plot_barplot(scores, teams, 'Teams', 'Number of goals', 'Most offensive teams in
La Liga','PRGn_r')
plt.savefig('offensiveteam.jpg', format='jpg', dpi=1000)
```

执行后的效果如图 4-6 所示。

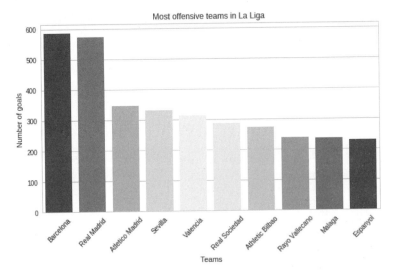

图 4-6 西甲联赛最具攻击力球队条形图

(2) 提取西甲联赛的射门数据，绘制西甲联赛攻击力较弱的球队条形图，代码如下：

```
grouping_by_offensive = df_events[df_events['league']=='La Liga']
[df_events['is_goal']==1].groupby('event_team')

grouping_by_offensive = grouping_by_offensive.count().sort_values(by='id_event',
ascending=True)[:10]
teams = grouping_by_offensive.index
scores = grouping_by_offensive['id_event']
```

```
plot_barplot(scores, teams, 'Teams', 'Number of goals', 'Less offensive teams in
La Liga', 'PRGn_r')
plt.savefig('lessoffensiveteam.jpg', format='jpg', dpi=1000)
```

执行后的效果如图 4-7 所示。

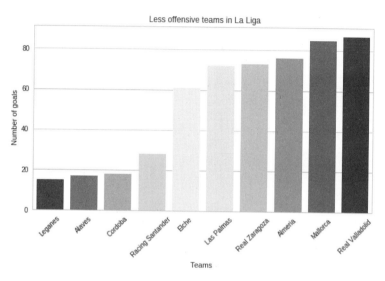

图 4-7　西甲联赛攻击力较弱的球队条形图

(3) 根据球员将进球数据分组，绘制西甲联赛最具攻击力的球员条形图，代码如下：

```
grouping_by_offensive_player = df_events[df_events['league']=='La Liga']
[df_events['is_goal']==1].groupby('player')

##根据球员的进球数排序，然后选出前10名
grouping_by_offensive_player =
grouping_by_offensive_player.count().sort_values(by='id_event',
ascending=False)[:10]
##提取球员名字
players = grouping_by_offensive_player.index
##提取进球数
scores = grouping_by_offensive_player['id_event']

##绘制图形
plot_barplot(scores, players, 'Players', 'Number of Goal', 'Most offensive players
in La Liga')
plt.savefig('offensiveteamplayer.jpg', format='jpg', dpi=1000)
```

执行后的效果如图 4-8 所示。

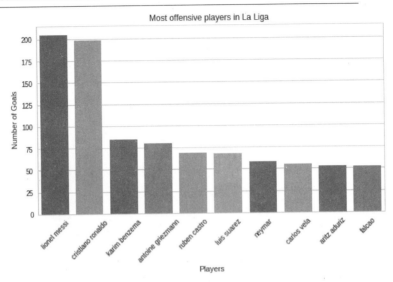

图 4-8 西甲联赛最具攻击力的球员条形图

(4) 根据 event_type 中的事件类型，绘制射门结果的饼形图，代码如下：

```
goal=df_events[df_events['event_team']=='Barcelona'][df_events['league']=='La Liga']
goal['shot_outcome'] = goal['shot_outcome'].map(shot_outcomes)
goal1=goal.copy()

plt.subplot(2,1,2)
plt.figure(figsize=(10,8))
data2=goal1.groupby(by=['shot_outcome'])['shot_outcome'].count()
colors=["green", "red","yellow", "pink"]
plt.pie(data2,autopct='%1.1f%%',labels=data2.index,startangle=60,explode=(0.1,0,0,0))
plt.axis('equal')
plt.title("Percentage of shot outcome",fontsize=15)
plt.legend(fontsize=12,loc='best')
plt.show()
```

执行后的效果如图 4-9 所示。

(5) 分别绘制巴塞罗那队进球情况和射门结果的饼形图，代码如下：

```
from matplotlib.gridspec import GridSpec

the_grid = GridSpec(1, 2)
goal = df_events[df_events['event_team']=='Barcelona'][df_events['league']==
'La Liga'][df_events['is_goal'] == 1]
goal['situation'] = goal['situation'].map(situations)
goal1=goal.copy()
plt.figure(figsize=(10,8))
```

```python
# plt.subplot(2,1,1)

data1=goal1.groupby(by=['situation'])['situation'].count()
# colors=["green", "red","yellow", "pink"]
plt.subplot(the_grid[0, 0], aspect=1)
plt.pie(data1,autopct='%1.1f%%',labels=data1.index,startangle=60,explode=(0,0,0,0.1))
plt.axis('equal')
plt.title("Percentage of goals situations for Barcelona",fontsize=15)
plt.legend(fontsize=12,loc='best')


goals = df_events[df_events['event_team']=='Barcelona'][df_events['league']=='La Liga']
goals['shot_outcome'] = goals['shot_outcome'].map(shot_outcomes)
goals1=goals.copy()
# plt.subplot(2,1,2)
# plt.figure(figsize=(10,8))
data2=goals1.groupby(by=['shot_outcome'])['shot_outcome'].count()
colors=["green", "red","yellow", "pink"]
plt.subplot(the_grid[0, 1], aspect=1)
plt.pie(data2,autopct='%1.1f%%',labels=data2.index,startangle=60,explode=(0,0,0,0.1))
plt.axis('equal')
plt.title("Percentage of shot outcome for Barcelona",fontsize=15)
plt.legend(fontsize=12,loc='best')
plt.savefig('barca.jpg', format='jpg', dpi=1000)
plt.show()
```

执行后的效果如图 4-10 所示。

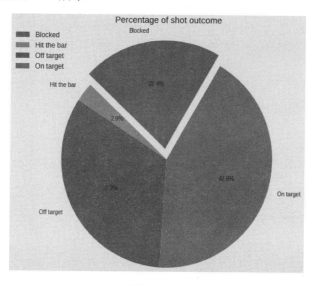

图 4-9　射门结果数据饼形图

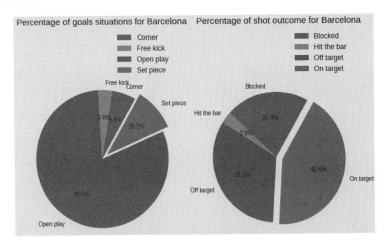

图 4-10 巴塞罗那队进球情况和射门结果的饼形图

(6) 分别绘制巴塞罗那队和皇家马德里队进球情况的饼形图，代码如下：

```
the_grid = GridSpec(1, 2)
goal = df_events[df_events['event_team']=='Barcelona'][df_events['league']==
'La Liga'][df_events['is_goal'] == 1]
goal['situation'] = goal['situation'].map(situations)
goal1=goal.copy()
plt.figure(figsize=(10,8))
# plt.subplot(2,1,1)

data1=goal1.groupby(by=['situation'])['situation'].count()
plt.subplot(the_grid[0, 0], aspect=1)
plt.pie(data1,autopct='%1.1f%%',labels=data1.index,startangle=60,explode=(0,0,0,0.1))
plt.axis('equal')
plt.title("Percentage of goals situations for Barcelona",fontsize=15)
plt.legend(fontsize=12,loc='best')

goal = df_events[df_events['event_team']=='Real Madrid'][df_events['league']==
'La Liga'][df_events['is_goal'] == 1]
goal['situation'] = goal['situation'].map(situations)
goal1=goal.copy()
data1=goal1.groupby(by=['situation'])['situation'].count()
plt.subplot(the_grid[0, 1], aspect=1)
plt.pie(data1,autopct='%1.1f%%',labels=data1.index,startangle=60,explode=(0,0,0,0.1))
plt.axis('equal')
plt.title("Percentage of goals situations for Real Madrid",fontsize=15)
plt.legend(fontsize=12,loc='best')
plt.savefig('barcavsrealmadrid.jpg', format='jpg', dpi=1000)
plt.show()
```

执行后的效果如图 4-11 所示。

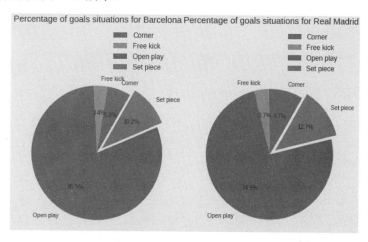

图 4-11 巴塞罗那队和皇家马德里队进球情况的饼形图

(7) 分别绘制皇家马德里队进球情况和射门成功率的饼形图，代码如下：

```
the_grid = GridSpec(1, 2)
goal = df_events[df_events['event_team']=='Real Madrid'][df_events['league']==
'La Liga'][df_events['is_goal'] == 1]
goal['situation'] = goal['situation'].map(situations)
goal1=goal.copy()
plt.figure(figsize=(10,8))

data1=goal1.groupby(by=['situation'])['situation'].count()
plt.subplot(the_grid[0, 0], aspect=1)
plt.pie(data1,autopct='%1.1f%%',labels=data1.index,startangle=60,explode=(0,0,0,0.1))
plt.axis('equal')
plt.title("Percentage of goals situations for Real Madrid",fontsize=15)
plt.legend(fontsize=12,loc='best')


goals = df_events[df_events['event_team']=='Real Madrid']
[df_events['league']=='La Liga']
goals['shot_outcome'] = goals['shot_outcome'].map(shot_outcomes)
goals1=goals.copy()
data2=goals1.groupby(by=['shot_outcome'])['shot_outcome'].count()
colors=["green", "red","yellow", "pink"]
plt.subplot(the_grid[0, 1], aspect=1)
plt.pie(data2,autopct='%1.1f%%',labels=data2.index,startangle=60,explode=(0,0,0,0.1))
plt.axis('equal')
plt.title("Percentage of shot outcome for Real Madrid",fontsize=15)
plt.legend(fontsize=12,loc='best')
```

```
plt.savefig('realmadrid.jpg', format='jpg', dpi=1000)
plt.show()
```

执行后的效果如图 4-12 所示。

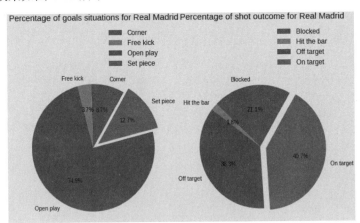

图 4-12　皇家马德里队进球情况和射门成功率的饼形图

### 4.4.4　联赛数据可视化

(1) 加载 events.csv 和 ginf.csv 中的数据,将日期、联赛、赛季和国家等信息添加到主数据集中,代码如下:

```
df_events1 = pd.read_csv("./events.csv")
df_ginf1 = pd.read_csv("./ginf.csv")
df_events1 = df_events1.merge(df_ginf1, how='left')
df_events1 = df_events1[['id_odsp', 'id_event', 'league', 'season', 'ht', 'at',
'event_team', 'is_goal']]
leagues = {'E0': 'Premier League', 'SP1': 'La Liga',
    'I1': 'Serie A', 'F1': 'League One', 'D1': 'Bundesliga'}
df_events1['league'] = df_events1['league'].map(leagues)
```

执行后会输出:

```
<class 'pandas.core.frame.DataFrame'>
Int64Index: 941009 entries, 0 to 941008
Data columns (total 8 columns):
id_odsp        941009 non-null object
id_event       941009 non-null object
league         941009 non-null object
season         941009 non-null int64
ht             941009 non-null object
at             941009 non-null object
```

```
event_team       941009 non-null object
is_goal          941009 non-null int64
dtypes: int64(2), object(6)
memory usage: 64.6+ MB
```

(2) 获取指定联赛的某赛季数据信息,例如获取意甲联赛数据,代码如下:

```
for league in Leagues[::-1]:
    if league == 'Premier League' :
        print('No details on {}'.format(league))
    else:
        for season in Seasons[::-1]:
            print('****Informations about {}'.format(league) + ' ' +'{} ****'.format(season))
            Stats = []
            Teams = df_events1[df_events1['league']== league][df_events1['season']
== season]['ht'].unique()
            Games = df_events1[df_events1['league']== league][df_events1['season']
== season]['id_odsp'].unique()
            for game in Games:
                Events = df_events1[df_events1['league']== league][df_events1['season']
== season][df_events1['id_odsp']==game]
                ht= Events.iloc[1,4]
                at= Events.iloc[1,5]
                Butat=0
                Butht=0
                for j in range(1,Events.shape[0]):
                    if Events.iloc[j,7] == 1 :
                        if Events.iloc[j,6] == ht:
                            Butat +=1
                        else:
                            Butht +=1
                item = [ht, at, Butht, Butat]
                Stats.append(item)
            Stats = np.array(Stats)
            df_Stats = pd.DataFrame({'Teamht': Stats[:,0], 'Butht':Stats[:,2],
'Teamat': Stats[:,1], 'But at':Stats[:,3] })

            results = []
            for team in Teams:
                data_team = df_Stats.loc[(df_Stats['Teamht'] ==
team)|(df_Stats['Teamat'] == team)]
                nbrgoals = 0
                for j in range(1, data_team.shape[0]):
                    if data_team.iloc[j,2]== team:
                        nbrgoals += int(data_team.iloc[j,0])
                    else:
                        nbrgoals += int(data_team.iloc[j,1])
                elem = [team, nbrgoals]
```

```
            results.append(elem)
    results = np.array(results)
    ids = np.argsort(results[:,1])[::-1]
    results[:,1] = results[:,1][ids]
    results[:,0] = results[:,0][ids]

    df_results = pd.DataFrame({'ATeam': results[:,0], 'ButsE': results[:,1]})
    print(df_results)
```

执行后会输出:

```
No details on Premier League
****Informations about Serie A 2017 ****
            ATeam ButsE Butsoo
0         Cagliari    42     43
1          Palermo    40     41
2       US Pescara    39     40
3          Crotone    35     36
4         Sassuolo    32     33
5    Chievo Verona    30     31
6           Torino    28     29
7            Genoa    28     29
8        Sampdoria    27     28
9           Empoli    25     26
10         Bologna    25     26
11         Udinese    24     25
12          Napoli    22     23
13      Fiorentina    21     22
14        AC Milan    21     22
15           Lazio    21     22
16        Atalanta    20     21
17   Internazionale    20     21
18         AS Roma    18     19
19        Juventus    15     16
****Informations about Serie A 2016 ****
            ATeam ButsE Butsoo
0        Frosinone    70     71
1          Palermo    61     62
2    Hellas Verona    59     60
3          Udinese    57     58
4        Sampdoria    54     55
5           Torino    52     53
6            Carpi    51     52
7            Lazio    47     48
8         Atalanta    45     46
9           Empoli    44     45
10           Genoa    44     45
```

```
11     Chievo Verona    44    45
12           Bologna    43    44
13        Fiorentina    40    41
14           AS Roma    39    40
15          AC Milan    38    39
16          Sassuolo    38    39
17     Internazionale   34    35
18            Napoli    28    29
19          Juventus    19    20
****Informations about Serie A 2015 ****
         ATeam  ButsE  Butsoo
0            Parma    74    75
1           Cesena    73    74
2         Cagliari    67    68
3    Hellas Verona    64    65
4          Udinese    56    57
5         Sassuolo    56    57
6         Atalanta    56    57
7          Palermo    54    55
8           Napoli    52    53
9           Empoli    50    51
10         AC Milan    49    50
11   Internazionale    47    48
12           Torino    45    46
13       Fiorentina    44    45
14            Genoa    44    45
15        Sampdoria    40    41
16    Chievo Verona    40    41
17            Lazio    35    36
18          AS Roma    31    32
19         Juventus    24    25
****Informations about Serie A 2014 ****
```

(3) 按球员分组统计何时进球，绘制不同时间段的进球数据条形图，代码如下：

```
goal = df_events[df_events['is_goal']==1]

plt.hist(goal.time, 100)
plt.xlabel("TIME (min)",fontsize=10)
plt.ylabel("Number of goals",fontsize=10)
plt.title("goal counts vs time",fontsize=15)
x=goal.groupby(by='time')['time'].count().sort_values(ascending=False).index[0]
y=goal.groupby(by='time')['time'].count().sort_values(ascending=False).iloc[0]
x1=goal.groupby(by='time')['time'].count().sort_values(ascending=False).index[1]
y1=goal.groupby(by='time')['time'].count().sort_values(ascending=False).iloc[1]
plt.text(x=x-10,y=y+10,s='time:'+str(x)+',max:'+str(y),fontsize=12,fontdict=
{'color':'red'})
```

```
plt.text(x=x1-10,y=y1+10,s='time:'+str(x1)+',the 2nd
max:'+str(y1),fontsize=12,fontdict={'color':'black'})
plt.savefig('goals.jpg', format='jpg', dpi=1000)
plt.show()
```

执行后的效果如图 4-13 所示。

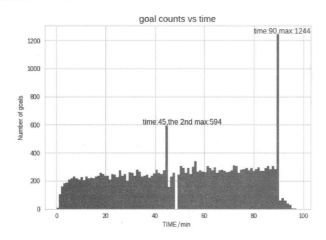

图 4-13　不同时间段的进球数据条形图

### 4.4.5　巴塞罗那队的进球数据饼形图

根据 event_type 中的事件类型，绘制巴塞罗那队的进球数据饼形图。On target 表示命中目标，Off target 表示偏离目标，Blocked 表示受阻挡，Hit the bar 表示几种横梁或门框。代码如下：

```
shot_outcomes = {1:'On target', 2:'Off target', 3:'Blocked', 4:'Hit the bar'}
assist_methods = {0:np.nan,1:'Pass', 2:'Cross', 3:'Headed pass', 4:'Through ball'}
situations = {1:'Open play', 2:'Set piece', 3:'Corner', 4:'Free kick'}
goal['shot_outcome'] = goal['shot_outcome'].map(shot_outcomes)
goal['assist_method'] =goal['assist_method'].map(assist_methods)
goal['situation'] = goal['situation'].map(situations)
goal1=goal.copy()
plt.subplot(2,1,1)
plt.figure(figsize=(10,8))
data1=goal1.groupby(by=['situation'])['situation'].count()
colors=["green", "red","yellow", "pink"]
plt.pie(data1,autopct='%1.1f%%',labels=data1.index,startangle=60,explode=(0,0,0,0.1))
plt.axis('equal')
plt.title("Percentage of situations for Barcelona",fontsize=15)
plt.legend(fontsize=12,loc='best')
plt.show()
```

执行后的效果如图 4-14 所示。

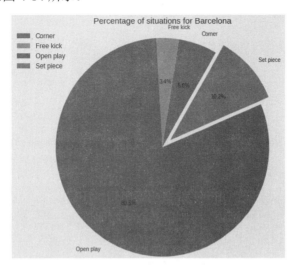

图 4-14　巴塞罗那队的进球数据饼形图

## 4.4.6　红牌和黄牌数据可视化

（1）绘制西甲联赛红牌数据条形图，代码如下：

```
redCards = df_events[df_events['league']=='La Liga'][df_events['event_type'] == 6]
['event_team']

##红牌事件发生次数
redCards_series = redCards.value_counts().sort_values(ascending=True)[:10]

## 绘制条形图
plot_barplot(redCards_series, redCards_series.index, "Event_team",
             "Number of Red Cards", "Red Cards per team in La Liga", 'gist_earth')
plt.savefig('redcard.jpg', format='jpg', dpi=1000)
```

执行后的效果如图 4-15 所示。

（2）绘制西甲联赛黄牌数据条形图，代码如下：

```
yellowCards = df_events[df_events['league']=='La Liga'][df_events['event_type']
== (4 or 5)]['event_team']

yellowCards_series = yellowCards.value_counts().sort_values(ascending=True)[:10]

plot_barplot(yellowCards_series, yellowCards_series.index, "Event_team",
             "Number of yellow Cards", "Yellow Cards per team", 'gist_earth')
```

执行后的效果如图 4-16 所示。

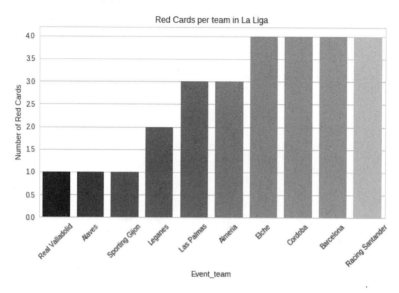

图 4-15　西甲联赛红牌数据条形图

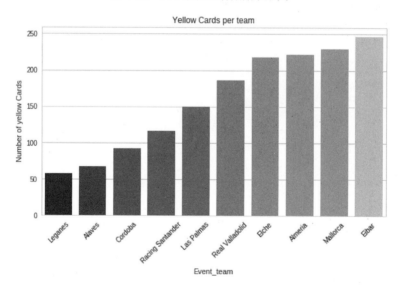

图 4-16　西甲联赛黄牌数据条形图

(3) 获取红牌和黄牌数据，然后绘制红牌条形图，展示最有可能收到红牌的时间。代码如下：

```
df_unique_events = df_events.drop_duplicates()

first_yellow_cards = df_unique_events [df_unique_events ['event_type'] == ('Yellow 
card')]     # 选择所有事件类型为"黄牌"的行
second_yellow_cards= df_unique_events [df_unique_events ['event_type'] == ('Second 
yellow card')]     # 选择所有事件类型为"第二张黄牌"的行
red_cards = df_unique_events [df_unique_events['event_type'] == ('Red card')]
# 选择所有事件类型为"红牌"的行
yellow_cards= df_unique_events [df_unique_events ['event_type'] == ('Yellow card' 
or 'Second yellow card')]

card_frames = [red_cards, yellow_cards]
all_cards = pd.concat(card_frames)
```

执行后的效果如图 4-17 所示。

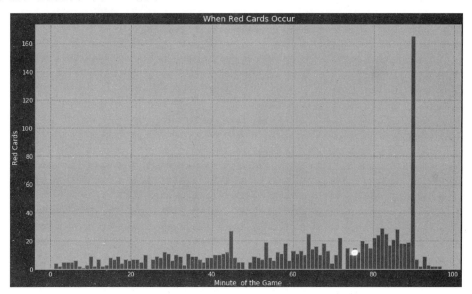

图 4-17  红牌数据条形图

(4) 绘制黄牌条形图,展示最有可能收到黄牌的时间,代码如下:

```
fig2 = plt.figure(figsize=(14,8))
plt.hist(first_yellow_cards.time, 100, color="yellow")
plt.xlabel("Minute of the Game")
plt.ylabel("First Yellow Cards")
plt.title("When First Yellow Cards Occur")
```

执行后的效果如图 4-18 所示。

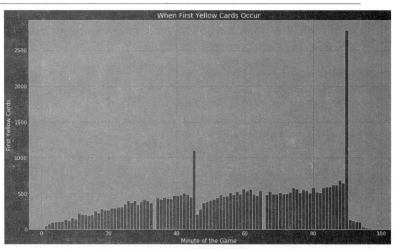

图 4-18　黄牌数据条形图

(5) 绘制第二张黄牌条形图,展示最有可能收到第二张黄牌的时间,代码如下:

```
fig3 = plt.figure(figsize=(14,8))
plt.hist(second_yellow_cards.time, 100, color="yellow")
plt.xlabel("Minute of the Game")
plt.ylabel("Second Yellow Cards")
plt.title("When Second Yellow Cards Occur")
```

执行后的效果如图 4-19 所示。

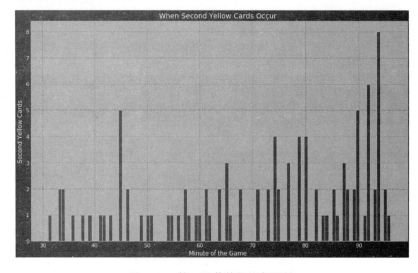

图 4-19　第二张黄牌数据条形图

(6) 绘制不同联赛的黄牌数据条形图，代码如下：

```
yellow_league = pd.crosstab(index=yellow_cards.event_type,
columns=yellow_cards.league)
yellow_league.plot(kind='bar', figsize=(14,14))
```

执行后的效果如图 4-20 所示。

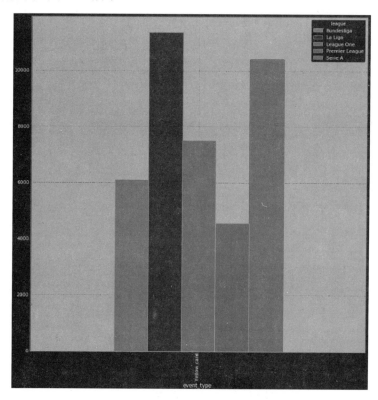

图 4-20　不同联赛的黄牌数据条形图

(7) 在拿到红牌、黄牌、第二张黄牌后，绘制影响球员在球场上表现的条形图，代码如下：

```
event_per_player = all_players_event_types.set_index('player')
f, ax = plt.subplots(figsize=(12, 12))
all_players_goals.goal_True
sns.distplot(all_players_goals.goal_False, color='g')
```

执行后的效果如图 4-21 所示。

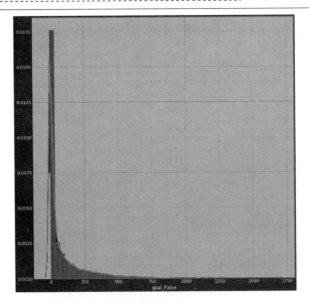

图 4-21 拿牌后影响球员表现的条形图

### 4.4.7 进球数据可视化

(1) 绘制尤文图斯队在各个时间段不同方式的进球数据条形图,代码如下:

```
team = 'Juventus'
Team_strategy(team)
```

执行后的效果如图 4-22 所示。

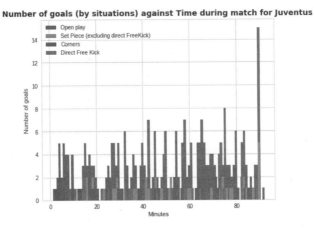

图 4-22 尤文图斯队在各个时间段不同方式的进球数据条形图

(2) 在数据中用 penalties 表示点球，通过下面的代码绘制点球数据饼图，可以展示出与点球相关的信息，例如中路、上中、左下、右下、没有命中等数据，代码如下：

```
penalties=df_events[df_events["location"]==14]

# 射门位置
for i in range(14):
    if sum(penalties["shot_place"]==i)==0:
        print(i)

top_left=sum(penalties["shot_place"]==12)
bot_left=sum(penalties["shot_place"]==3)
top_right=sum(penalties["shot_place"]==13)
bot_right=sum(penalties["shot_place"]==4)
centre=sum(penalties["shot_place"]==5)+sum(penalties["shot_place"]==11)
missed=sum(penalties["shot_place"]==1)+sum(penalties["shot_place"]==6)+sum(penalties["shot_place"]==7)+sum(penalties["shot_place"]==8)+sum(penalties["shot_place"]==9)+sum(penalties["shot_place"]==10)

labels_pen=["top left","bottom left","centre","top right","bottom right","missed"]
num_pen=[top_left,bot_left,centre,top_right,bot_right,missed]
colors_pen=["red", "aqua","royalblue","yellow","violet","m"]
plt.pie(num_pen,labels=labels_pen,colors=colors_pen,autopct='%1.1f%%',startangle=60,explode=(0,0,0,0,0,0.2))
plt.axis('equal')
plt.title("Percentage of each placement of penalties",fontsize=18,fontweight="bold")
fig=plt.gcf()
fig.set_size_inches(8,6)
plt.show()
```

执行后的效果如图 4-23 所示。

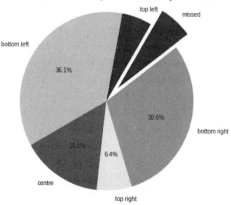

图 4-23 点球数据饼图

(3) 绘制命中点球时的左、右脚统计图,代码如下:

```
scored_pen=penalties[penalties["is_goal"]==1]
pen_rightfoot=scored_pen[scored_pen["bodypart"]==1].shape[0]
pen_leftfoot=scored_pen[scored_pen["bodypart"]==2].shape[0]

penalty_combi=pd.DataFrame({"right foot":pen_rightfoot,"left foot":pen_leftfoot},
index=["Scored"])
penalty_combi.plot(kind="bar")
plt.title("Penalties scored (Right/Left foot)",fontsize=14,fontweight="bold")
penalty_combi
```

执行后的效果如图 4-24 所示。

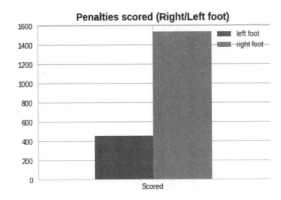

图 4-24　命中点球时的左、右脚统计图

(4) 分别绘制比赛前 15 分钟和后 15 分钟的进球数据条形图,代码如下:

```
first_15 = df_events[df_events['time'] <= 15]
last_15 = df_events[(df_events['time'] >= 75) & (df_events['time'] <= 90)]

top_10_scorer_first_15 = first_15[first_15['is_goal'] == 1].
groupby('event_team').count().sort_values(by='id_event', ascending=False)

teams = top_10_scorer_first_15.index[:10]
scores = top_10_scorer_first_15['id_event'][:10]

sns.set_style("whitegrid")
fig, axs = plt.subplots(ncols=2, figsize=(15, 6))
ax = sns.barplot(x = [j for j in range(0, len(scores))], y=scores.values, ax=axs[0])
ax.set_xticks([j for j in range(0, len(scores))])
ax.set_xticklabels(teams, rotation=45)
```

```
ax.set(xlabel = 'Teams', ylabel = 'Number of goals', title = 'Goals scored in the
1st 15 minutes');

top_10_scorer_last_15 = last_15[last_15['is_goal'] ==
1].groupby('event_team').count().sort_values(by='id_event',
ascending=False)[:10]

teams_last_15 = top_10_scorer_last_15.index[:10]
scores_last_15 = top_10_scorer_last_15['id_event'][:10]

ax = sns.barplot(x = [j for j in range(0, len(scores_last_15))],
y=scores_last_15.values, ax=axs[1])
ax.set_xticks([j for j in range(0, len(scores_last_15))])
ax.set_xticklabels(teams_last_15, rotation=45)
ax.set(xlabel = 'Teams', ylabel = 'Number of goals', title = 'Goals scored in the
last 15 minutes');
plt.savefig('Lastminutewinners.jpg', format='jpg', dpi=1000)
fig.tight_layout()
```

执行后的效果如图 4-25 所示。

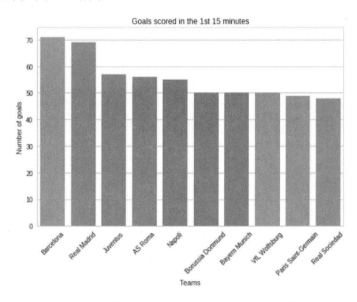

图 4-25　比赛前 15 分钟和后 15 分钟的进球数据条形图

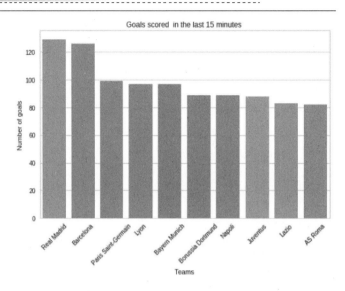

图 4-25  比赛前 15 分钟和后 15 分钟的进球数据条形图(续)

(5) 统计不同球队在各个时间段不同方式的进球数据，例如，绘制切尔西队的进球数据条形图，代码如下：

```
def Team_strategy(team):
    goal = df_events[df_events['is_goal']==1][df_events['event_team'] == team]
    plt.hist(goal[goal["situation"]==1]["time"],width=1,bins=100,label="Open play")
    plt.hist(goal[goal["situation"]==2]["time"],width=1,bins=100,label="Set Piece (excluding direct FreeKick)")

plt.hist(goal[goal["situation"]==3]["time"],width=1,bins=100,label="Corners")
    plt.hist(goal[goal["situation"]==4]["time"],width=1,bins=100,label="Direct Free Kick")
    plt.xlabel("Minutes")
    plt.ylabel("Number of goals")
    plt.legend()
    plt.title("Number of goals (by situations) against Time during match for {}".format(team),fontname="Times New Roman Bold",fontsize=14,fontweight="bold")
    plt.tight_layout()

team = 'Chelsea'
Team_strategy(team)
```

执行后的效果如图 4-26 所示。

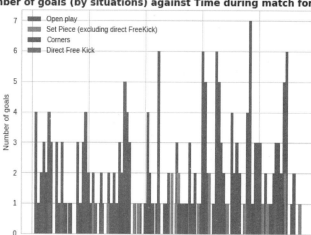

图 4-26　切尔西队在各个时间段不同方式的进球数据条形图

(6) 绘制各个球队的射门效率条形图，需要先创建包含总尝试次数和总目标的数据帧，代码如下：

```
result_df = pd.DataFrame({'total': total.dropna(), 'is_goal':
grouped_by_player_is_goal_filtered['id_event']})

result_df.sort_values('total', ascending=False, inplace=True)
sns.set(style="darkgrid")

f, ax = plt.subplots(figsize=(10, 6))

## 绘制条形图
sns.set_color_codes("pastel")
sns.barplot(x="total",
        y=result_df.index,
        data=result_df,
        label="# of attempts", color="b")

sns.set_color_codes("muted")
sns.barplot(x='is_goal',
        y=result_df.index,
        data=result_df,
        label="# of goals", color="b")

ax.legend(ncol=2, loc="lower right", frameon=True)
ax.set(ylabel="Teams",
    xlabel="Number of goals x attempts", title='Shooting Accuracy')
```

```
each = result_df['is_goal'].values
the_total = result_df['total'].values
x_position = 50

for i in range(len(ax.patches[:30])):
    ax.text(ax.patches[i].get_width() - x_position, ax.patches[i].get_y() +.50,
            str(round((each[i]/the_total[i])*100, 2))+'%')

sns.despine(left=True, bottom=True)
f.tight_layout()
plt.savefig('ShootingAccuracy.jpg', format='jpg', dpi=1000)
```

执行后的效果如图4-27所示。

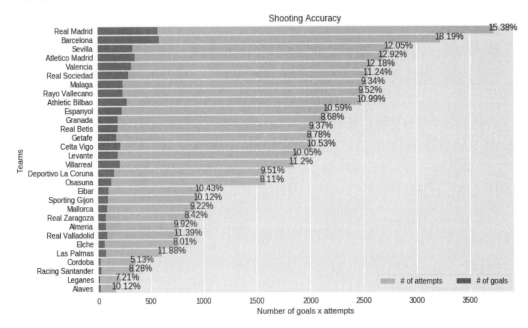

图4-27 各球队的射门效率条形图

(7) 绘制进球时间条形图，分析什么时候最有可能进球，代码如下：

```
goals=df_unique_events[df_unique_events["is_goal"]==1]

fig4=plt.figure(figsize=(14,8))
plt.hist(goals.time,width=1,bins=100,color="green")    #绘制进球时间直方图，设置每个
#柱状图的宽度为1分钟，总共有100个柱状图，即每分钟一个柱状图，颜色设置为绿色
plt.xlabel("Minutes")
plt.ylabel("Number of goals")
plt.title("Number of goals against Time during match")
```

执行后的效果如图 4-28 所示。

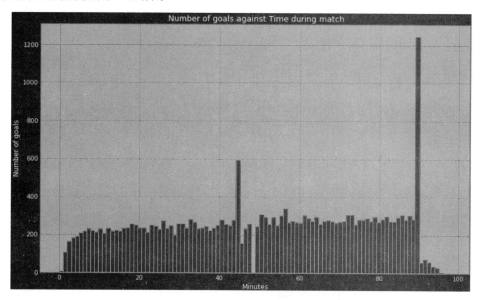

图 4-28 进球时间条形图

(8) 绘制五大联赛进球数据对比条形图,代码如下:

```
no_goal=df_unique_events[df_unique_events["is_goal"]==0]
goal=df_unique_events[df_unique_events["is_goal"]==1]

goals=pd.concat([no_goal, goal])

player_no_goal = no_goal[['player','is_goal']]
player_no_goals = player_no_goal.groupby('player').count()

player_no_goals.columns = ['goal_False']

player_goal = goal[['player','is_goal']]
player_goals = player_goal.groupby('player').count()
player_goals.columns = ['goal_True']

plt.figure(figsize=(14,8))
sns.countplot(x='league', hue='is_goal', data=goals)
```

执行后的效果如图 4-29 所示。

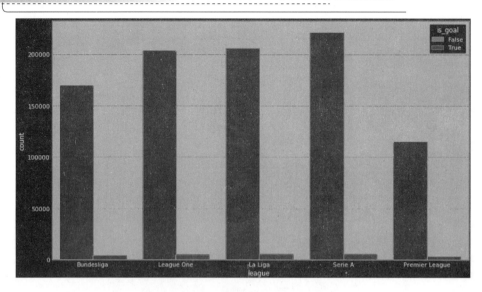

图 4-29 五大联赛进球数据对比条形图

### 4.4.8 梅西和 C 罗的数据可视化

(1) 分别统计梅西和 C 罗的射门数据，代码如下：

```
def pen_full_stats(player):
    player_pen=penalties[penalties["player"]==player]
    scored_pen=player_pen[player_pen["is_goal"]==1]
    missed_pen=player_pen[player_pen["is_goal"]==0]

    top_left_rightfoot=scored_pen[scored_pen["shot_place"]==12] [scored_pen
["bodypart"]==1].shape[0]
    top_left_leftfoot=scored_pen[scored_pen["shot_place"]==12] [scored_pen
["bodypart"]==2].shape[0]
    bot_left_rightfoot=scored_pen[scored_pen["shot_place"]==3] [scored_pen
["bodypart"]==1].shape[0]
    bot_left_leftfoot=scored_pen[scored_pen["shot_place"]==3] [scored_pen
["bodypart"]==2].shape[0]
    top_right_rightfoot=scored_pen[scored_pen["shot_place"]==13] [scored_pen
["bodypart"]==1].shape[0]
    top_right_leftfoot=scored_pen[scored_pen["shot_place"]==13] [scored_pen
["bodypart"]==2].shape[0]
    bot_right_rightfoot=scored_pen[scored_pen["shot_place"]==4] [scored_pen
["bodypart"]==1].shape[0]
    bot_right_leftfoot=scored_pen[scored_pen["shot_place"]==4] [scored_pen
["bodypart"]==2].shape[0]
```

```
    centre_rightfoot=scored_pen[scored_pen["shot_place"]==5][scored_pen
["bodypart"]==1].shape[0]+scored_pen[scored_pen["shot_place"]==11][scored_pen
["bodypart"]==1].shape[0]
    centre_leftfoot=scored_pen[scored_pen["shot_place"]==5][scored_pen
["bodypart"]==2].shape[0]+scored_pen[scored_pen["shot_place"]==11][scored_pen
["bodypart"]==2].shape[0]
    scored_without_recorded_loc_rightfoot=scored_pen[scored_pen["shot_place"]
.isnull()][scored_pen["bodypart"]==1].shape[0]
    scored_without_recorded_loc_leftfoot=scored_pen[scored_pen["shot_place"]
.isnull()][scored_pen["bodypart"]==2].shape[0]
    missed_rightfoot=missed_pen[missed_pen["bodypart"]==1].shape[0]
    missed_leftfoot=missed_pen[missed_pen["bodypart"]==2].shape[0]

    right_foot=pd.DataFrame({"Top Left Corner":top_left_rightfoot,"Bottom Left
Corner":bot_left_rightfoot,"Top Right Corner":top_right_rightfoot,"Bottom Right
Corner":bot_right_rightfoot,"Centre":centre_rightfoot,"Unrecorded placement":
scored_without_recorded_loc_rightfoot,"Missed":missed_rightfoot},index=["Right
Foot attempt"])
    left_foot=pd.DataFrame({"Top Left Corner":top_left_leftfoot,"Bottom Left
Corner":bot_left_leftfoot,"Top Right Corner":top_right_leftfoot,"Bottom Right
Corner":bot_right_leftfoot,"Centre":centre_leftfoot,"Unrecorded placement":
scored_without_recorded_loc_leftfoot,"Missed":missed_leftfoot},index=["Left
Foot attempt"])

    fullstats=right_foot.append(left_foot)
    fullstats=fullstats[["Top Right Corner","Bottom Right Corner","Top Left
Corner","Bottom Left Corner","Centre","Unrecorded placement","Missed"]]
    return fullstats
pen_full_stats("lionel messi")
pen_full_stats("cristiano ronaldo")
```

执行后会输出：

|  | Top Right Corner | Bottom Right Corner | Top Left Corner | Bottom Left Corner | Centre | Unrecorded placement | Missed |
| --- | --- | --- | --- | --- | --- | --- | --- |
| Right Foot attempt | 0 | 0 | 0 | 0 | 0 | 0 | 0 |
| Left Foot attempt | 7 | 8 | 3 | 6 | 5 | 1 | 7 |
|  | Top Right Corner | Bottom Right Corner | Top Left Corner | Bottom Left Corner | Centre | Unrecorded placement | Missed |
| Right Foot attempt | 3 | 14 | 2 | 19 | 3 | 2 | 8 |
| Left Foot attempt | 0 | 0 | 0 | 0 | 0 | 0 | 0 |

(2) 分别统计梅西和 C 罗的进球数据，代码如下：

```
def stats(player):
    player_pen=df_events[df_events["player"]==player]
    right_attempt=player_pen[player_pen["bodypart"]==1]
    right_attempt_scored=right_attempt[right_attempt["is_goal"]==1].shape[0]
```

```
        right_attempt_missed=right_attempt[right_attempt["is_goal"]==0].shape[0]
        left_attempt=player_pen[player_pen["bodypart"]==2]
        left_attempt_scored=left_attempt[left_attempt["is_goal"]==1].shape[0]
        left_attempt_missed=left_attempt[left_attempt["is_goal"]==0].shape[0]
        head_attempt=player_pen[player_pen["bodypart"]==3]
        head_attempt_scored=head_attempt[head_attempt["is_goal"]==1].shape[0]
        head_attempt_missed=head_attempt[head_attempt["is_goal"]==0].shape[0]
        scored=pd.DataFrame({"right foot":right_attempt_scored,"left
foot":left_attempt_scored, "head": head_attempt_scored},index=["Scored"])
        missed=pd.DataFrame({"right foot":right_attempt_missed,"left
foot":left_attempt_missed, "head": head_attempt_missed},index=["Missed"])
        combi=scored.append(missed)
        return combi

stats('lionel messi')

stats("cristiano ronaldo")
```

执行后会输出：

|  | head | left foot | right foot |
|---|---|---|---|
| Scored | 8 | 167 | 30 |
| Missed | 45 | 585 | 79 |

|  | head | left foot | right foot |
|---|---|---|---|
| Scored | 36 | 32 | 130 |
| Missed | 123 | 205 | 664 |

（3）绘制其他多名球员的射门效率条形图，代码如下：

```
#最有效的球员
##对于事件类型为射门(event_type==1)的记录，按球员分组并计数
grouped_by_player = df_events[df_events['event_type'] == 1].groupby('player').count()

##对于事件类型为射门(event_type==1)且成功进球(is_goal==1)的记录，按球员分组并计数
grouped_by_player_goals = df_events[(df_events['event_type'] == 1) &
                        (df_events['is_goal'] == 1)].groupby('player').count()

##当不是目标且尝试==1时，按球员分组
grouped_by_player_not_goals = df_events[(df_events['event_type'] == 1) &
                        (df_events['is_goal'] == 0)].groupby('player').count()

##设置一个阈值，以过滤少量尝试的球员，这可能导致最终结果缺乏一致性
threshold = grouped_by_player['id_event'].std()
```

```
grouped_by_player_is_goal = df_events[df_events['is_goal'] == 1].groupby('player').count()
##过滤至少有超过平均次数的球员
##例如，有两次尝试和 1 个进球的球员具有非常高的效率
##这位球员没有为他的球队创造多少机会
grouped_by_player_is_goal_filtered =
grouped_by_player_goals[grouped_by_player_goals['id_event'] > threshold]
grouped_by_players_not_goal_filtered =
grouped_by_player_not_goals[grouped_by_player_not_goals['id_event'] > threshold]

##射门总数
total = grouped_by_players_not_goal_filtered['id_event'] +
grouped_by_player_is_goal_filtered['id_event']

result = total/grouped_by_player_is_goal_filtered['id_event']
result.dropna(inplace=True)
sorted_results = result.sort_values(ascending=True)
players = sorted_results[:10].index
effectiveness = sorted_results[:10]
plot_barplot(effectiveness, players, 'Players',
        '# of shots needed to score a goal',
        'Most effective players',
        'RdBu_r', 8, 6)
```

执行后的效果如图 4-30 所示。

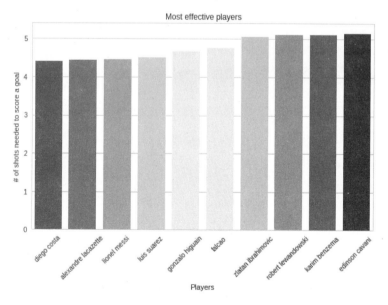

图 4-30　多名球员的射门效率条形图

### 4.4.9 五大联赛的球员数量可视化

绘制 Venn 图,展示五大联赛的球员数量,代码如下:

```
AA = goals.player.groupby(goals.league)
names = list(AA.groups.keys())
labels = venn.get_labels([set(AA.get_group(names[i])) for i in range(5)],
fill=['number', 'logic'])

fig, ax = venn.venn5(labels, names=names, figsize=(20,20), fontsize=14)
fig.show()
```

执行后的效果如图 4-31 所示。

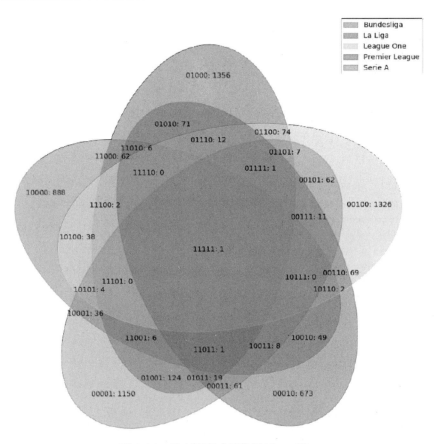

图 4-31 五大联赛球员数量 Venn 图

## 4.5 比赛预测

接下来，我们将根据射门的位置、射手、联赛、助攻方式、比赛状态、球场上的球员数量、时间等因素，来预测射门成功进球的概率，并以此来预测整场比赛的结果。

扫码看视频

### 4.5.1 读取数据

(1) 编写函数 read_merge()，功能是读取文件 events.csv 和 ginf.csv 合并后的数据，代码如下：

```
def read_merge():
    df_events = pd.read_csv("../events.csv")
    df_game_info = pd.read_csv("../ginf.csv")

    #手动将dictionary.txt转换为Python dicts
    event_types = {1:'Attempt', 2:'Corner', 3:'Foul', 4:'Yellow card', 5:'Second yellow card', 6:'Red card', 7:'Substitution', 8:'Free kick won', 9:'Offside', 10:'Hand ball', 11:'Penalty conceded'}
    event_types2 = {12:'Key Pass', 13:'Failed through ball', 14:'Sending off', 15:'Own goal'}
    sides = {1:'Home', 2:'Away'}
    shot_places = {1:'Bit too high', 2:'Blocked', 3:'Bottom left corner', 4:'Bottom right corner', 5:'Centre of the goal', 6:'High and wide', 7:'Hits the bar', 8:'Misses to the left', 9:'Misses to the right', 10:'Too high', 11:'Top centre of the goal', 12:'Top left corner', 13:'Top right corner'}
    shot_outcomes = {1:'On target', 2:'Off target', 3:'Blocked', 4:'Hit the bar'}
    locations = {1:'Attacking half', 2:'Defensive half', 3:'Centre of the box', 4:'Left wing', 5:'Right wing', 6:'Difficult angle and long range', 7:'Difficult angle on the left', 8:'Difficult angle on the right', 9:'Left side of the box', 10:'Left side of the six yard box', 11:'Right side of the box', 12:'Right side of the six yard box', 13:'Very close range', 14:'Penalty spot', 15:'Outside the box', 16:'Long range', 17:'More than 35 yards', 18:'More than 40 yards', 19:'Not recorded'}
    bodyparts = {1:'right foot', 2:'left foot', 3:'head'}
    assist_methods = {0:np.nan, 1:'Pass', 2:'Cross', 3:'Headed pass', 4:'Through ball'}
    situations = {1:'Open play', 2:'Set piece', 3:'Corner', 4:'Free kick'}

    #将dict映射到events(事件)dataframe
    df_events['event_type'] = df_events['event_type'].map(event_types)
    df_events['event_type2'] = df_events['event_type2'].map(event_types2)
    df_events['side'] = df_events['side'].map(sides)
    df_events['shot_place'] = df_events['shot_place'].map(shot_places)
```

```
df_events['shot_outcome']= df_events['shot_outcome'].map(shot_outcomes)
df_events['location'] = df_events['location'].map(locations)
df_events['bodypart'] = df_events['bodypart'].map(bodyparts)
df_events['assist_method']= df_events['assist_method'].map(assist_methods)
df_events['situation'] = df_events['situation'].map(situations)

##为了清晰起见，我们使用主流的联赛名称来标识各个联赛
leagues = {'E0': 'Premier League', 'SP1': 'La Liga',
           'I1': 'Serie A', 'F1': 'League 1', 'D1': 'Bundesliga'}

## 映射到 events
df_game_info.league = df_game_info.league.map(leagues)

#合并到一个表中(合并其他数据集，需要包含国家、联赛、日期和赛季等信息)
df_events = df_events.merge(df_game_info ,how = 'left')

return df_events
```

(2) 编写函数 fill_unk() 填充数据，用 UNK 填充 df 中的分类特征：shot_place、player、shot_outcome、bodypart、location 和 assist_method。代码如下：

```
def fill_unk(df):
    df.shot_place.fillna('UNK',inplace = True)
    df.player.fillna('UNK',inplace = True)
    df.shot_outcome.fillna('UNK',inplace = True)
    df.bodypart.fillna('UNK',inplace = True)
    df.location.fillna('UNK',inplace = True)
    df.assist_method.fillna('UNK',inplace = True)
    df.situation.fillna('UNK',inplace = True)
    df.location.replace('Not recorded', 'UNK', inplace= True)
```

(3) 编写函数 missing_values_table()，功能是计算数据框中每列的缺失值，代码如下：

```
def missing_values_table(df):

    #缺失值总计
    mis_val = df.isnull().sum()

    #缺失值的百分比
    mis_val_percent = 100 * df.isnull().sum() / len(df)

    #制作结果表
    mis_val_table = pd.concat([mis_val, mis_val_percent], axis=1)

    #重命名列
    mis_val_table_ren_columns = mis_val_table.rename(
    columns = {0 : 'Missing Values', 1 : '% of Total Values'})
```

```
#按缺少的降序百分比对表格进行排序
mis_val_table_ren_columns = mis_val_table_ren_columns[
    mis_val_table_ren_columns.iloc[:,1] != 0].sort_values(
'% of Total Values', ascending=False).round(1)

#打印一些摘要信息
print ("Your selected dataframe has " + str(df.shape[1]) + " columns.\n"
    "There are " + str(mis_val_table_ren_columns.shape[0]) +
    " columns that have missing values.")

#返回缺少信息的数据帧
return mis_val_table_ren_columns
```

### 4.5.2 清洗数据

接下来,我们需要处理缺失值,用 UNK 填充分类特征,在分类列中用新类 UNK 替换任意空值。代码如下:

```
df = read_merge()
df[df.location == 'Not recorded'].location.count()
df.dtypes
#检查每个功能中缺少的值
missing_values_table(df)

#手动功能选择
feat_cols = ['odd_h', 'odd_d', 'odd_a',
        'assist_method', 'location',
        'side', 'shot_place', 'situation',
        'bodypart', 'time', 'is_goal']

df_feats = select_feats(df, feat_cols)
df_feats['first_half'] = df_feats.time <= 45
```

浏览数据并绘制可视化条形图。

```
#查找与 is_goal 的相关性
correlations = df_feats.corr()['is_goal'].sort_values()
correlations

import seaborn as sns
%matplotlib inline

sns.barplot(df_feats.situation, df_feats.is_goal);
```

执行效果如图 4-32 所示。

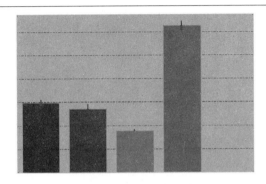

图 4-32　绘制的可视化条形图

### 4.5.3　逻辑回归算法

使用逻辑回归算法进行预测分析，并绘制出逻辑回归对比图。代码如下：

```
from sklearn import metrics
from sklearn.linear_model import LogisticRegression

model = LogisticRegression()
model.fit(X_res_u, y_res_u)

print(model)

#做出预测
expected = y_val
predicted = model.predict(x_val)

#总结模型的适合性
print(metrics.classification_report(expected, predicted))
print(metrics.confusion_matrix(expected, predicted))

print('Accuracy: ', metrics.accuracy_score(expected, predicted))
print('Balanced Acc: ', metrics.balanced_accuracy_score(expected, predicted))

# 查找残差
pred = predicted.astype('int8')
y = y_val.as_matrix().astype('int8')
residuals = abs(pred - y)
x = x_val.as_matrix()
# 提取最错误的预测
wrong = x[np.argmax(residuals), :]

from lime import lime_tabular
```

```
# 创建 LIME 解释器对象
explainer = lime_tabular.LimeTabularExplainer(training_data = X_res_u,
                        mode = 'classification',
                        training_labels = y_res_u,
                        feature_names = df_feats_dumm.columns.tolist())

exp = explainer.explain_instance(data_row = wrong,
                        predict_fn = model.predict_proba, labels=(0,1))

# 绘制预测解释
exp.as_pyplot_figure(label=0);
```

执行效果如图 4-33 所示。

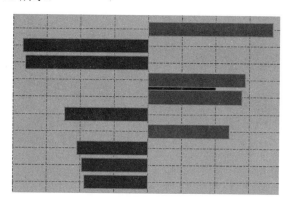

图 4-33 逻辑回归对比图

## 4.5.4 创建梯度提升模型

创建梯度提升模型(gradient boosting model),并评估该模型。代码如下:

```
from sklearn.ensemble import GradientBoostingClassifier

gradient_boosted = GradientBoostingClassifier()

#根据训练数据拟合模型
gradient_boosted.fit(X_res_u, y_res_u)

#对测试数据进行预测
predictions = gradient_boosted.predict(x_val)

#总结模型的适合性
print(metrics.classification_report(y_val, predictions))
print(metrics.confusion_matrix(y_val, predictions))
```

#评估模型
```
print('Accuracy: ', metrics.accuracy_score(y_val, predictions))
print('Balanced Acc: ', metrics.balanced_accuracy_score(y_val, predictions))
```

执行后输出：

```
              precision    recall   f1-score   support

      False       1.00      0.94      0.97      91631
       True       0.32      1.00      0.49       2470

  micro avg       0.94      0.94      0.94      94101
  macro avg       0.66      0.97      0.73      94101
weighted avg      0.98      0.94      0.96      94101

[[86439  5192]
 [    1  2469]]
Accuracy:  0.9448146140848662
Balanced Acc:  0.9714665475065742
```

### 4.5.5 创建随机森林分类器模型

创建随机森林分类器模型，提取项目中 15 个重要的特征，并绘制出对应的可视化条形图。代码如下：

```
def plot_feature_importances(df):
    #根据重要性对特征进行排序
    df = df.sort_values('importance', ascending = False).reset_index()
    
    #标准化特征
    df['importance_normalized'] = df['importance'] / df['importance'].sum()
    
    #制作特征重要性的水平条形图
    plt.figure(figsize = (10, 6))
    ax = plt.subplot()
    
    #需要反转索引，以在顶部绘制最重要的内容
    ax.barh(list(reversed(list(df.index[:15]))),
            df['importance_normalized'].head(15),
            align = 'center', edgecolor = 'k')
    
    # 设置yticks
    ax.set_yticks(list(reversed(list(df.index[:15]))))
    ax.set_yticklabels(df['feature'].head(15))
    
    #绘图标签
    plt.xlabel('Normalized Importance'); plt.title('Feature Importances')
    plt.show()
```

```
    return df
from sklearn.ensemble import RandomForestClassifier

#制作随机森林分类器
random_forest = RandomForestClassifier(n_estimators = 100, random_state = 12,
verbose = 1, n_jobs = -1)
#根据训练数据拟合模型
random_forest.fit(X_res_u, y_res_u);

features = list(df_feats_dumm.columns)

#提取重要特征
feature_importance_values = random_forest.feature_importances_
feature_importances = pd.DataFrame({'feature': features, 'importance':
feature_importance_values})

#对测试数据进行预测
predictions = random_forest.predict(x_val)

#总结模型
print(metrics.classification_report(y_val, predictions))
print(metrics.confusion_matrix(y_val, predictions))

print('Accuracy: ', metrics.accuracy_score(y_val, predictions))
print('Balanced Acc: ', metrics.balanced_accuracy_score(y_val, predictions))

#显示默认功能
feature_importances_sorted = plot_feature_importances(feature_importances)
```

执行效果如图 4-34 所示。

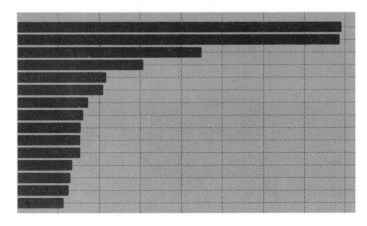

图 4-34　15 个特征的可视化条形图

### 4.5.6 不平衡处理

接下来，我们将使用 UNK 填充空值，然后用 Imbalanced-Learn 中的 SMOTENC 方法解决采样数据不平衡的问题。

(1) 首先查看原始数据集的 shape 属性，代码如下：

```
from imblearn.over_sampling import SMOTENC
from collections import Counter
print('Original dataset shape %s' % Counter(y_train))
```

(2) 在删除空值后，保存使用 SMOTE 重新采样的数据集，代码如下：

```
x_train.iloc[0,1]
num_feats = len(x_train.columns)
x_train.head()
cat_indices = np.arange(3, num_feats)
df_X_smote = pd.DataFrame(X_res, columns=df_feats_dumm.columns)
df_y_smote = pd.DataFrame(y_res, columns=['is_goal'])
df_X_smote.to_csv('../X_res_SMOTE_UNK.csv')
df_y_smote.to_csv('../y_res_SMOTE_UNK.csv')
print('Resampled dataset shape %s' % Counter(y_res))
```

执行后输出：

```
Resampled dataset shape Counter({False: 824932, True: 824932})
```

(3) 重新实现逻辑回归模型，代码如下：

```
from sklearn import metrics
from sklearn.linear_model import LogisticRegression

model = LogisticRegression()
model.fit(X_res, y_res)

expected = y_val
predicted = model.predict(x_val)

print(metrics.classification_report(expected, predicted))
print(metrics.confusion_matrix(expected, predicted))

print('Accuracy: ', metrics.accuracy_score(expected, predicted))
print('Balanced Acc: ', metrics.balanced_accuracy_score(expected, predicted))
```

执行后输出：

```
              precision    recall  f1-score   support

       False       1.00      0.95      0.97     91631
        True       0.34      0.99      0.51      2470

   micro avg       0.95      0.95      0.95     94101
   macro avg       0.67      0.97      0.74     94101
weighted avg       0.98      0.95      0.96     94101

[[86870  4761]
 [   21  2449]]
Accuracy:  0.9491822616125227
Balanced Acc:  0.9697697886749339
```

(4) 实现随机森林分类器，代码如下：

```
random_forest = RandomForestClassifier(n_estimators = 100, random_state = 12, 
verbose = 1, n_jobs = -1)
#训练数据
random_forest.fit(X_res, y_res);

#功能名称
features = list(df_feats_dumm.columns)

#提取功能重要性
feature_importance_values = random_forest.feature_importances_
feature_importances = pd.DataFrame({'feature': features, 'importance': 
feature_importance_values})

#对测试数据进行预测
predictions = random_forest.predict(x_val)

print(metrics.classification_report(y_val, predictions))
print(metrics.confusion_matrix(y_val, predictions))

print('Accuracy: ', metrics.accuracy_score(y_val, predictions))
print('Balanced Acc: ', metrics.balanced_accuracy_score(y_val, predictions))
feature_importances_sorted = plot_feature_importances(feature_importances)
```

执行后输出：

```
              precision    recall  f1-score   support

       False       0.99      0.98      0.99     91631
        True       0.49      0.79      0.61      2470

   micro avg       0.97      0.97      0.97     94101
   macro avg       0.74      0.88      0.80     94101
weighted avg       0.98      0.97      0.98     94101
```

```
[[89600  2031]
 [  514  1956]]
Accuracy: 0.9729545913433438
Balanced Acc: 0.8848689230882341
```

执行效果如图 4-35 所示。

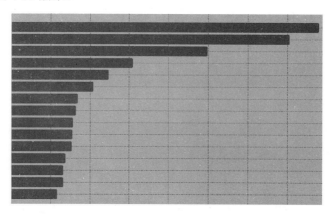

图 4-35 二次处理后 15 个特征的可视化条形图

## 4.6 进球预测

进球预测也就是比分预测,是本项目中的一个重要功能,我们将根据以往比赛的相关数据来预测比分。在本节的内容中,将详细讲解进球预测的实现过程。

扫码看视频

### 4.6.1 预处理

(1) 加载文件 events.csv 和 ginf.csv 中的数据,代码如下:

```
df_events = pd.read_csv("events.csv")
df_game_info = pd.read_csv("ginf.csv")
```

(2) 将数据处理成训练和测试模型时所需的标签(即目标变量),代码如下:

```
df_game_info_select=df_game_info[['fthg','ftag']].fillna(-1)
l1=np.array(df_game_info_select.values, dtype=int)
l1=l1
l1.shape
```

执行后输出:

(10112, 2)

(3) 将二维数组 ll 中的每个进球数(即主队和客队的进球数)转换为二进制表示,并存储到新的二维数组 coded_ll 中。代码如下:

```
coded_ll=np.empty((ll.shape[0],2))
for j in range (ll.shape[0]):
        coded_ll[j,0]=int(np.binary_repr(ll[j,0]))
        coded_ll[j,1]=int(np.binary_repr(ll[j,1]))
coded_ll.shape
```

执行后输出:

(10112, 2)

(4) 从原始的 df_events DataFrame 中选取特定的列,并处理缺失值。代码如下:

```
df_events_select=df_events[['event_type','event_type2','shot_place','shot_outcome',
'location','side']]
#put NAn = -1
df_events_select=df_events_select.fillna(-1)
colums=['id_odsp','event_type','event_type2','shot_place','shot_outcome','location',
'side',]
```

(5) 将 Pandas DataFrame 转换为 NumPy 数组,代码如下:

```
event_np=df_events_select.values
event_np=np.array(event_np,dtype=int)
event_np.shape
```

执行后输出:

(941009, 6)

(6) 将一个包含足球比赛事件数据的 NumPy 数组 event_np 转换成一个二进制表示的 NumPy 数组 coded_events,代码如下:

```
coded_events=np.ones((event_np.shape[0],6))
for i in range (event_np.shape[0]):
   for j in range (event_np.shape[1]):
        coded_events[i,j]=int(np.binary_repr(event_np[i,j]))
coded_events.shape
```

执行后输出:

(941009, 6)

(7) 将 events 分为主场和客场比赛,根据一个二维数组 full 中的特定列(第 7 列,索引为 6)的值,将数组中的行分为两组:一组是当第 7 列的值为 1 时的行;另一组是当第 7 列的值为 2 时的行。然后,将这两组行转换成 NumPy 数组,并且只保留每行的前 6 个特征。

代码如下:

```
full_home=[]
full_away=[]
for i in range (full.shape[0]):
    if full[i,6]==1:
        full_home.append(full[i])
    if full[i,6]==2:
        full_away.append(full[i])
full_home=np.array(full_home)[:,:6]
full_away=np.array(full_away)[:,:6]
print('full_home.shape',full_home.shape)
print('full_away.shape',full_away.shape)
```

执行后输出:

```
full_home.shape (488224, 6)
full_away.shape (452785, 6)
```

(8) 将 codedevents 分为主场和客场，从二维数组 coded_full 中根据第 7 列的值将数据分为两组：一组是当第 7 列的值为 1 时的行；另一组是当第 7 列的值为 10 时的行。这表明已成功按第 7 列的值分为主场和客场数据。代码如下：

```
coded_full_home=[]
coded_full_away=[]
for i in range (coded_full.shape[0]):
    if coded_full[i,6]==1:
        coded_full_home.append(coded_full[i])
    if coded_full[i,6]==10:
        coded_full_away.append(coded_full[i])
coded_full_home=np.array(coded_full_home)[:,:6]
coded_full_away=np.array(coded_full_away)[:,:6]
print('coded_full_home.shape',coded_full_home.shape)
print('coded_full_away.shape',coded_full_away.shape)
```

执行后输出:

```
coded_full_home.shape (488224, 6)
coded_full_away.shape (452785, 6)
```

(9) 创建主场比赛事件的三维 NumPy 数组，代码如下：

```
a3=-1*np.ones((9074,180,5),dtype=int)  #每场比赛最多包含180个事件
j=0
c=0
for i in range (full_home.shape[0]-1):
    if full_home[i,0]==full_home[i+1,0]:
        a3[c,j]=full_home[i,1:]
        j+=1
```

```
        else:
            j=0
            c+=1
print('c',c)
full1=a3
full1.shape
```

执行后输出:

```
(9074, 180, 5)
```

(10) 创建客场比赛事件的三维 NumPy 数组，代码如下:

```
a3=-1*np.ones((9074,180,5),dtype=int)  #每场比赛最多包含180个事件
j=0
c1=0
for i in range (full_away.shape[0]-1):
    if full_away[i,0]==full_away[i+1,0]:
        a3[c1,j]=full_away[i,1:]
        j+=1
    else:
        j=0
        c1+=1
print('c1',c1)
full2=a3
full2.shape
```

执行后输出:

```
(9074, 180, 5)
```

(11) 分别创建 coded 主场和客场比赛事件的三维 NumPy 数组，代码如下:

```
a3=-1*np.ones((9074,180,5),dtype=int)
j=0
c=0
for i in range (coded_full_home.shape[0]-1):
    if coded_full_home[i,0]==coded_full_home[i+1,0]:
        a3[c,j]=coded_full_home[i,1:]
        j+=1
    else:
        j=0
        c+=1
print('c',c)
full3=a3
full3.shape

a3=-1*np.ones((9074,180,5),dtype=int)
j=0
c=0
```

```
for i in range (coded_full_away.shape[0]-1):
    if coded_full_away[i,0]==coded_full_away[i+1,0]:
        a3[c,j]=coded_full_away[i,1:]
        j+=1
    else:
        j=0
        c+=1
print('c',c)
full4=a3
full4.shape
```

执行后输出：

```
(9074, 180, 5)
(9074, 180, 5)
```

(12) 将数据从(180*5)向量化到(1900)，代码如下：

```
training_data1=np.empty((9074,900))
for i in range (full1.shape[0]):
    training_data1[i]=full1[i].reshape(-1)
training_data1.shape

training_data2=np.empty((9074,900))
for i in range (full2.shape[0]):
    training_data2[i]=full2[i].reshape(-1)
training_data2.shape

training_data3=np.empty((9074,900))
for i in range (full3.shape[0]):
    training_data3[i]=full3[i].reshape(-1)
training_data3.shape

training_data4=np.empty((9074,900))
for i in range (full4.shape[0]):
    training_data4[i]=full4[i].reshape(-1)
training_data4.shape
```

执行后会输出：

```
(9074, 900)
(9074, 900)
(9074, 900)
```

## 4.6.2 创建循环神经网络

(1) 创建循环神经网络(RNN)，分别创建训练函数和测试函数，主要代码如下：

```python
class SimpleRNN(nn.Module):
    def __init__(self, input_size, hidden_size, output_size):
        super().__init__()
        self.rnn = torch.nn.RNN(input_size, hidden_size, nonlinearity='relu', batch_first=True)
        self.linear = torch.nn.Linear(hidden_size, output_size)

    def forward(self, x):
        x, _ = self.rnn(x)
        x = self.linear(x)
        return x

class SimpleLSTM(nn.Module):
    def __init__(self, input_size, hidden_size, output_size):
        super().__init__()
        self.lstm = torch.nn.LSTM(input_size, hidden_size, batch_first=True)
        self.linear = torch.nn.Linear(hidden_size, output_size)

    def forward(self, x):
        x, _ = self.lstm(x)
        x = self.linear(x)
        return x

def train(model, train_data_gen, labels, criterion, optimizer):
    device = torch.device('cuda:0' if torch.cuda.is_available() else 'cpu')
    model.train()
    num_correct = 0

    for batch_idx in range(len(train_data_gen)):
        optimizer.zero_grad()

        data, target = train_data_gen[batch_idx], labels[batch_idx]
        data=data.view(1,1,900)
        target=target.view(1,1,1)
        data, target = data.float().to(device), target.long().to(device)

        # 模型的前向传播
        output = model(data)
        loss = criterion(output.float(), target.float())
        loss.backward()

        optimizer.step()

        y_pred = output
        if abs(y_pred - target.float()) <= 0.5:
            num_correct+=1

    return num_correct, loss.item()
```

```python
def test(model, test_data_gen,labels_test, criterion):
    device = torch.device('cuda:0' if torch.cuda.is_available() else 'cpu')
    model.eval()
    
    num_correct = 0
    
    with torch.no_grad():
        for batch_idx in range(len(test_data_gen)):
            data, target = test_data_gen[batch_idx],labels_test[batch_idx]
            data=data.view(1,1,900)
            target=target.view(1,1,1)
            
            data, target = (data).float().to(device), (target).long().to(device)
            
            output = model(data)
            
            loss = criterion(output.float(), target.float())
            y_pred = output
            if abs(y_pred - target.float()) <= 0.5:
                num_correct+=1
    return num_correct, loss.item()
```

(2) 获取没有经过 One-hot 编码的主场训练数据，主要代码如下：

```
train_d=training_data1[0:2000]
train_d=torch.from_numpy(train_d)
print('train_d.shape',train_d.shape)
train_label=l1[0:2000,0].reshape(-1,1)
train_label=torch.from_numpy(train_label)
print('train_label.shape',train_label.shape)
test_d=training_data1[2000:2100]
test_d=torch.from_numpy(test_d)
print('test_d.shape',test_d.shape)
test_label=l1[2000:2100,0]
test_label=torch.from_numpy(test_label)
print('test_label.shape',test_label.shape)

batch_size = 100
train_data_gen = train_d
test_data_gen = test_d
labels = train_label
labels_test=test_label

input_size = 900
hidden_size = 500
output_size = 1
model = SimpleRNN(input_size, hidden_size, output_size)
criterion = torch.nn.MSELoss()
optimizer =torch.optim.Adam(model.parameters(),lr=0.001,weight_decay=1e-3)
```

```python
max_epochs = 100
device = torch.device('cuda:0' if torch.cuda.is_available() else 'cpu')
start=time.perf_counter()
model = train_and_test(model, train_data_gen,labels, test_data_gen,labels_test,
criterion, optimizer, max_epochs)
print(model)
print('Total time is {} Second'.format(time.perf_counter()-start))
```

执行后会输出:

```
[Epoch 1 / 100] loss: 2.213989496231079, acc 25.8%
[Epoch 2 / 100] loss: 2.2713334560394287, acc 24.999999999999996%
[Epoch 3 / 100] loss: 1.7270112037658691, acc 30.0%
[Epoch 4 / 100] loss: 0.44491833448410034, acc 31.05%
[Epoch 5 / 100] loss: 1.29188072681427, acc 34.25%
[Epoch 6 / 100] loss: 0.2968190312385559, acc 33.800000000000004%
[Epoch 7 / 100] loss: 0.1385488212108612, acc 35.25%
[Epoch 8 / 100] loss: 0.32342100143432617, acc 35.449999999999996%
[Epoch 9 / 100] loss: 0.2702680230140686, acc 37.55%
///省略部分结果
[Epoch 88 / 100] loss: 0.16087624430656433, acc 65.9%
[Epoch 89 / 100] loss: 0.32201728224754333, acc 66.4%
[Epoch 90 / 100] loss: 0.12332797050476074, acc 66.25%
[Epoch 91 / 100] loss: 0.14182744920253754, acc 64.45%
[Epoch 92 / 100] loss: 0.13937370479106903, acc 64.14999999999999%
[Epoch 93 / 100] loss: 0.14549803733825684, acc 64.5%
[Epoch 94 / 100] loss: 0.13153932988643646, acc 66.9%
[Epoch 95 / 100] loss: 0.13768810033798218, acc 67.2%
[Epoch 96 / 100] loss: 0.10590659081935883, acc 67.2%
[Epoch 97 / 100] loss: 0.02342037670314312, acc 66.9%
[Epoch 98 / 100] loss: 0.12119510769844055, acc 66.60000000000001%
[Epoch 99 / 100] loss: 0.022589420899748802, acc 67.95%
[Epoch 100 / 100] loss: 0.04556592181324959, acc 69.25%
SimpleRNN(
  (rnn): RNN(900, 500, batch_first=True)
  (linear): Linear(in_features=500, out_features=1, bias=True)
)
Total time is 4984.555050394993 Second
```

(3) 获取没有经过 One-hot 编码的客场训练数据，主要代码如下:

```python
train_d=training_data2[0:2000]
train_d=torch.from_numpy(train_d)
print('train_d.shape',train_d.shape)
train_label=l1[0:2000,1].reshape(-1,1)
train_label=torch.from_numpy(train_label)
print('train_label.shape',train_label.shape)
test_d=training_data2[2000:2100]
```

```
test_d=torch.from_numpy(test_d)
print('test_d.shape',test_d.shape)
test_label=l1[2000:2100,1]
test_label=torch.from_numpy(test_label)
print('test_label.shape',test_label.shape)

batch_size = 100
train_data_gen = train_d
test_data_gen = test_d
labels = train_label
labels_test=test_label

input_size = 900
hidden_size = 500
output_size = 1
model = SimpleRNN(input_size, hidden_size, output_size)
criterion = torch.nn.MSELoss()
optimizer =torch.optim.Adam(model.parameters(),lr=0.001,weight_decay=1e-3)
max_epochs = 100
device = torch.device('cuda:0' if torch.cuda.is_available() else 'cpu')
start=time.perf_counter()
model = train_and_test(model, train_data_gen,labels, test_data_gen,labels_test,
    criterion, optimizer, max_epochs)

print(model)
print('Total time is {} Second'.format(time.perf_counter()-start))
```

预测比赛的进球数如图 4-36 所示。

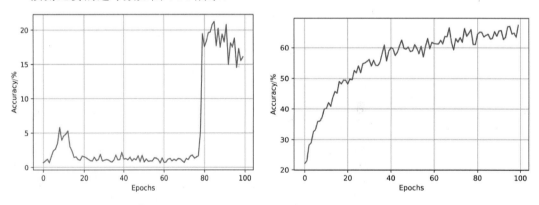

图 4-36 预测比赛的进球数

为了节省本书的篇幅，深度学习方面的知识不再继续讲解，本项目的详细实现过程请参阅本书提供的素材源码和视频讲解。

# 第 5 章 智能素描绘图系统

> 智能绘图系统在人工智能领域比较常见，很多人认为人工智能技术的出现会替代绘画者和设计师。本章介绍了一个基于图像处理和绘图技术的系统，旨在实现自动化图像绘制。该系统使用了一系列算法和技术，包括边缘增强、图像量化和调色以及笔画绘制等。通过这些步骤，系统可以将输入的图像转换为以笔画形式绘制的作品，具体流程由 PyTorch+OpenCV+Matplotlib+NumPy+Scikit–image 实现。

## 5.1 项目介绍

本项目是一款经典的智能素描绘图软件,旨在提供一个有趣的、富有创造性的平台,让用户能够轻松地绘制各种素描图像。这款绘图软件具有以下特点。

扫码看视频

- 绘画工具:软件提供了多种绘画工具,包括不同颜色的画笔、油漆、纹理和装饰品。用户可以根据自己的喜好选择合适的工具,绘制出个性化的素描图像。
- 多种绘画模式:提供了多种绘画模式,如自由绘画模式、填色模式和涂鸦模式。用户可以根据自己的创意选择不同的模式,创作出独特的作品。
- 保存和分享:用户可以将绘制的图像保存到本地,也可以分享给朋友、社交媒体或打印成实体作品。
- 用户界面友好:软件具有简洁直观的用户界面,方便操作和导航,为用户提供更好的体验。

## 5.2 需求分析

1) 功能需求

扫码看视频

- 图形绘制:允许用户选择需要绘制的图形类型(如人物、景色、猫、狗、树等),并指定相应的参数(如大小、颜色、位置等),以绘制该图形并显示在屏幕上。
- 图形编辑:允许用户选择已绘制的图形,并进行编辑操作,如修改颜色、调整大小、移动位置等。
- 图形删除:允许用户选择已绘制的图形,并将其从画布上删除。
- 图形保存:允许用户将绘制的图形保存为图像文件,以便后续使用或分享。
- 图形加载:允许用户从图像文件中加载已保存的图形,并显示在画布上进行编辑或查看。

2) 用户界面需求

- 图形选择界面:提供一个用户界面,展示可供选择的图形类型列表,用户可以从中选择要绘制的图形。
- 参数输入界面:在选定图形后,显示参数输入界面,允许用户输入图形的参数值,如大小、颜色、位置等。
- 画布显示界面:提供一个画布区域,用于显示绘制的图形,并可以进行编辑和删除操作。
- 文件操作界面：提供文件操作选项，允许用户保存绘制的图形为图像文件，或加载已保存的图像文件。

3) 性能需求
- 响应时间：用户输入命令或进行操作后软件应能够快速响应，并及时显示绘制的图形或执行相应的操作。
- 稳定性：能够处理各种异常情况和错误输入，并为用户提供相应的提示和处理方式。
- 可扩展性：具备良好的可扩展性，方便用户后续增加更多图形类型和功能。

4) 安全性需求
- 用户身份验证：提供用户身份验证机制，确保只有授权用户可以进行绘制和编辑操作。
- 数据保护：确保用户输入的参数和绘制的图形数据得到适当的保护，防止数据泄露或篡改。

5) 兼容性需求
- 平台兼容性：应能够兼容多个操作系统平台，如 Windows、Mac OS 和 Linux 等。
- 文件格式兼容性：支持常见的图像文件格式，如 PNG、JPEG 等，以方便用户保存和加载文件。

## 5.3 功能模块

本项目的具体结构如图 5-1 所示。

扫码看视频

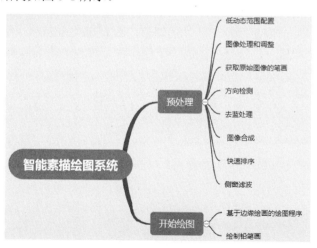

图 5-1　项目结构

## 5.4 预处理

在正式绘图之前,需要预先对图像进行处理,主要包括低动态范围配置、图像处理和调整、获取原始图像的笔画、方向检测、去蓝处理、图像合成、快速排序和侧窗滤波等。在本节中,将详细介绍上述预处理功能的实现过程。

扫码看视频

### 5.4.1 低动态范围配置

编写文件 LDR.py,实现图像的低动态范围配置,LDR 常见的格式包括 png 和 jpg 等,例如一张 8 位图,RGB 三个通道灰度值变化范围为 0~255,即 8 位图可以表示 $256^3$ 种颜色,虽然看似很多,但实际上远远不够来描述我们身处真实世界中的色彩。

文件 LDR.py 的具体实现流程如下。

(1) 编写方法 LDR(),其功能是将输入的图像转换为浮点数格式,将像素值划分为 n 个区间,然后进行量化,并将像素值重新映射回原始范围。对应的实现代码如下:

```
def LDR(img, n):
    Interval = 255.0/n
    img = np.float32(img)
    img = np.uint8(img/Interval)
    img = np.clip(img,0,n-1)
    img = np.uint8((img+0.5)*Interval)
    return img
```

(2) 编写方法 HistogramEqualization(),其功能是调用函数 createCLAHE()生成自适应均衡化图像,参数 clipLimit 表示颜色对比度的阈值,参数 tileGridSize 用于设置像素均衡化的网格大小,即在多个网格下进行直方图的均衡化操作。该方法的具体实现代码如下:

```
def HistogramEqualization(img,clipLimit=2, tileGridSize=(10,10)):
    # 创建一个 CLAHE 对象(参数可选)
    clahe = cv2.createCLAHE(clipLimit=clipLimit, tileGridSize=tileGridSize)
    img = clahe.apply(img)
    return img
```

(3) 在测试主函数中加载一个灰度图像,对图像进行直方图均衡化处理,具体实现代码如下:

```
if __name__ == '__main__':
    img_path = './input/jiangwen/010s.jpg'
    img = cv2.imread(img_path, cv2.IMREAD_GRAYSCALE)
```

```
img = HistogramEqualization(img)

s = SideWindowFilter(radius=1, iteration=1)
img_tensor = F.to_tensor(img).unsqueeze(0)

res = s.forward(img_tensor).squeeze().detach().numpy()

for n in range(8, 11):
    img_ldr = LDR(res, n)
    cv2.imwrite("LDR{}.jpg".format(n), img_ldr)

print("done")
```

上述代码的具体说明如下。

(1) 创建一个 SideWindowFilter 对象(s = SideWindowFilter(radius=1, iteration=1))。

(2) 将图像转换为 PyTorch 张量(img_tensor = F.to_tensor(img).unsqueeze(0))。

(3) 使用 SideWindowFilter 对象对图像进行滤波处理(res = s.forward(img_tensor).squeeze().detach().numpy())。

(4) 对处理后的图像进行多次 LDR 处理，并将结果保存为不同的图像文件(for n in range(8, 11): img_ldr = LDR(res, n) cv2.imwrite("LDR{}.jpg".format(n), img_ldr))。

(5) 打印输出 done 表示处理完成。

## 5.4.2 图像处理和调整

编写文件 tone.py，用于对图像进行处理和调整，主要实现图像色调调整和单通道低动态范围处理的功能。文件 tone.py 的具体实现流程如下。

(1) 导入所需的库，包括 cv2、NumPy 和 math，具体实现代码如下：

```
import cv2
import numpy as np
import math
```

(2) 编写函数 transferTone()，用于调整图像的色调。首先，计算输入图像的直方图，并对直方图进行归一化处理。然后，根据给定的权重和概率密度函数，计算出每个像素点的新值，以实现色调调整。最后，显示调整后的图像。函数 transferTone()的具体实现代码如下：

```
def transferTone(img):
    ho, _ = np.histogram(img.flatten(), bins=256, range=[0, 256], density=True)
    ho_cumulative = np.cumsum(ho)

    omiga1 = 76
    omiga2 = 22
```

```
omiga3 = 2
p1 = lambda x: (1 / 9.0) * np.exp(-(255 - x) / 9.0)
p2 = lambda x: (1.0 / (225 - 105)) * ((x >= 105) & (x <= 225))
p3 = lambda x: (1.0 / np.sqrt(2 * math.pi * 11)) * np.exp(-((x - 90) ** 2) / float((2 * (11 ** 2))))
p = lambda x: (omiga1 * p1(x) + omiga2 * p2(x) + omiga3 * p3(x)) * 0.01

prob = np.array([p(i) for i in range(256)])
prob /= np.sum(prob)
histo_cumulative = np.cumsum(prob)

Iadjusted = np.interp(ho_cumulative, histo_cumulative, np.arange(256))
Iadjusted = np.uint8(Iadjusted.reshape(img.shape))

cv2.imshow('adjust tone', Iadjusted)
cv2.waitKey(0)
J = cv2.blur(Iadjusted, (3, 3))
cv2.imshow('blurred adjust tone', J)
cv2.waitKey(1)
return J
```

(3) 编写函数 LDR_single()，用于对图像进行单通道低动态范围处理。首先将输入图像分为 n 个区间，然后对每个区间的像素值进行量化，并将像素值映射为对应的色调。最后，将每个区间的色调图像保存到指定路径下。函数 LDR_single() 的具体实现代码如下：

```
def LDR_single(img, n, output_pathos):
    Interval = 250.0 / n
    img = np.float32(img)
    img = np.uint8(img / Interval)
    img = np.clip(img, 0, n - 1)

    tones = [(i + 0.5) * Interval for i in range(n)]
    tones = np.array(tones)
    mask = np.zeros(img.shape, dtype=bool)
    for i in range(n):
        mask = (img == i)
        tone = np.uint8(tones[i] * mask + (1 - mask) * 255)
        cv2.imwrite(output_pathos + "/tone{}.png".format(i), tone)

    return
```

(4) 编写函数 LDR_single_add()，用于对图像进行单通道的累积低动态范围处理。它与 LDR_single() 函数类似，将输入图像分为 n 个区间，并根据累积的像素值计算每个像素点的色调值。此外，还将每个区间的二值掩码和累积的掩码保存到指定路径下。函数 LDR_single_add() 的具体实现代码如下：

```python
def LDR_single_add(img, n, output_pathos):
    Interval = 250.0 / n
    img = np.float32(img)
    img = np.uint8(img / Interval)
    img = np.clip(img, 0, n - 1)

    mask_add = np.zeros(img.shape, dtype=bool)
    for i in range(n):
        mask = (img == i)
        mask_add += mask
        cv2.imwrite(output_pathos + "/mask/mask{}.png".format(i),
np.uint8(mask_add * 255))
        tone = np.uint8((i + 0.5) * Interval * mask_add + (1 - mask_add) * 255)
        cv2.imwrite(output_pathos + "/mask/tone_cumulate{}.png".format(i), tone)

    return
```

（5）加载一个灰度图像，调用 LDR_single() 函数对图像进行单通道的 LDR 处理，并将结果保存到指定路径下。调用 LDR_single_add() 函数对图像进行单通道的累积 LDR 处理，并将结果保存到指定路径下。具体实现代码如下：

```python
if __name__ == '__main__':
    img_path = './input/jw.png'
    img = cv2.imread(img_path, cv2.IMREAD_GRAYSCALE)

    # img = transferTone(img)
    # cv2.imwrite("./input/jiangwen/transferTone.png", img)

    LDR_single(img, 10, output_pathos="./output")
    LDR_single_add(img, 10, output_pathos="./output")
    print("done")
```

## 5.4.3 获取原始图像的笔画

在计算机视觉和图像处理领域，stroke 通常用来指代图像中的笔画或轮廓。获取图像的笔画可以使用边缘检测、轮廓提取或图像分割等技术。这些技术可以帮助我们找到图像中物体的边界或轮廓，并进一步分析和处理这些边界信息。要计算和获取图像的笔画，可以使用各种图像处理库和算法，常用的主要有 OpenCV、Pillow、Scikit-image 等，具体的实现方法可能会根据应用的需求和所使用的库而有所不同。

在本项目中，编写文件 Stroke_origin.py，用于计算和获取原始图像的笔画(stroke)。通过计算图像的梯度信息和滤波操作，可以生成具有线条效果的笔画图像，并对图像进行后处理以增强视觉效果。文件 Stroke_origin.py 的具体实现流程如下。

(1) 编写函数 zhuanImg()，其功能是对输入的图像进行旋转操作，返回旋转后的图像。具体实现代码如下：

```
def zhuanImg(img, angle):
    # 将图像按照给定角度旋转
    row, col = img.shape
    M = cv2.getRotationMatrix2D((row / 2, col / 2), angle, 1)
    res = cv2.warpAffine(img, M, (row, col))
    return res
```

(2) 编写函数 genStroke()，其功能是接收一个灰度图像和 dirNum 参数作为输入，生成图像的笔画效果。具体实现代码如下：

```
# 计算并获取原始图像的笔画
def genStroke(img, dirNum, verbose=False):
    height, width = img.shape[0], img.shape[1]
    img = np.float32(img) / 255.0
    print("输入图像高度：%d, 宽度：%d" % (height, width))

    print("图像预处理，去噪...")
    img = cv2.medianBlur(img, 3)

    print("生成梯度图像...")
    imX = np.append(np.absolute(img[:, 0:width - 1] - img[:, 1:width]), np.zeros((height, 1)), axis=1)
    imY = np.append(np.absolute(img[0:height - 1, :] - img[1:height, :]), np.zeros((1, width)), axis=0)
    ###############################################################
    #####   有许多方法来生成渐变效果   #####
    ###############################################################
    img_gradient = np.sqrt((imX ** 2 + imY ** 2))
    img_gradient = imX + imY
    if verbose == True:
        cv2.imshow('梯度图像', np.uint8(255 - img_gradient * 255))
        cv2.imwrite('output/grad.jpg', np.uint8(255 - img_gradient * 255))
        cv2.waitKey(0)

    # 滤波核大小
    tempsize = 0
    if height > width:
        tempsize = width
    else:
        tempsize = height
    tempsize /= 30
    ###############################################################
```

```python
#核大小是边长的1/30
###################################################################
halfKsize = int(tempsize / 2)
if halfKsize < 1:
    halfKsize = 1
if halfKsize > 9:
    halfKsize = 9
kernalsize = halfKsize * 2 + 1
print("核大小 = %s" % (kernalsize))

###################################################################
############### 这里生成滤波核 #################
###################################################################
kernel = np.zeros((dirNum, kernalsize, kernalsize))
kernel[0, halfKsize, :] = 1.0
for i in range(0, dirNum):
    kernel[i, :, :] = temp = zhuanImg(kernel[0, :, :], i * 180 / dirNum)
    kernel[i, :, :] *= kernalsize / np.sum(kernel[i])
    if verbose == True:
        title = '线性滤波核 %d' % i
        cv2.imshow(title, np.uint8(temp * 255))
        cv2.waitKey(0)

# 在不同方向上滤波梯度图像
print("在不同方向上滤波梯度图像...")
response = np.zeros((dirNum, height, width))
for i in range(dirNum):
    ker = kernel[i, :, :]
    response[i, :, :] = cv2.filter2D(img_gradient, -1, ker)
if verbose == True:
    for i in range(dirNum):
        title = '响应 %d' % i
        cv2.imshow(title, np.uint8(response[i, :, :] * 255))
        cv2.waitKey(0)

# 将梯度图像分割成不同的子图
print("计算梯度分类...")
Cs = np.zeros((dirNum, height, width))
for x in range(width):
    for y in range(height):
        i = np.argmax(response[:, y, x])
        Cs[i, y, x] = img_gradient[y, x]
if verbose == True:
    for i in range(dirNum):
        title = '最大响应 %d' % i
```

```
            cv2.imshow(title, np.uint8(Cs[i, :, :] * 255))
            cv2.waitKey(0)

    # 生成线条形状
    print("生成线条形状...")
    spn = np.zeros((dirNum, height, width))
    for i in range(dirNum):
        ker = kernel[i, :, :]
        spn[i, :, :] = cv2.filter2D(Cs[i], -1, ker)
    sp = np.sum(spn, axis=0)

    sp = sp * np.power(img_gradient, 0.4)
    sp = (sp - np.min(sp)) / (np.max(sp) - np.min(sp))
    S = 1 - sp
    return S
```

上述代码的执行过程如下。

① 对输入图像进行预处理,包括去噪声操作(使用 cv2.medianBlur()函数)。

② 生成图像的梯度信息(imX 和 imY),计算梯度幅值(img_gradient)。

③ 生成卷积核(kernel),并对其进行旋转和归一化处理。

④ 使用生成的核对图像的梯度进行滤波操作,得到不同方向上的响应值(response)。

⑤ 将梯度图像根据响应值进行分类,得到不同方向上的梯度子图(Cs)。

⑥ 通过对子图进行滤波操作得到线条形状(spn),并将所有子图求和得到最终的笔画图像(sp)。

⑦ 对笔画图像进行后处理,调整其亮度和对比度,并进行归一化处理得到最终的笔画图像(S)。

(3) 加载输入图像并调用函数 genStroke()生成笔画图像,对生成的笔画图像进行处理,包括指数变换和归一化处理,将最终的笔画图像保存到文件中并进行显示。具体实现代码如下:

```
if __name__ == '__main__':

    img_path = './input/1.jpg'
    img = cv2.imread(img_path, cv2.IMREAD_GRAYSCALE)
    stroke = genStroke(img, 18, False)
    # stroke = stroke*(np.exp(stroke)-np.exp(1)+1)
    stroke = np.power(stroke, 3)
    # stroke=(stroke - np.min(stroke)) / (np.max(stroke) - np.min(stroke))  # 加深边缘
    stroke = np.uint8(stroke * 255)

    cv2.imwrite('output/edge.jpg', stroke)
```

```
cv2.imshow('笔画', stroke)
cv2.waitKey(0)
```

执行后会得到图像 cat.jpg 的笔画图像，如图 5-2 所示。

图 5-2　图像 cat.jpg 的笔画图像

## 5.4.4　方向检测

图像的方向检测在计算机视觉和图像处理领域中应用得很广泛，下面列举了一些常见的用法。

- 图像分类与识别：在图像分类与识别任务中，识别图像的主要方向对于理解图像内容和结构至关重要。主要方向的识别不仅有助于特征提取、特征匹配和模式识别等任务，而且可以显著提高分类和识别的准确性。
- 边缘检测与分割：通过检测图像中的边缘和线条，可以更好地理解图像中的结构和边界。方向信息可以用于优化边缘检测算法和分割方法，从而提高结果的准确性和质量。
- 图像增强与滤波：根据图像主要方向的信息，可以应用方向性滤波器或增强算法，以增强图像特定方向上的特征和细节，从而增强图像的视觉效果。
- 视觉导航与目标跟踪：在计算机视觉中，方向信息对于视觉导航和目标跟踪至关重要。通过检测和跟踪图像中的主要方向，可以帮助机器视觉系统进行定位、导航和目标追踪，实现自主导航、目标跟踪和场景理解等任务。

在本项目中，编写文件 tools.py，用于对输入的图像进行方向检测，并输出图像的主要方向。文件 tools.py 的具体实现流程如下。

（1）编写函数 get_start_end()，根据给定的二值掩码图像获取图像中连续区间的起始和结束位置。函数 get_start_end()的具体实现代码如下：

```
def get_start_end(mask):
    lines=[]
    Flag = True   # 没有新的间隔
    for i in range(mask.shape[0]):
        if Flag == True:
            if mask[i]==1:
                if len(lines)>0 and i-lines[-1][1]<=1:   ####### 太接近了
                    Flag = False
                    continue
                else:
                    lines.append([i,i])
                    Flag = False
            else:
                continue
        else:
            if mask[i]==1:
                continue
            else:
                lines[-1][1]=i
                Flag = True

    if Flag == False:
        lines[-1][1]=i

    return lines
```

(2) 编写函数 zhuanImg()，其功能是对给定图像进行旋转，旋转角度由参数 angle 指定。函数 zhuanImg()的具体实现代码如下：

```
def zhuanImg(img, angle):
    row, col = img.shape
    M = cv2.getRotationMatrix2D((row / 2 , col / 2 ), angle, 1)
    res = cv2.warpAffine(img, M, (row, col))
    return res
```

(3) 编写函数 get_directions()，其功能是在给定的图像中检测出主要的方向。其中，Num_choose 表示要选择的主要方向数量，dirNum 表示在方向空间中划分的方向数量，img 是输入的灰度图像。在函数 get_directions()中，首先，对输入图像计算梯度图像，并根据梯度图像生成掩码。然后，构建滤波核，并将滤波核应用于梯度图像，得到在不同方向上滤波后的响应图像。接着，根据响应图像的最大值确定每个像素点的主要方向，并统计各个方向的像素值。最后，根据像素值选择主要的方向，并返回这些方向的角度值。函数 get_directions()的具体实现代码如下：

```python
def get_directions(Num_choose, dirNum, img):
    height,width = img.shape
    img = np.float32(img)/255.0
    # print("输入图像高度：%d，宽度：%d" % (height,width))
    
    imX = np.append(np.absolute(img[:, 0 : width - 1] - img[:, 1 : width]),
np.zeros((height, 1)), axis = 1)
    imY = np.append(np.absolute(img[0 : height - 1, :] - img[1 : height, :]),
np.zeros((1, width)), axis = 0)
    
    img_gradient = np.sqrt((imX ** 2 + imY ** 2))
    mask = (img_gradient-0.02)>0
    cv2.imshow('mask',np.uint8(mask*255))
    # img_gradient = imX + imY
    
    # 滤波核大小
    tempsize = 0
    if height > width:
        tempsize = width
    else:
        tempsize = height
    tempsize /= 30   # 核大小是边长的1/30
    
    halfKsize = int(tempsize / 2)
    if halfKsize < 1:
        halfKsize = 1
    if halfKsize > 9:
        halfKsize = 9
    kernalsize = halfKsize * 2 + 1
    # print("核大小 = %s" %(kernalsize))

###############################################################
############## 这里生成滤波核 #################
###############################################################
    kernel = np.zeros((dirNum, kernalsize, kernalsize))
    kernel [0,halfKsize,:] = 1.0
    for i in range(0,dirNum):
        kernel[i,:,:] = temp = zhuanImg(kernel[0,:,:], i * 180 / dirNum)
        kernel[i,:,:] *= kernalsize/np.sum(kernel[i])
    
    # 在不同方向上滤波梯度图像
    print("在不同方向上滤波梯度图像 ...")
    response = np.zeros((dirNum, height, width))
    for i in range(dirNum):
```

```
        ker = kernel[i,:,:];
        response[i, :, :] = cv2.filter2D(img_gradient, -1, ker)

cv2.waitKey(0)

# 将梯度图像分割成不同的子图
print("计算方向分类 ...")
direction = np.zeros(( height, width))
for x in range(width):
    for y in range(height):
        direction[y, x] = np.argmax(response[:,y,x])
# direction = direction*mask

dirs = np.zeros(dirNum)
for i in range (dirNum):
    dirs[i]=np.sum((direction-i)==0)
sort_dirs = np.sort(dirs,axis=0)
print(dirs,sort_dirs)
angles = []
for i in range(Num_choose):
    for j in range (dirNum):
        if sort_dirs[-1-i]==dirs[j]:
            angles.append(j*180/dirNum)
            continue
return angles
```

(4) 读取指定的灰度图像 cat.jpg，然后调用函数 get_directions()检测图像中的主要方向，并将结果保存在 angles 变量中，具体实现代码如下：

```
if __name__ == '__main__':
    input_pathos = './input/cat.jpg'
    img = cv2.imread(input_pathos, cv2.IMREAD_GRAYSCALE)
    angles = get_directions(4,12,img)
```

执行后会对图像 cat.jpg 进行方向检测，并输出图像的主要方向：

```
在不同方向上滤波梯度图像 ...
计算方向分类 ...
[13849. 2705. 3283. 2941. 3641. 1400. 6787. 1122. 3174. 3480. 3416. 4827.]
[ 1122. 1400. 2705. 2941. 3174. 3283. 3416. 3480. 3641. 4827. 6787. 13849.]
```

### 5.4.5 去蓝处理

去蓝处理是一种图像处理技术，主要目的是减少或消除图像中的蓝色成分，这种处理可以应用于不同的场景，主要用途如下。

- ❏ 色彩校正：某些图像可能受到蓝色光线的干扰或偏色，但通过去除图像中的蓝色成分，可以校正图像的色彩偏差，使其更加真实和准确。
- ❏ 特定效果：去蓝处理可以产生一种老化或怀旧的效果，通过减少蓝色成分，使图像呈现出暖色调的效果，增强了图像的复古感。
- ❏ 图像增强：在某些情况下，蓝色成分可能会对图像的细节和对比度产生负面影响。通过去除蓝色成分，可以减少图像中的噪声或干扰，提高图像的清晰度和细节。
- ❏ 图像分析：对于一些特定的图像分析任务，蓝色成分可能会对目标或特征的识别造成干扰。通过去除蓝色成分，可以减少干扰，提高对目标或特征的检测和分析准确性。

需要注意的是，在具体应用中是否需要去蓝处理取决于实际需求和图像的特点。有些情况下可能需要保留蓝色成分，有些情况下则需要去除它。去蓝处理作为一种图像处理技术，可根据具体需求进行调整和应用。

在本项目中编写文件 dellblue.py，用于实现去蓝处理，具体实现流程如下所示。

(1) 编写函数 dellblue()，用于对图像进行去蓝处理，在函数中，首先将输入的灰度图像转换为彩色图像(BGR 格式)。接下来，将彩色图像转换为 HSV 颜色空间。利用高斯函数模拟生成两个高斯噪声图像，分别用于修改 HSV 图像的色相(H 通道)和饱和度(S 通道)。然后将修改后的 HSV 图像转换回 BGR 颜色空间，并将其转换为 8 位无符号整数类型，将处理后的图像保存到指定路径下的文件 aging.jpg 中。函数 dellblue()的具体实现代码如下：

```python
def dellblue(img, output_pathos):
    BGR = cv2.cvtColor(img, cv2.COLOR_GRAY2BGR)
    HSV = cv2.cvtColor(BGR, cv2.COLOR_BGR2HSV)

    size = img.shape

    HSV[:,:,0] = Gassian(size, mean=15, var=1)
    HSV[:,:,1] = Gassian(size, mean=20, var=2)

    result = cv2.cvtColor(HSV, cv2.COLOR_HSV2BGR)
    result = np.uint8(result)

    cv2.imwrite(output_pathos+'/aging.jpg', result)
    cv2.imshow("aging",result)
    cv2.waitKey(0)
```

(2) 读取输入图像的路径，然后调用函数 dellblue()对输入图像进行去蓝处理，并将处理后的图像保存到指定路径下的文件中。具体实现代码如下：

```python
if __name__ == '__main__':
    input_pathos = './output/draw.png'
```

```
output_pathos = './output'
input_img = cv2.imread(input_pathos, cv2.IMREAD_GRAYSCALE)
dellblue(input_img, output_pathos)
```

## 5.4.6 图像合成

在前面的去蓝处理文件 dellblue.py 中调用了文件 simulate.py 的内容，文件 simulate.py 实现了一个简单的图像合成功能，通过生成随机的平行直线并在画布上叠加，可以生成一幅具有一定曲线和衰减效果的图像。文件 simulate.py 的具体实现流程如下。

(1) 编写函数 Gassian()，功能是生成一个服从高斯分布的随机矩阵，用于添加噪声。具体实现代码如下：

```
constant_length = 1000

def Gassian(size, mean = 0, var = 0):
    norm = np.random.randn(*size)
    denorm = norm * np.sqrt(var) + mean
    return np.uint8(np.round(np.clip(denorm,0,255)))
```

在上述代码中，constant_length = 1000 是一个常数，用于判断直线的长度是否达到阈值。在代码中，当直线的长度小于 constant_length 时，直线是对齐的，即直线的起点和终点在同一水平位置上。当直线的长度大于或等于 constant_length 时，直线不再对齐，即直线的起点和终点不在同一水平位置上。这个阈值的设置主要用于确定直线的对齐方式。如果长度较短，直线对齐可以使整体效果更加规整；而长度较长时，不对齐的直线可以产生一种更加自由和随机的效果。

(2) 编写函数 Getline()，功能是根据给定的长度和分布参数生成一条随机直线。根据直线长度 constant_length 的大小，选择对齐和不对齐的方式生成直线。这个阈值的设定可以根据具体需求进行调整，以获得期望的直线效果。具体实现代码如下：

```
def Getline(distribution, length):
    linewidth = distribution.shape[0]
    if length < constant_length:  # 如果长度太短，线条将被对齐
        patch = Gassian((2*linewidth, length), mean=250, var=3)
        for i in range(linewidth):
            patch[i] = Gassian((1, length), mean=distribution[i, 0], var=distribution[i, 1])
        begin, end = 0, 1
    else:  # if length is't too short, lines is't Aligned
        patch = Gassian((2*linewidth, length+4*linewidth), mean=250, var=3)
        begin = np.clip(np.round(2.0 * linewidth), 0, 4 * linewidth).astype(np.uint8)
        end = np.clip(np.round(2.0 * linewidth), 1, 4 * linewidth + 1).astype(np.uint8)
        real_length = length + 4 * linewidth - end - begin
```

```
       patch[:linewidth, begin:-end] = np.array([Gassian((1, real_length),
mean=distribution[i, 0], var=distribution[i, 1]) for i in range(linewidth)])

    patch = Attenuation(patch, linewidth=linewidth, distribution=distribution,
begin=begin, end=end)
    patch = Distortion(patch, begin=begin, end=end)

    return np.clip(patch, 0, 255).astype(np.uint8)
```

（3）编写函数 Attenuation()，功能是对输入的图像进行衰减处理，使图像在水平方向上逐渐变淡。具体实现代码如下：

```
def Attenuation(patch, linewidth, distribution, begin, end):
    order = int((patch.shape[1]-begin-end)/2)+1
    radius = (linewidth-1)/2
    canvas = Gassian((patch.shape[0], patch.shape[1]), mean=250, var=3)
    patch = np.float32(patch)
    canvas = np.float32(canvas)
    for i in range(begin, patch.shape[1]-end+1):
        for j in range(linewidth):
            a = np.abs((1.0-(i-begin)/order)**2)/3
            b = np.abs((1.0-j/radius)**2)*1
            patch[j,i] += (canvas[j,i]-patch[j,i])*np.sqrt(a+b)/1.5
            # patch[j,i] += 0.75*(canvas[j,i]-patch[j,i]) *
(np.abs((1.0-(i-begin)/order)**2))**0.5

    return np.uint8(np.round(np.clip(patch,0,255)))
```

（4）编写函数 Distortion()，功能是对输入的图像进行扭曲处理，使图像呈现一定的曲线形状。具体实现代码如下：

```
def Distortion(patch, begin, end):
    height = patch.shape[0] // 2
    length = patch.shape[1]
    patch = patch.astype(np.float32)
    patch_copy = patch.copy()

    central = (length - begin - end) / 2 + begin
    if length > 100:
        radius = length ** 2 / (4 * height)
    else:
        radius = length ** 2 / (2 * height)

    offset_vals = ((central - np.arange(length)) ** 2) / (2 * radius)
    int_offsets = offset_vals.astype(int)
    decimal_offsets = offset_vals - int_offsets
```

```python
    for i in range(length):
        int_offset = int_offsets[i]
        decimal_offset = decimal_offsets[i]
        for j in range(height):
            if j > int_offset:
                patch[j, i] = int(decimal_offset * patch_copy[j - 1 - int_offset, i]
+ (1 - decimal_offset) * patch_copy[j - int_offset, i])
            else:
                patch[j, i] = np.random.randn() * np.sqrt(3) + 250

    patch_copy = patch.copy()
    if length > 100:
        for i in range(length):
            int_offset = int_offsets[i]
            decimal_offset = decimal_offsets[i]
            for j in range(patch.shape[0]):
                if j > int_offset:
                    patch[j, i] = int(decimal_offset * patch_copy[j - 1 - int_offset,
 i] + (1 - decimal_offset) * patch_copy[j - int_offset, i])
                else:
                    patch[j, i] = np.random.randn() * np.sqrt(3) + 250

    return np.clip(patch, 0, 255).astype(np.uint8)
```

(5) 编写函数 GetParallel()，功能是根据给定的分布参数、高度、长度和线宽生成一组平行的随机直线。具体实现代码如下：

```python
def GetParallel(distribution, height, length, linewidth):
    if length<constant_length: # constant length
        canvas = Gassian((height+2*linewidth,length), mean=250, var = 3)
    else: # variable length
        canvas = Gassian((height+2*linewidth,length+4*linewidth), mean=250, var = 3)

    distensce = Gassian((1,int(height/linewidth)+2), mean = linewidth, var = linewidth/5)
    # distensce = Gassian((1,int(height/linewidth)+1), mean = linewidth, var = 0)
    distensce = np.uint8(np.round(np.clip(distensce, linewidth*0.8,linewidth*1.25)))

    begin = 0
    for i in np.squeeze(distensce).tolist():
        newline = Getline(distribution=distribution, length=length)
        h,w = newline.shape
        # cv2.imshow('line', newline)
        # cv2.waitKey(0)
        # cv2.imwrite("D:/ECCV2020/simu_patch/Line3.jpg",newline)
```

```
        if begin < height:
            m = np.minimum(canvas[begin:(begin + h),:], newline)
            canvas[begin:(begin + h),:] = m
            begin += i
        else:
            break

    return canvas[:height,:]
```

（6）编写函数 ChooseDistribution()，功能是根据线宽和灰度值选择合适的分布参数，用于生成直线的灰度分布。具体实现代码如下：

```
def ChooseDistribution(linewidth, Grayscale):
    distribution = np.zeros((linewidth,2))
    c = linewidth/2.0
    difference = 250-Grayscale
    for i in range(distribution.shape[0]):
        distribution[i][0] = Grayscale + difference*abs(i-c)/c
        distribution[i][1] = np.cos((i-c)/c*(0.5*3.1415929))*difference

    return np.abs(distribution)
```

（7）在主程序中，先生成一个基础画布 canvas，然后根据给定的参数调用函数来生成随机的平行直线，并将生成的直线放置在画布上。最后将显示生成的图像，并将其保存为 maomao.png。具体实现代码如下：

```
if __name__ == '__main__':
    np.random.seed(100)
    canvas = Gassian((500,500), mean=250, var = 3)

    # distribution =
np.array([[245,31],[238,27],[218,48],[205,33],[214,38],[234,24],[240,42]])
    linewidth = 7
    Grayscale = 128
    H,L = (100,200)
    distribution = ChooseDistribution(linewidth=linewidth, Grayscale=Grayscale)
    print(distribution)
    patch = GetParallel(distribution=distribution, height=H, length=L, linewidth=linewidth)
    (h,w) = patch.shape
    # patch = GetOffsetParallel(offset=4, distribution=distribution,
patch_size=(40,200), linewidth_mean=distribution.shape[0], linewidth_var=1)
    # (h,w) = patch.shape
    canvas[250-int(h/2):250-int(h/2)+h,250-int(w/2):250-int(w/2)+w] = patch
```

```
    # cv2.imshow('Parallel', patch[:,
2*distribution.shape[0]:w-2*distribution.shape[0]])
    cv2.imshow('Parallel', canvas)
    cv2.waitKey(0)
    cv2.imwrite("maomao.png",patch)
    print("done")
```

### 5.4.7 快速排序

编写文件 quicksort.py,实现快速排序算法,用于对给定数组进行排序,具体实现流程如下。

(1) 编写函数 partition(),功能是根据选取的随机基准元素将数组分割为两部分,并返回基准元素的索引位置。函数内部使用了双指针的方式进行元素交换,将小于基准的元素放在基准的左边,大于基准的元素放在基准的右边。具体实现代码如下:

```
def partition(arr, low, high):
    # 随机选择基准元素
    random_index = random.randint(low, high)
    arr[random_index], arr[high] = arr[high], arr[random_index]

    pivot = arr[high]
    i = low - 1

    for j in range(low, high):
        if arr[j]['importance'] >= pivot['importance']:
            i = i + 1
            arr[i], arr[j] = arr[j], arr[i]

    arr[i+1], arr[high] = arr[high], arr[i+1]
    return i + 1
```

(2) 编写函数 quickSort(),功能是递归地对子数组进行快速排序。函数首先检查子数组的长度是否小于或等于阈值,若是,则使用插入排序进行排序。若子数组长度大于阈值,则选择一个基准元素,将子数组划分为两部分,并递归地对这两部分进行排序。为了优化性能,在递归调用中,会对较短的子数组进行尾递归优化,以减少递归深度。具体实现代码如下:

```
def quickSort(arr, low, high):
    # 设置阈值,当子数组长度小于或等于该阈值时,使用插入排序
    threshold = 10

    while low < high:
        # 当子数组长度小于或等于阈值时,使用插入排序
```

```
        if high - low + 1 <= threshold:
            insertionSort(arr, low, high)
            return

        pi = partition(arr, low, high)

        # 对较短的子数组进行尾递归优化
        if pi - low < high - pi:
            quickSort(arr, low, pi - 1)
            low = pi + 1
        else:
            quickSort(arr, pi + 1, high)
            high = pi - 1
```

(3) 编写函数 insertionSort(),功能是实现插入排序算法,用于在子数组长度小于或等于阈值时进行排序。该函数通过比较元素的重要性(importance)进行排序,将重要性较高的元素向前移动,以实现递减排序。具体实现代码如下:

```
def insertionSort(arr, low, high):
    for i in range(low + 1, high + 1):
        key = arr[i]
        j = i - 1
        while j >= low and arr[j]['importance'] < key['importance']:
            arr[j + 1] = arr[j]
            j -= 1
        arr[j + 1] = key
```

## 5.4.8  侧窗滤波

编写文件 SideWindowFilter.py,定义一个名为 SideWindowFilter 的 PyTorch 模型,用于实现侧窗滤波。侧窗滤波是一种图像处理技术,通过在每个像素点的邻域内进行滤波操作来改变图像的外观。文件 SideWindowFilter.py 的具体实现代码如下:

```
import torch
import torch.nn as nn
import torch.nn.functional as F

class SideWindowFilter(nn.Module):

    def __init__(self, radius, iteration, filter='box'):
        super(SideWindowFilter, self).__init__()
        self.radius = radius
        self.iteration = iteration
        self.kernel_size = 2 * self.radius + 1
        self.filter = filter
```

```python
def forward(self, im):
    b, c, h, w = im.size()

    d = torch.zeros(b, 8, h, w, dtype=torch.float)
    res = im.clone()

    if self.filter.lower() == 'box':
        filter = torch.ones(1, 1, self.kernel_size, self.kernel_size)
        L, R, U, D = [filter.clone() for _ in range(4)]

        L[:, :, :, self.radius + 1:] = 0
        R[:, :, :, 0: self.radius] = 0
        U[:, :, self.radius + 1:, :] = 0
        D[:, :, 0: self.radius, :] = 0

        NW, NE, SW, SE = U.clone(), U.clone(), D.clone(), D.clone()

        L, R, U, D = L / ((self.radius + 1) * self.kernel_size), R / ((self.radius + 1) * self.kernel_size), \
                     U / ((self.radius + 1) * self.kernel_size), D / ((self.radius + 1) * self.kernel_size)

        NW[:, :, :, self.radius + 1:] = 0
        NE[:, :, :, 0: self.radius] = 0
        SW[:, :, :, self.radius + 1:] = 0
        SE[:, :, :, 0: self.radius] = 0

        NW, NE, SW, SE = NW / ((self.radius + 1) ** 2), NE / ((self.radius + 1) ** 2), \
                        SW / ((self.radius + 1) ** 2), SE / ((self.radius + 1) ** 2)

        # sum = self.kernel_size * self.kernel_size
        # sum_L, sum_R, sum_U, sum_D, sum_NW, sum_NE, sum_SW, sum_SE = \
        #     (self.radius + 1) * self.kernel_size, (self.radius + 1) * self.kernel_size, \
        #     (self.radius + 1) * self.kernel_size, (self.radius + 1) * self.kernel_size, \
        #     (self.radius + 1) ** 2, (self.radius + 1) ** 2, (self.radius + 1) ** 2, (self.radius + 1) ** 2

        print('L:', L)
        print('R:', R)
        print('U:', U)
        print('D:', D)
        print('NW:', NW)
        print('NE:', NE)
        print('SW:', SW)
```

```
            print('SE:', SE)

        for ch in range(c):
            im_ch = im[:, ch, ::].clone().view(b, 1, h, w)
            # print('im size in each channel:', im_ch.size())

            for i in range(self.iteration):
                # print('###', (F.conv2d(input=im_ch, weight=L, padding=(self.radius,
self.radius)) / sum_L -
                # im_ch).size(), d[:, 0,::].size())
                d[:, 0, ::] = F.conv2d(input=im_ch, weight=L, padding=(self.radius,
self.radius)) - im_ch
                d[:, 1, ::] = F.conv2d(input=im_ch, weight=R, padding=(self.radius,
self.radius)) - im_ch
                d[:, 2, ::] = F.conv2d(input=im_ch, weight=U, padding=(self.radius,
self.radius)) - im_ch
                d[:, 3, ::] = F.conv2d(input=im_ch, weight=D, padding=(self.radius,
self.radius)) - im_ch
                d[:, 4, ::] = F.conv2d(input=im_ch, weight=NW, padding=(self.radius,
self.radius)) - im_ch
                d[:, 5, ::] = F.conv2d(input=im_ch, weight=NE, padding=(self.radius,
self.radius)) - im_ch
                d[:, 6, ::] = F.conv2d(input=im_ch, weight=SW, padding=(self.radius,
self.radius)) - im_ch
                d[:, 7, ::] = F.conv2d(input=im_ch, weight=SE, padding=(self.radius,
self.radius)) - im_ch

                d_abs = torch.abs(d)
                print('im_ch', im_ch)
                print('dm = ', d_abs.shape, d_abs)
                mask_min = torch.argmin(d_abs, dim=1, keepdim=True)
                print('mask min = ', mask_min.shape, mask_min)
                dm = torch.gather(input=d, dim=1, index=mask_min)
                im_ch = dm + im_ch

            res[:, ch, ::] = im_ch
        return res
```

该模型的构造函数\_\_init\_\_()接收参数 radius、iteration 和 filter，分别用于设置滤波器的半径、迭代次数和滤波类型。函数 forward()定义了模型的前向传播过程，接收输入图像 im，并根据设置的滤波类型进行侧窗滤波操作。

在上述代码中还使用了库 torch 的一些功能，如 torch.zeros、torch.ones、torch.clone、torch.view、torch.abs、torch.argmin 和 torch.gather 等，以及使用了 torch.nn.functional 中的 F.conv2d()函数进行卷积操作。另外，还使用了 PIL 和 OpenCV 库来加载和显示图像，以及使用 NumPy 进行数据处理。

总之，这段代码实现了侧窗滤波的功能，通过定义模型和实现模型的前向传播过程，对输入图像进行侧窗滤波操作，并输出滤波后的图像。

## 5.5 开始绘图

经过前面的预处理工作之后，接下来开始正式的绘图工作。在本节的内容中，将详细讲解两种绘图方案的实现过程。

### 5.5.1 基于边缘绘画的绘图程序

扫码看视频

编写文件 cat.py，功能是实现了一个基于边缘绘画的图像艺术生成算法，将输入的图像转换为具有艺术效果的图像。该文件通过组合边缘细化、量化、笔触绘制和颜色处理等步骤，将输入图像转换为艺术效果的图像，具体实现流程如下。

（1）设置一系列参数，用于控制整个图像处理和绘制的过程。通过对这些参数的设置，可以影响最终生成的手绘风格图像的效果和绘制过程的展示方式。通过调整这些参数，可以控制线条的样式、绘制顺序，以及图像处理的方式，从而获得不同的艺术化效果。具体实现代码如下：

```
input_pathos = './input/cat_up.png'
output_pathos = './output'

np.random.seed(1)
n = 6
linewidth = 4
direction = 10
Freq = 100
deepen = 1
transTone = False
kernel_radius = 3
iter_time = 15
background_dir = 45
CLAHE = True
edge_CLAHE = True
draw_new = True
random_order = False
ETF_order = True
process_visible = True
```

每个参数的功能说明如下。

❑ input_pathos：输入图像的路径。

- output_pathos：输出结果的路径。
- np.random.seed(1)：设置随机数种子，用于保持随机结果的可重复性。
- n：量化顺序，用于将灰度值分为多个级别。
- linewidth：线条的宽度。
- direction：设置绘制线条时使用的方向数量。
- Freq：保存绘制结果的频率，表示每绘制多少条线条保存一次结果。
- deepen：用于边缘加深的参数。
- transTone：是否进行色调转换。
- kernel_radius：边缘切向流(ETF)算法中的核半径。
- iter_time：ETF 算法的迭代次数。
- background_dir：ETF 算法中的背景方向。
- CLAHE：是否进行直方图自适应均衡化。
- edge_CLAHE：是否对边缘图像进行直方图自适应均衡化。
- draw_new：是否重新绘制线条。
- random_order：是否对线条绘制顺序进行随机排序。
- ETF_order：是否按照 ETF 算法计算结果的顺序绘制线条。
- process_visible：是否显示绘制过程的中间结果。

(2) 根据输入路径中的文件名创建相应的输出文件夹，并在输出文件夹中创建两个子文件夹：mask 和 process，用于存储相关的结果和中间过程。这样可以更好地组织和保存处理后的图像及其相关数据。具体实现代码如下：

```
file_name = os.path.basename(input_pathos)
file_name = file_name.split('.')[0]
print(file_name)
output_pathos = output_pathos+"/"+file_name
if not os.path.exists(output_pathos):
    os.makedirs(output_pathos)
    os.makedirs(output_pathos+"/mask")
    os.makedirs(output_pathos+"/process")
```

对上述代码的具体说明如下。

- file_name = os.path.basename(input_pathos)：从输入路径中获取文件名，包括扩展名。
- file_name = file_name.split('.')[0]：将文件名按照扩展名进行分割，只保留文件名部分(去除扩展名)。
- print(file_name)：打印文件名，这一步是为了在控制台显示文件名，供用户查看。
- output_pathos = output_pathos+"/"+file_name：将输出路径设置为输出文件夹路径和

文件名的组合。

- if not os.path.exists(output_pathos)：检查输出文件夹是否存在，如果不存在则执行以下操作。
- os.makedirs(output_pathos)：创建输出文件夹。
- os.makedirs(output_pathos+"/mask")：在输出文件夹中创建一个名为 mask 的子文件夹。
- os.makedirs(output_pathos+"/process")：在输出文件夹中创建一个名为 process 的子文件夹。

(3) 执行 ETF 滤波器的操作，生成 ETF 滤波后的图像，并保存输入图像的灰度版本供后续使用。具体实现代码如下：

```
time_start=time.time()
ETF_filter = ETF(input_pathos=input_pathos, output_pathos=output_pathos+'/mask',\
    dir_num=direction, kernel_radius=kernel_radius, iter_time=iter_time,
background_dir=background_dir)
ETF_filter.forward()
print('ETF done')

input_img = cv2.imread(input_pathos, cv2.IMREAD_GRAYSCALE)
(h0,w0) = input_img.shape
cv2.imwrite(output_pathos + "/input_gray.jpg", input_img)
```

上述代码的主要功能是进行图像处理中的 ETF(edge tangent flow)操作，以及保存输入图像的灰度版本。具体说明如下：

- time_start=time.time()：记录开始时间，用于计算 ETF 操作的执行时间。
- ETF_filter = ETF(input_pathos=input_pathos, output_pathos=output_pathos+'/mask',\ dir_num=direction, kernel_radius=kernel_radius, iter_time=iter_time, background_dir=background_dir)：创建一个名为 ETF_filter 的 ETF 滤波器对象，其中传入的参数包括输入路径、输出路径(包括子文件夹 mask)、方向数(dir_num)、核半径(kernel_radius)、迭代次数(iter_time)和背景方向(background_dir)。
- ETF_filter.forward()：执行 ETF 滤波器的前向操作，对输入图像进行 ETF 处理。
- print('ETF done')：在控制台打印 ETF done，表示 ETF 操作已完成。
- input_img = cv2.imread(input_pathos, cv2.IMREAD_GRAYSCALE)：使用 OpenCV 读取输入图像的灰度版本，以便后续处理。
- (h0,w0) = input_img.shape：获取灰度图像的高度和宽度。
- cv2.imwrite(output_pathos + "/input_gray.jpg", input_img)：将灰度图像保存到输出文件夹中，命名为 input_gray.jpg。

(4)对输入图像进行色调转换(如果 transTone 为 True)，创建绘制图像所需的初始变量和数据结构，并将输入图像按不同的角度进行旋转，以准备进行后续的绘制操作，具体实现代码如下：

```
if transTone == True:
    input_img = transferTone(input_img)
now_ = np.uint8(np.ones((h0,w0)))*255
step = 0
if draw_new==True:
    time_start=time.time()
    stroke_sequence=[]
    stroke_temp={'angle':None, 'grayscale':None, 'row':None, 'begin':None, 'end':None}
    for dirs in range(direction):
        angle = -90+dirs*180/direction
        print('angle:', angle)
        stroke_temp['angle'] = angle
        img,_ = rotate(input_img, -angle)
```

对上述代码的具体说明如下。

- if transTone == True：如果变量 transTone 的值为 True，则执行下面的代码块。
- input_img = transferTone(input_img)：调用 transferTone() 函数，对 input_img 进行色调转换。
- now_ = np.uint8(np.ones((h0,w0)))*255：创建一个大小为 (h0, w0) 的二维数组 now_，并用值 255 填充。这个数组将用于存储绘制图像的笔画信息。
- step = 0：设置变量 step 的初始值为 0。
- if draw_new==True：如果变量 draw_new 的值为 True，则执行下面的代码块。
- time_start=time.time()：记录开始时间，用于计算绘制图像的执行时间。
- stroke_sequence=[]：创建一个空列表 stroke_sequence，用于存储笔画序列信息。
- stroke_temp={'angle':None, 'grayscale':None, 'row':None, 'begin':None, 'end':None}：创建一个字典 stroke_temp，用于存储每个笔画的角度、灰度、行号、起始点和终止点。
- for dirs in range(direction)：对于 0 ～ direction-1 的每个值，执行循环。
- angle = -90+dirs*180/direction：计算当前方向的角度值。
- print('angle:', angle)：打印当前角度值。
- stroke_temp['angle'] = angle：将当前角度值存储到 stroke_temp 字典中的 angle 键。
- img,_ = rotate(input_img, -angle)：调用 rotate() 函数，对 input_img 进行逆时针旋转 -angle 度，并将旋转后的图像存储到 img 变量中。

(5) 进行直方图均衡化操作,这样可以提高图像的对比度和亮度分布。如果 CLAHE 变量为 True,则对旋转后的图像 img 进行直方图均衡化,并打印一条消息表示操作已完成。具体实现代码如下:

```
if CLAHE==True:
    img = HistogramEqualization(img)
print('HistogramEqualization done')
```

(6) 进行梯度计算和归一化操作,具体实现代码如下:

```
img_pad = cv2.copyMakeBorder(img, 2*linewidth, 2*linewidth, 2*linewidth,
2*linewidth, cv2.BORDER_REPLICATE)
img_normal = cv2.normalize(img_pad.astype("float32"), None, 0.0, 1.0, cv2.NORM_MINMAX)

x_der = cv2.Sobel(img_normal, cv2.CV_32FC1, 1, 0, ksize=5)
y_der = cv2.Sobel(img_normal, cv2.CV_32FC1, 0, 1, ksize=5)

x_der = torch.from_numpy(x_der) + 1e-12
y_der = torch.from_numpy(y_der) + 1e-12

gradient_magnitude = torch.sqrt(x_der**2.0 + y_der**2.0)
gradient_norm = gradient_magnitude/gradient_magnitude.max()
```

对上述代码的具体说明如下。

- img_pad=cv2.copyMakeBorder(img, 2*linewidth, 2*linewidth, 2*linewidth, 2*linewidth, cv2.BORDER_REPLICATE):将图像 img 进行边界填充,边界宽度为 2*linewidth,填充方式为复制边界像素值。

- img_normal=cv2.normalize(img_pad.astype("float32"), None, 0.0, 1.0, cv2.NORM_MINMAX):将填充后的图像 img_pad 转换为浮点型,并进行归一化处理,像素值范围从原始值域映射到 [0.0, 1.0]。

- x_der = cv2.Sobel(img_normal, cv2.CV_32FC1, 1, 0, ksize=5):对归一化后的图像 img_normal 在 x 方向上应用 Sobel 算子,计算 x 方向上的梯度。

- y_der = cv2.Sobel(img_normal, cv2.CV_32FC1, 0, 1, ksize=5):对归一化后的图像 img_normal 在 y 方向上应用 Sobel 算子,计算 y 方向上的梯度。

- x_der = torch.from_numpy(x_der) + 1e-12 和 y_der = torch.from_numpy(y_der) + 1e-12:将计算得到的 x 和 y 方向上的梯度转换为 PyTorch 张量,并加上一个小的常数 1e-12,用于避免除零错误。

- gradient_magnitude = torch.sqrt(x_der**2.0 + y_der**2.0):计算梯度幅值,即 x 和 y 方向上梯度的欧氏距离。

- gradient_norm = gradient_magnitude/gradient_magnitude.max()：将梯度幅值进行归一化，除以最大幅值，使得梯度的范围在[0, 1]。

(7) 进行图像的量化操作，具体实现代码如下：

```
ldr = LDR(img, n)
cv2.imshow('Quantization', ldr)
cv2.waitKey(0)
cv2.imwrite(output_pathos + "/Quantization.png", ldr)

LDR_single(ldr,n,output_pathos) # debug
```

对上述代码的具体说明如下：

- ldr = LDR(img, n)：调用函数 LDR() 对梯度归一化图像 img 进行量化操作，将图像分为 n 个不同的灰度级别。
- cv2.imshow('Quantization', ldr) 和 cv2.waitKey(0)：显示量化图像。
- cv2.imwrite(output_pathos + "/Quantization.png", ldr)：将量化后的图像 ldr 保存为文件，文件路径为"output_pathos+"/Quantization.png"，即输出路径下的 Quantization.png 文件。
- LDR_single(ldr,n,output_pathos)：是一个被注释掉的函数调用，可能用于调试目的，根据注释内容看，与单个灰度级别的量化有关。

(8) 使用 for 循环在每个灰度级别下生成笔画序列，首先设置当前迭代的灰度级别为 j*256/n；再读取对应灰度级别的掩码图像，并进行归一化处理；继续读取方向掩码图像，并对其进行旋转和二值化处理；然后生成高斯分布，用于计算笔画的长度；接下来初始化笔画的起始行位置，遍历高斯分布中的每个值，表示笔画的长度。具体实现代码如下：

```
LDR_single_add(ldr,n,output_pathos)
print('Quantization done')


# get tone
(h,w) = ldr.shape
canvas = Gassian((h+4*linewidth,w+4*linewidth), mean=250, var = 3)


for j in range(n):
   # print('tone:',j)
   # distribution = ChooseDistribution(linewidth=linewidth,Grayscale=j*256/n)
   stroke_temp['grayscale'] = j*256/n
   mask = cv2.imread(output_pathos +
'/mask/mask{}.png'.format(j),cv2.IMREAD_GRAYSCALE)/255
     dir_mask = cv2.imread(output_pathos +
'/mask/dir_mask{}.png'.format(dirs),cv2.IMREAD_GRAYSCALE)
```

```
        # if angle==0:
        #     dir_mask[::] = 255
        dir_mask,_ = rotate(dir_mask, -angle, pad_color=0)
        dir_mask[dir_mask<128]=0
        dir_mask[dir_mask>127]=1

        distensce = Gassian((1,int(h/linewidth)+4), mean = linewidth, var = 1)
        distensce = np.uint8(np.round(np.clip(distensce, linewidth*0.8, linewidth*1.25)))
        raw = -int(linewidth/2)

        for i in np.squeeze(distensce).tolist():
            if raw < h:
                y = raw + 2*linewidth # y < h+2*linewidth
                raw += i
                for interval in get_start_end(mask[y-2*linewidth]*dir_mask
[y-2*linewidth]):

                    begin = interval[0]
                    end = interval[1]

                    # length = end - begin

                    begin -= 2*linewidth
                    end += 2*linewidth

                    length = end - begin
                    stroke_temp['begin'] = begin
                    stroke_temp['end'] = end
                    stroke_temp['row'] = y-int(linewidth/2)
                    print(gradient_norm[y,interval[0]+2*linewidth:interval[1]+
2*linewidth])
                    stroke_temp['importance'] = (255-stroke_temp
['grayscale'])*torch.sum(gradient_norm[y:y+linewidth,interval[0]+2*linewidth:
interval[1]+2*linewidth]).numpy()

                    stroke_sequence.append(stroke_temp.copy())
```

对上述代码的具体说明如下。

- LDR_single_add(ldr,n,output_pathos)：调用函数 LDR_single_add()将每个灰度级别的量化图像进行累积操作，将结果保存到输出路径中。
- print('Quantization done')：输出提示信息，表示量化操作已完成。
- canvas = Gassian((h+4*linewidth,w+4*linewidth), mean=250, var = 3)：创建一个高斯噪声背景图像，尺寸为 (h+4*linewidth,w+4*linewidth)，均值为 250，方差为 3。
- for j in range(n)：遍历每个灰度级别。

(9) 根据笔画序列逐步绘制图像,生成最终的绘画结果。在绘制过程中,可以选择显示每个步骤的图像,以及保存中间过程的图像帧用于生成绘画动画。具体实现代码如下:

```
time_end=time.time()
print('total time',time_end-time_start)
print('stoke number',len(stroke_sequence))
# cv2.imwrite(output_pathos + "/draw.png", now_)
# cv2.imshow('draw', now_)
# cv2.waitKey(0)

if random_order == True:
    random.shuffle(stroke_sequence)

 if ETF_order == True:
    random.shuffle(stroke_sequence)
    quickSort(stroke_sequence,0,len(stroke_sequence)-1)
result = Gassian((h0,w0), mean=250, var = 3)
canvases = []

for dirs in range(direction):
    angle = -90+dirs*180/direction
    canvas,_ = rotate(result, -angle)
    # (h,w) = canvas.shape
    canvas = np.pad(canvas, pad_width=2*linewidth, mode='constant',
constant_values=(255,255))
    canvases.append(canvas)

for stroke_temp in stroke_sequence:
    angle = stroke_temp['angle']
    dirs = int((angle+90)*direction/180)
    grayscale = stroke_temp['grayscale']
    distribution = ChooseDistribution(linewidth=linewidth,Grayscale=grayscale)
    row = stroke_temp['row']
    begin = stroke_temp['begin']
    end = stroke_temp['end']
    length = end - begin

    newline = Getline(distribution=distribution, length=length)

    canvas = canvases[dirs]

    if length<1000 or begin == -2*linewidth or end == w-1+2*linewidth:
temp = canvas[row:row+2*linewidth,2*linewidth+ begin:2*linewidth+end]
        m = np.minimum(temp, newline[:,:temp.shape[1]])
        canvas[row:row+2*linewidth,2*linewidth+begin:2*linewidth+end] = m
```

```
    # else:
    #     temp = canvas[row:row+2*linewidth,2*linewidth+begin-2*linewidth:
2*linewidth+end+2*linewidth]
    #     m = np.minimum(temp, newline)
    #     canvas[row:row+2*linewidth,2*linewidth+begin- 2*linewidth:
2*linewidth+end+2*linewidth] = m

    now,_ = rotate(canvas[2*linewidth:-2*linewidth,2*linewidth:-2*linewidth], angle)
    (H,W) = now.shape
    now = now[int((H-h0)/2):int((H-h0)/2)+h0, int((W-w0)/2):int((W-w0)/2)+w0]
    result = np.minimum(now,result)
    if process_visible == True:
        cv2.imshow('step', result)
        cv2.waitKey(1)

    step += 1
    if step % Freq == 0:
        cv2.imwrite(output_pathos + "/process/{0:04d}.jpg".format(int(step/Freq)), result)
if step % Freq != 0:
    step = int(step/Freq)+1
    cv2.imwrite(output_pathos + "/process/{0:04d}.jpg".format(step), result)

cv2.destroyAllWindows()
time_end=time.time()
print('total time',time_end-time_start)
print('stoke number',len(stroke_sequence))
```

在上述代码中，stroke_sequence 是一个存储笔画信息的列表。在绘制图像的过程中，通过分析图像的边缘和灰度信息，将图像分解为一系列的笔画(strokes)。每个笔画由以下信息组成。

- angle：笔画的角度(方向)。
- grayscale：笔画的灰度值(亮度)。
- row：笔画的起始行位置。
- begin：笔画的起始列位置。
- end：笔画的结束列位置。
- importance：笔画的重要性(根据灰度值和边缘信息计算得出)。

列表 stroke_sequence 按照一定的规则存储了所有的笔画信息。在绘制图像时，会根据笔画的重要性和其他属性来确定绘制的顺序和方式。快速排序函数 quickSort()用于根据笔画的重要性对列表 stroke_sequence 中的内容进行排序，以便在绘制过程中先绘制重要性较高的笔画，从而实现更精确的绘制效果。

(10) 将生成的边缘图像与之前的绘画结果图像进行合并，生成最终的结果图像。然后对结果图像进行颜色处理，去除蓝色成分，并将结果图像的亮度通道替换为合成的图像，从而生成最终的彩色图像。具体实现代码如下：

```python
edge = genStroke(input_img,18)
edge = np.power(edge, deepen)
edge = np.uint8(edge*255)
if edge_CLAHE==True:
    edge = HistogramEqualization(edge)

cv2.imwrite(output_pathos + '/edge.jpg', edge)
cv2.imshow("edge",edge)


############# merge #############
edge = np.float32(edge)
now_ = cv2.imread(output_pathos + "/draw.jpg", cv2.IMREAD_GRAYSCALE)
result = res_cross= np.float32(now_)

result[1:,1:] = np.uint8(edge[:-1,:-1] * res_cross[1:,1:]/255)
result[0] = np.uint8(edge[0] * res_cross[0]/255)
result[:,0] = np.uint8(edge[:,0] * res_cross[:,0]/255)
result = edge*res_cross/255
result=np.uint8(result)

cv2.imwrite(output_pathos + '/result.jpg', result)
# cv2.imwrite(output_pathos + "/process/{0:04d}.png".format(step+1), result)
cv2.imshow("result",result)


# 调用函数dellblue()
dellblue(result, output_pathos)

# RGB
img_rgb_original = cv2.imread(input_pathos, cv2.IMREAD_COLOR)
cv2.imwrite(output_pathos + "/input.jpg", img_rgb_original)
img_yuv = cv2.cvtColor(img_rgb_original, cv2.COLOR_BGR2YUV)
img_yuv[:,:,0] = result
img_rgb = cv2.cvtColor(img_yuv, cv2.COLOR_YUV2BGR)

cv2.imshow("RGB",img_rgb)
cv2.waitKey(0)
cv2.imwrite(output_pathos + "/result_RGB.jpg",img_rgb)
```

上述代码的实现流程如下。

① 生成边缘图像。使用 genStroke() 函数生成输入图像的边缘图像，将其深化，并转换为灰度图像。

② 如果需要，对边缘图像进行直方图均衡化。
③ 将边缘图像保存为文件，并显示边缘图像。
④ 将边缘图像和之前生成的绘画结果图像进行合并，生成最终的结果图像。
⑤ 将结果图像保存为文件，并显示结果图像。
⑥ 进行去蓝色处理，删除结果图像中的蓝色成分。
⑦ 读取输入图像的彩色原始图像，并将结果图像的亮度通道替换为结果图像。
⑧ 将结果图像转换回彩色图像，并保存为文件。

执行效果如图 5-3 所示。

图 5-3　执行效果

## 5.5.2　绘制铅笔画

编写文件 process_order.py，功能是使用一系列的图像处理函数和方法绘制指定的铅笔画，包含的图像处理方法有 ETF 滤波、直方图均衡化、梯度计算、量化、绘制线条等。与前面的文件 cat.py 相比，保存绘图结果的路径不同：文件 process_order.py 将处理结果保存在指定的路径下的不同文件夹中；而文件 cat.py 直接显示结果图像，不保存文件。文件 process_order.py 的具体实现流程如下。

(1) 导入所需的库和模块。
(2) 设置程序运行所需的参数。
(3) 创建输出文件夹和子文件夹。
(4) 进行边缘流动场(ETF)滤波。
(5) 读取输入图像，并进行一些预处理操作。
(6) 根据给定的方向数量，生成一系列笔画序列。
(7) 根据笔画序列，将笔画逐步添加到画布上，形成最终的效果图。

(8) 生成边缘图像。

(9) 将边缘图像与最终效果图进行合并。

(10) 进行图像后处理操作，如去除蓝色。

(11) 将结果转换为 RGB 格式。

(12) 保存生成的结果图像。

总之，该代码通过将图像进行量化、边缘检测和笔画生成等步骤，实现将输入图像转换为类似铅笔画效果的输出图像。文件 process_order.py 的主要实现代码如下：

```python
edge = genStroke(input_img,18)
edge = np.power(edge, deepen)
edge = np.uint8(edge*255)
if edge_CLAHE==True:
    edge = HistogramEqualization(edge)

cv2.imwrite(output_pathos + '/edge.jpg', edge)
cv2.imshow("edge",edge)


############# merge #############
edge = np.float32(edge)
now_ = cv2.imread(output_pathos + "/draw.jpg", cv2.IMREAD_GRAYSCALE)
result = res_cross= np.float32(now_)

result[1:,1:] = np.uint8(edge[:-1,:-1] * res_cross[1:,1:]/255)
result[0] = np.uint8(edge[0] * res_cross[0]/255)
result[:,0] = np.uint8(edge[:,0] * res_cross[:,0]/255)
result = edge*res_cross/255
result=np.uint8(result)

cv2.imwrite(output_pathos + '/result.jpg', result)
# cv2.imwrite(output_pathos + "/process/{0:04d}.png".format(step+1), result)
cv2.imshow("result",result)


# dellblue
dellblue(result, output_pathos)

# RGB
img_rgb_original = cv2.imread(input_pathos, cv2.IMREAD_COLOR)
cv2.imwrite(output_pathos + "/input.jpg", img_rgb_original)
img_yuv = cv2.cvtColor(img_rgb_original, cv2.COLOR_BGR2YUV)
img_yuv[:,:,0] = result
img_rgb = cv2.cvtColor(img_yuv, cv2.COLOR_YUV2BGR)
```

```
cv2.imshow("RGB",img_rgb)
cv2.waitKey(0)
cv2.imwrite(output_pathos + "/result_RGB.jpg",img_rgb)
```

执行效果如图 5-4 所示。

图 5-4　执行效果

# 第 6 章
## ChatGPT 微信客服机器人

ChatGPT(Chat Generative Pre-trained Transformer)是美国 OpenAI 公司研发的聊天机器人程序，于 2022 年 11 月 30 日发布。ChatGPT 是人工智能技术驱动的自然语言处理工具，它能够通过理解和学习人类的语言来进行对话，还能根据聊天的上下文进行互动，真正像人类一样来聊天交流，甚至能完成撰写邮件、视频脚本、文案、代码、论文等任务。在本章的内容中，将详细讲解使用 ChatGPT 开发一个智能机器人系统的过程。

## 6.1 ChatGPT 概述

ChatGPT 是一个基于大规模预训练语言模型的对话系统，由 OpenAI 公司开发。ChatGPT 的核心技术是 GPT(generative pre-trained transformer)模型，它是一种基于深度学习的自然语言处理技术。GPT 模型采用 Transformer 架构，利用无监督学习从大规模语料库中学习语言知识，具有强大的语言理解和生成能力。ChatGPT 将 GPT 模型应用于对话生成，可以进行自然流畅的对话，具有人类般的语言交互能力，本质上是一个聊天工具。

扫码看视频

ChatGPT 的训练数据来自大规模的互联网文本语料库，例如维基百科和 BookCorpus。通过大规模的训练，ChatGPT 可以学习到丰富的语言知识和语言模式，从而能够在各种自然语言处理任务中表现出色。

ChatGPT 的应用非常广泛，例如在智能客服中，ChatGPT 可以帮助自动回复用户的问题，提高客户服务效率；在机器翻译中，ChatGPT 可以帮助实现更准确、流畅的翻译结果；在问答系统中，ChatGPT 可以回答用户的问题，从而实现人机交互。

### 6.1.1 ChatGPT 的发展历程

2015 年 12 月，OpenAI 公司在美国旧金山成立。值得一提的是，特斯拉的创始人马斯克也是该公司创始人之一。

2017 年，谷歌大脑团队推出了用于自然语言处理的 Transformer 模型，成为当时最先进的大型语言模型(large language model)。自诞生之日起，Transformer 模型就深刻地影响了接下来几年人工智能发展的各个领域，而 OpenAI 公司就是专注于研究 Transformer 模型的众多团队之一。

2018 年，Transformer 模型诞生不到一年，OpenAI 公司就推出了具有 1.17 亿个参数的 GPT-1 模型。

2019 年，OpenAI 公司公布了 GPT-2，具有 15 亿个参数。该模型架构与 GPT-1 原理相同，主要区别是比 GPT-2 的规模更大。

2020 年，这个创业团队推出了最新的 GPT-3 模型——这时它具有 1750 亿个参数。面对如此庞大的 GPT-3 模型，用户提供小样本的提示语或直接询问，就能获得符合要求的高质量答案。经过早期测试结束后，OpenAI 公司对 GPT-3 模型进行了商业化。

2020 年 9 月，微软公司获得了 GPT-3 模型的独占许可，意味着微软公司可以独家接触到 GPT-3 的源代码。

2022 年 3 月，OpenAI 公司推出了 InstructGPT 模型，该模型为 GPT-3 的微调版。

2022 年 11 月底，人工智能对话聊天机器人 ChatGPT 推出，使用的模型是 GPT-3.5。短短几个月时间，ChatGPT 在 2023 年 1 月份的月活跃用户数已达 1 亿，这使其成为史上用户数增长最快的消费者应用。

2023 年 3 月推出 GPT-4，模型系统的能力已经能够把美国的模拟律师资格考试考到前 10%了，并且顺利地在美国高考题(SAT 考试)中拿到了进入哈佛大学的成绩。有专家认为，GPT-4 的专业和学术水平接近人类。

而尚未公布的 GPT-5 已经看完了人类网络上所有的视频(大约 2000PB 的容量)，可以瞬间标记出所有它看过的视频中的一切声光信息，并且可准确到每一秒。有专家称，GPT-5 的智商也许接近天才级别。照这样的演进速度，看起来人工智能的能力将加快超越人类。

现在的 ChatGPT 可以扮演生活中各种各样的角色，如医生、翻译员、办公助手、程序员、历史学家、情感分析师、心理咨询师、写作润色师等。它可以根据用户的需求和背景，生成各种类型和风格的文本，并提供有用的信息和建议。

## 6.1.2 GPT 系列的演变

- GPT-1：结合有监督学习和无监督学习，更接近处理特定语言任务的专家模型，而非通用的语言模型。
- GPT-2：构建了泛化能力更强的语言模型，使语言模型的通用性得到了更充分的展现。
- GPT-3：参数量相较于 GPT-2 提升了两个数量级，达到了 1750 亿，成为真正意义上的超大语言模型。
- ChatGPT：通过引入人类反馈强化学习等新的训练方式，语言生成能力大幅提升，并且展现出了思维链和逻辑推理等多种能力。
- GPT-4：在推理能力、文本生成能力、对话能力等方面进一步提升的同时，实现了从大语言模型向多模态模型进化的第一步。

## 6.1.3 ChatGPT 的主要特点

OpenAI 使用人类反馈强化学习(reinforcement learning from human feedbac，RLHF) 技术对 ChatGPT 进行了训练，且加入了更多人工监督进行微调。此外，ChatGPT 还具有以下特点。

(1) 可以主动承认自身错误：若用户指出其错误，模型会听取意见并优化答案。

(2) ChatGPT 可以质疑不正确的问题。例如，被询问"哥伦布 2015 年来到美国的情景"的问题时，机器人会说明哥伦布不属于这一时代并调整输出结果。

(3) ChatGPT 可以承认自身的无知，承认对专业技术的不了解。

(4) 支持连续多轮对话。与大家在生活中用到的各类智能音箱和"人工智障"不同，ChatGPT 在对话过程中会记忆先前使用者的对话信息，即上下文理解，以回答某些假设性的问题。ChatGPT 可实现连续对话，极大地提升了对话交互模式下的用户体验。

## 6.2 系统介绍

在本项目中，将使用 ChatGPT 的对话模型把我们的微信或微信公众号打造成一个智能机器人，能够实时聊天，聊天功能都是通过 ChatGPT 实现的。本项目的具体功能如下。

扫码看视频

- 基础对话：私聊及群聊的消息智能回复，支持多轮会话上下文记忆，支持 GPT-3、GPT-3.5、GPT-4 模型等。
- 语音识别：可识别语音消息，通过文字或语音回复，支持 Azure、baidu、google、OpenAI 等多种语音模型。
- 图片生成：支持图片生成和照片修复，可选择 Dell-E、Stable Diffusion、Replicate 模型。
- 丰富插件：支持个性化插件扩展，已实现多角色切换、文字冒险、敏感词过滤、聊天记录总结等插件。
- Tool 工具：与操作系统和互联网交互，支持最新信息搜索、数学计算、天气和资讯查询、网页总结，基于 chatgpt-tool-hub 实现。
- 不仅支持个人微信和微信公众号，还可以接入更多应用和添加新的插件。

## 6.3 项目结构

本项目的具体结构如图 6-1 所示。

扫码看视频

第 6 章　ChatGPT 微信客服机器人

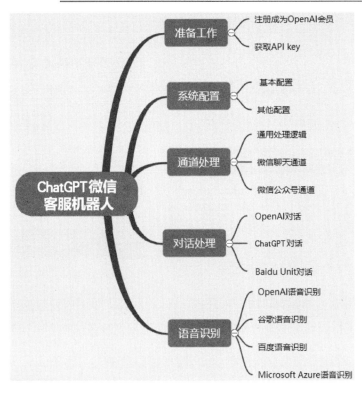

图 6-1　项目结构

## 6.4　准备工作

在开发本项目之前需要先注册成为 OpenAI 会员，登录后获取一个 API key。在本节的内容中，将详细讲解注册成为 OpenAI 会员并获取 API key 的过程。

扫码看视频

### 6.4.1　注册成为 OpenAI 会员

（1）在浏览器中输入 https://chat.openai.com/auth/login，进入 ChatGPT 会话登录页面，如图 6-2 所示。

（2）单击 Sign up 按钮来到注册页面，如图 6-3 所示。在文本框中输入邮箱地址，或者直接使用 Gmail 邮箱地址。

（3）单击 Continue 按钮来到设置密码页面，在文本框中

图 6-2　ChatGPT 会话登录页面

177

设置注册账号的密码，如图 6-4 所示。

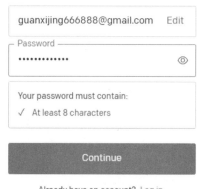

图 6-3　输入邮箱地址　　　　　图 6-4　设置密码

（4）单击 Continue 按钮来到注册成功页面，需要分别验证邮箱和电话号码，如图 6-5 和图 6-6 所示。

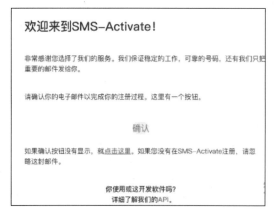

图 6-5　验证邮箱　　　　　图 6-6　验证电话号码

## 6.4.2　获取 API key

API key 也就是 API 密钥，是我们在程序中调用并识别 ChatGPT API 请求的唯一标识，获取 API key 的流程如下。

(1) 在浏览器中输入 https://chat.openai.com/auth/login，来到 ChatGPT 会话登录页面，输入自己的账号信息，单击 Continue 按钮登录系统。登录成功后的页面效果如图 6-7 所示。

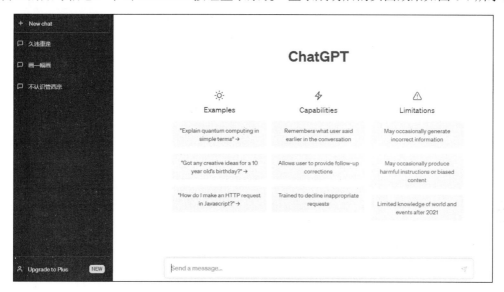

图 6-7　ChatGPT 会话页面

(2) 用户可以在下方的文本框中输入聊天信息与 ChatGPT 对话，例如，在文本框中输入"go 语言有几种循环语句"，ChatGPT 的回答非常完美，具体如图 6-8 所示。

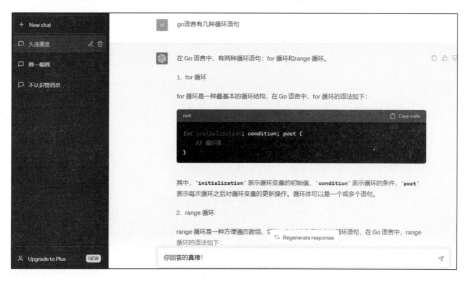

图 6-8　与 ChatGPT 对话

(3) 输入 https://platform.openai.com/account/usage，来到个人中心页面，如图 6-9 所示。

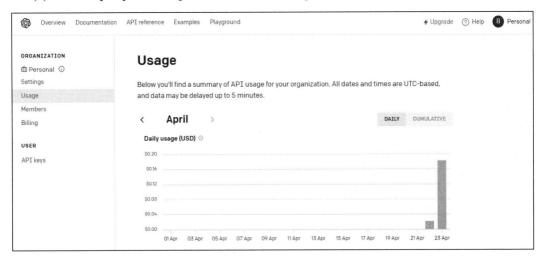

图 6-9　个人中心页面

(4) 单击图 6-9 中左侧列表中的 API keys 链接，来到个人 API keys 页面，在此页面中列出了当前会员的所有 API keys 信息，如图 6-10 所示。

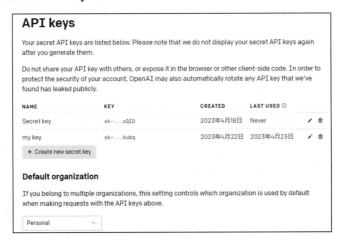

图 6-10　个人 API keys 页面

(5) 在 API keys 列表中可以删除或修改已经存在的 API keys，也可以单击 Create new secret key 按钮，创建一个新的 API key，在弹出的页面中设置 key 的名字，如图 6-11 所示。

(6) 单击 Create secret key 按钮后会显示创建的 key，如图 6-12 所示。单击 Done 按钮，完成 API keys 的创建工作，这个 key 的值将在项目中用到。

## 第 6 章　ChatGPT 微信客服机器人

图 6-11　设置 key 的名字

图 6-12　创建的 key

注意：ChatGPT 的 API keys 十分重要，获取后请用户妥善保管，不要泄露。

## 6.5　系统配置

在本项目中，需要利用前面申请的 API key，将 ChatGPT 功能应用在微信或公众号的聊天模块中。在具体编码之前，需要先设置配置信息。

扫码看视频

### 6.5.1　基本配置

编写文件 config-template.json，功能是设置申请的 API key 和代理地址，并分别设置启动 ChatGPT 机器人的参数。文件 config-template.json 的具体实现代码如下：

```
{
  "open_ai_api_key": "申请的 API Key",
  "model": "gpt-3.5-turbo",
  "proxy": "http://127.0.0.1:10809",
```

```json
"single_chat_prefix": [
  "bot",
  "@bot"
],
"single_chat_reply_prefix": "[bot] ",
"group_chat_prefix": [
  "@bot"
],
"group_name_white_list": [
  "足球聊天群",
  "炒股群"
],
"group_chat_in_one_session": [
  "ChatGPT 测试群"
],
"image_create_prefix": [
  "画",
  "看",
  "找"
],
"speech_recognition": false,
"group_speech_recognition": false,
"voice_reply_voice": false,
"conversation_max_tokens": 1000,
"expires_in_seconds": 3600,
"character_desc": "你是 ChatGPT,一个由 OpenAI 训练的大型语言模型,你旨在回答并解决人们的任何问题,并且可以使用多种语言与人交流。"
}
```

在上述代码中分别设置了使用的模型信息、个人聊天、群组聊天参数、图像处理参数、声音识别参数,本项目默认关闭声音识别。

## 6.5.2 其他配置

除了在前面文件 config-template.json 中介绍的配置信息外,还可以配置其他相关信息,例如公众号机器人、百度语音识别、代理信息等参数。编写文件 config.py 实现其他信息配置功能,具体实现代码如下:

```python
# 将所有可用的配置项写在字典里,请使用小写字母
available_setting = {
    # OpenAI API 配置
    "open_ai_api_key": " openai api key ",
    # 当use_azure_chatgpt 为 true 时,需要设置对应的 api base
    "open_ai_api_base": "https://api.openai.com/v1",
    "proxy": "http://127.0.0.1:10809",  # OpenAI 使用的代理
```

```
    # ChatGPT 模型, 当use_azure_chatgpt 为true 时, 其名称为Azure 上model deployment 名称
    "model": "gpt-3.5-turbo",
    "use_azure_chatgpt": False,        # 是否使用 azure 的 ChatGPT
    "azure_deployment_id": "",         # azure 模型部署名称
    # Bot 触发配置
    "single_chat_prefix": ["bot", "@bot"],   # 私聊时文本需要包含该前缀才触发机器人回复
    "single_chat_reply_prefix": ["bot"],     # 私聊时自动回复的前缀,用于区分真人
    "group_chat_prefix": ["@bot"],           # 群聊时包含该前缀则会触发机器人回复
    "group_chat_reply_prefix": "",           # 群聊时自动回复的前缀
    "group_chat_keyword": [],                # 群聊时包含该关键词则会触发机器人回复
    "group_at_off": False,                   # 是否关闭群聊时@bot 的触发
    "group_name_white_list": ["足球群", "炒股交流群"],   # 开启自动回复的群名称列表
    "group_name_keyword_white_list": [],     # 开启自动回复的群名称关键词列表
    "group_chat_in_one_session": ["Python 训练营群 2"],  # 支持会话上下文共享的群名称
    "trigger_by_self": False,                # 是否允许机器人触发
    "image_create_prefix": ["画", "看", "找"],  # 开启图片回复的前缀
    "concurrency_in_session": 1,             # 同一会话最多有多少条消息在处理中,大于 1 可能乱序
    "image_create_size": "256x256",          # 图片大小,可选尺寸有 256×256、512×512、1024×1024
    # ChatGPT 会话参数
    "expires_in_seconds": 3600,              # 无操作会话的过期时间
    "character_desc": "你是 ChatGPT, 一个由 OpenAI 训练的大型语言模型, 你旨在回答并解决人们
的任何问题, 并且可以使用多种语言与人交流。",   # 人格描述
    "conversation_max_tokens": 1000,         # 支持上下文记忆的最多字符数
    # ChatGPT 限流配置
    "rate_limit_chatgpt": 20,                # ChatGPT 的调用频率限制
    "rate_limit_dalle": 50,                  # OpenAI dalle 的调用频率限制
    # ChatGPT API 参数参考 https://platform.openai.com/docs/api-reference/chat/create
    "temperature": 0.9,
    "top_p": 1,
    "frequency_penalty": 0,
    "presence_penalty": 0,
    "request_timeout": 60,                   # ChatGPT 请求超时时间,OpenAI 接口默认设置为 600
                                             # 对于难问题一般需要较长时间
    "timeout": 120,                          # ChatGPT 重试超时时间,在这个时间内,将会自动重试
                                             # 语音设置
    "speech_recognition": False,             # 是否开启语音识别
    "group_speech_recognition": False,       # 是否开启群组语音识别
    "voice_reply_voice": False,              # 是否使用语音回复,需要设置对应语音合成引擎的 API key
    "always_reply_voice": False,             # 是否一直使用语音回复
    "voice_to_text": "openai",               # 语音识别引擎,支持 openai、baidu、google、azure
    "text_to_voice": "baidu",    # 语音合成引擎,支持 baidu、google、pytts(offline)、azure
    # baidu 语音 api 配置,   使用百度语音识别和语音合成时需要
    "baidu_app_id": "",
    "baidu_api_key": "",
    "baidu_secret_key": "",
    # 1536 普通话(支持简单的英文识别) 1737 英语 1637 粤语 1837 四川话 1936 普通话远场
    "baidu_dev_pid": "1536",
```

```python
    # azure 语音 api 配置，使用 azure 语音识别和语音合成时需要
    "azure_voice_api_key": "",
    "azure_voice_region": "japaneast",
    # 服务时间限制，目前支持 itchat
    "chat_time_module": False,            # 是否开启服务时间限制
    "chat_start_time": "00:00",           # 服务开始时间
    "chat_stop_time": "24:00",            # 服务结束时间
    # itchat 的配置
    "hot_reload": False,                  # 是否开启热重载
    # wechaty 的配置
    "wechaty_puppet_service_token": "",   # wechaty 的 token
    # wechatmp 的配置
    "wechatmp_token": "",                 # 微信公众平台的 Token
    "wechatmp_port": 8080,                # 微信公众平台的端口号，需将其端口转发到 80 或 443
    "wechatmp_app_id": "",                # 微信公众平台的 appID
    "wechatmp_app_secret": "",            # 微信公众平台的 appsecret
    "wechatmp_aes_key": "",               # 微信公众平台的 EncodingAESKey, 加密模式需要
    # ChatGPT 指令自定义触发词
    "clear_memory_commands": ["#清除记忆"],  # 重置会话指令，必须以"#"开头
    # channel 配置
    "channel_type": "wx",                 # 通道类型，支持类型有: {wx,wxy,terminal,wechatmp,
                                          #   wechatmp_service}
    "debug": False,                       # 是否开启 debug 模式，开启后会打印更多日志
    "appdata_dir": "",                    # 数据目录
    # 插件配置
    "plugin_trigger_prefix": "$",         # 规范插件提供聊天相关指令的前缀，建议不要和管理员指令
                                          # 前缀"#"冲突
}

class Config(dict):
    def __init__(self, d: dict = {}):
        super().__init__(d)
        # user_datas 为用户数据，key 为用户名，value 为用户数据，也是 dict
        self.user_datas = {}

    def __getitem__(self, key):
        if key not in available_setting:
            raise Exception("key {} not in available_setting".format(key))
        return super().__getitem__(key)

    def __setitem__(self, key, value):
        if key not in available_setting:
            raise Exception("key {} not in available_setting".format(key))
        return super().__setitem__(key, value)

    def get(self, key, default=None):
        try:
```

```python
            return self[key]
        except KeyError as e:
            return default
        except Exception as e:
            raise e

    def get_user_data(self, user) -> dict:
        if self.user_datas.get(user) is None:
            self.user_datas[user] = {}
        return self.user_datas[user]

    def load_user_datas(self):
        try:
            with open(os.path.join(get_appdata_dir(), "user_datas.pkl"), "rb") as f:
                self.user_datas = pickle.load(f)
                logger.info("[Config] User datas loaded.")
        except FileNotFoundError as e:
            logger.info("[Config] User datas file not found, ignore.")
        except Exception as e:
            logger.info("[Config] User datas error: {}".format(e))
            self.user_datas = {}

    def save_user_datas(self):
        try:
            with open(os.path.join(get_appdata_dir(), "user_datas.pkl"), "wb") as f:
                pickle.dump(self.user_datas, f)
                logger.info("[Config] User datas saved.")
        except Exception as e:
            logger.info("[Config] User datas error: {}".format(e))
config = Config()

def load_config():
    global config
    config_path = "./config.json"
    if not os.path.exists(config_path):
        logger.info("配置文件不存在，将使用 config-template.json 模板")
        config_path = "./config-template.json"
    config_str = read_file(config_path)
    logger.debug("[INIT] config str: {}".format(config_str))

    # 将 json 字符串反序列化为 dict 类型
    config = Config(json.loads(config_str))
    for name, value in os.environ.items():
        name = name.lower()
        if name in available_setting:
            logger.info("[INIT] override config by environ args: {}={}".format(name, value))
            try:
```

```
                config[name] = eval(value)
            except:
                if value == "false":
                    config[name] = False
                elif value == "true":
                    config[name] = True
                else:
                    config[name] = value
    if config.get("debug", False):
        logger.setLevel(logging.DEBUG)
        logger.debug("[INIT] set log level to DEBUG")
    logger.info("[INIT] load config: {}".format(config))
    config.load_user_datas()

def get_root():
    return os.path.dirname(os.path.abspath(__file__))

def read_file(path):
    with open(path, mode="r", encoding="utf-8") as f:
        return f.read()

def conf():
    return config

def get_appdata_dir():
    data_path = os.path.join(get_root(), conf().get("appdata_dir", ""))
    if not os.path.exists(data_path):
        logger.info("[INIT] data path not exists, create it: {}".format(data_path))
        os.makedirs(data_path)
    return data_path
```

各个配置参数的具体说明如下。

1) 个人聊天
- 在个人聊天中,需要以 bot 或@bot 为开头的内容触发机器人,对应的配置项为 single_chat_prefix (如果不需要以前缀触发,可以填写"single_chat_prefix": [""])。
- 机器人回复的内容会以 [bot] 作为前缀,以区分真人,对应的配置项为 single_chat_reply_prefix (如果不需要前缀,可以填写"single_chat_reply_prefix": "")。

2) 群组聊天
- 在群组聊天中,群名称配置在 group_name_white_list 中,才能开启群聊自动回复。如果想对所有群聊生效,可以直接填写"group_name_white_list": ["ALL_GROUP"]。
- 默认只要被人@就会触发机器人自动回复。在群聊天中,只要检测到以@bot 开头的内容,同样会自动回复(方便自己触发),对应的配置项为 group_chat_prefix。
- group_name_keyword_white_list 配置项:支持模糊匹配群名称,group_chat_keyword

配置项则支持模糊匹配群消息内容，用法与上述两个配置项相同。
- group_chat_in_one_session：使群聊共享一个会话上下文，配置项为["ALL_GROUP"]，则作用于所有群聊。

3) 语音识别
- 将 speech_recognition 值设置为 true 后将会开启语音识别，默认使用 OpenAI 的 whisper 模型识别为文字，同时以文字回复，该参数仅支持私聊(注意，由于语音消息无法匹配前缀，一旦开启将对所有语音自动回复，支持语音触发画图)。
- 添加 group_speech_recognition：true 将开启群组语音识别，默认使用 OpenAI 的 whisper 模型识别为文字，同时以文字回复，参数仅支持群聊(会匹配 group_chat_prefix 和 group_chat_keyword，支持语音触发画图)。
- 添加 voice_reply_voice：true 将开启语音回复功能(同时作用于私聊和群聊)，但是需要配置对应语音合成平台的 key，由于 itchat 协议的限制，只能发送语音 MP3 文件，若使用 wechaty，则回复的是微信语音。

4) 其他配置
- model：模型名称，目前支持 gpt-3.5-turbo、text-davinci-003、gpt-4、gpt-4-32k (其中 gpt-4 api 暂未开放)。
- temperature,frequency_penalty,presence_penalty：Chat API 接口参数，详情参考 OpenAI 官方文档。
- proxy：由于目前 OpenAI 接口国内无法访问，需配置代理客户端的地址。
- 图像生成：在满足个人或群组触发条件外，还需要额外的关键词前缀来触发，对应的配置项为 image_create_prefix。
- OpenAI 对话及图片接口的参数(内容自由度、回复字数限制、图片大小等)配置。可以在对话接口和图像接口文件 bot/openai/open_ai_bot.py 中进行调整。
- conversation_max_tokens：表示能够记忆的上下文最大字数(一问一答为一组对话，如果累积的对话字数超出限制，就会优先移除最早的一组对话)。
- rate_limit_chatgpt/rate_limit_dalle：每分钟最高问答速率、画图速率，超速后排队按序处理。
- clear_memory_commands：对话内指令，主动清空前文记忆，字符串数组可自定义指令别名。
- hot_reload：程序退出后，暂存微信扫码状态，默认为关闭状态。
- character_desc：保存了你对机器人说的一段话，它会记住这段话并作为它的设定，你可以为它定制任何人格。

## 6.6 通道处理

为了便于系统维护，为不同的聊天类型(微信聊天、微信公众号聊天)提供不同的聊天通道，通过不同的通道传输不同类型的聊天信息。

扫码看视频

### 6.6.1 通用处理逻辑

编写文件 channel/chat_channel.py 实现通道抽象类，在此文件中包含了与消息通道无关的通用处理逻辑。文件 chat_channel.py 的具体实现流程如下。

(1) 创建通道抽象类 ChatChannel，设置相关的通道参数 name、user_id、sessions 等。对应的实现代码如下：

```python
class ChatChannel(Channel):
    name = None         # 登录的用户名
    user_id = None      # 登录的用户 id
    # 记录每个 session_id 提交到线程池的 future 对象，用于重置会话时把没执行的 future 对象取消
    # 正在执行的 future 对象不会被取消
    futures = {}
    sessions = {}       # 用于控制并发，每个 session_id 同时只能有一个 context 在处理
    lock = threading.Lock()  # 用于控制对 sessions 的访问
    handler_pool = ThreadPoolExecutor(max_workers=8)  # 处理消息的线程池
```

(2) 编写方法_compose_context()，功能是根据获取的消息构造 context，将与消息内容相关的触发项写在这里。对应的实现代码如下：

```python
def _compose_context(self, ctype: ContextType, content, **kwargs):
    context = Context(ctype, content)
    context.kwargs = kwargs
    # context 首次传入时，origin_ctype 是 None，引入的起因是：当输入语音时会嵌套生成两个 context
    # 第一步语音转文本，第二步通过文本生成文字回复
    # origin_ctype 用于第二步文本回复时，判断是否需要匹配前缀，如果是私聊的语音，就不需要匹配前缀
    if "origin_ctype" not in context:
        context["origin_ctype"] = ctype
    # context 首次传入时，receiver 是 None，根据类型设置 receiver
    first_in = "receiver" not in context
    # 群名匹配过程，设置 session_id 和 receiver
    if first_in:  # context 首次传入时，receiver 是 None，根据类型设置 receiver
        config = conf()
        cmsg = context["msg"]
        if context.get("isgroup", False):
            group_name = cmsg.other_user_nickname
            group_id = cmsg.other_user_id
```

```python
            group_name_white_list = config.get("group_name_white_list", [])
            group_name_keyword_white_list =
                config.get("group_name_keyword_white_list", [])
            if any(
                [
                    group_name in group_name_white_list,
                    "ALL_GROUP" in group_name_white_list,
                    check_contain(group_name, group_name_keyword_white_list),
                ]
            ):
                group_chat_in_one_session =
                    conf().get("group_chat_in_one_session", [])
                session_id = cmsg.actual_user_id
                if any(
                    [
                        group_name in group_chat_in_one_session,
                        "ALL_GROUP" in group_chat_in_one_session,
                    ]
                ):
                    session_id = group_id
            else:
                return None
            context["session_id"] = session_id
            context["receiver"] = group_id
        else:
            context["session_id"] = cmsg.other_user_id
            context["receiver"] = cmsg.other_user_id
        e_context = PluginManager().emit_event(EventContext(Event.ON_RECEIVE_MESSAGE,
{"channel": self, "context": context}))
        context = e_context["context"]
        if e_context.is_pass() or context is None:
            return context
        if cmsg.from_user_id == self.user_id and not config.get("trigger_by_self", True):
            logger.debug("[WX]self message skipped")
            return None
    # 消息内容匹配过程,并处理content
    if ctype == ContextType.TEXT:
        if first_in and "」\n- - - - - - -" in content:  # 初次匹配过滤引用消息
            logger.debug("[WX]reference query skipped")
            return None
        if context.get("isgroup", False):  # 群聊
            # 校验关键字
            match_prefix = check_prefix(content, conf().get("group_chat_prefix"))
            match_contain = check_contain(content, conf().get("group_chat_keyword"))
            flag = False
            if match_prefix is not None or match_contain is not None:
                flag = True
                if match_prefix:
```

```python
                    content = content.replace(match_prefix, "", 1).strip()
                if context["msg"].is_at:
                    logger.info("[WX]receive group at")
                    if not conf().get("group_at_off", False):
                        flag = True
                    pattern = f"@{re.escape(self.name)}(\u2005|\u0020)"
                    content = re.sub(pattern, r"", content)
                if not flag:
                    if context["origin_ctype"] == ContextType.VOICE:
                        logger.info("[WX]receive group voice, but checkprefix didn't match")
                    return None
            else:  # 单聊
                match_prefix = check_prefix(content, conf().get("single_chat_prefix", [""]))
                if match_prefix is not None:
                    # 判断如果匹配到自定义前缀,则返回过滤掉前缀+空格后的内容
                    content = content.replace(match_prefix, "", 1).strip()
                elif context["origin_ctype"] == ContextType.VOICE:
                    # 如果源消息是私聊的语音消息,允许不匹配前缀,放宽条件
                    pass
                else:
                    return None
            img_match_prefix = check_prefix(content, conf().get("image_create_prefix"))
            if img_match_prefix:
                content = content.replace(img_match_prefix, "", 1)
                context.type = ContextType.IMAGE_CREATE
            else:
                context.type = ContextType.TEXT
            context.content = content.strip()
            if "desire_rtype" not in context and conf().get("always_reply_voice") and \
                    ReplyType.VOICE not in self.NOT_SUPPORT_REPLYTYPE:
                context["desire_rtype"] = ReplyType.VOICE
        elif context.type == ContextType.VOICE:
            if "desire_rtype" not in context and conf().get("voice_reply_voice") and \
                    ReplyType.VOICE not in self.NOT_SUPPORT_REPLYTYPE:
                context["desire_rtype"] = ReplyType.VOICE

    return context
```

(3) 编写处理方法_handle(),功能是根据获取的信息进行预回复处理。对应的实现代码如下:

```python
def _handle(self, context: Context):
    if context is None or not context.content:
        return
    logger.debug("[WX] ready to handle context: {}".format(context))
    # reply 的构建步骤
    reply = self._generate_reply(context)
```

```
    logger.debug("[WX] ready to decorate reply: {}".format(reply))
# reply 的包装步骤
reply = self._decorate_reply(context, reply)
# reply 的发送步骤
self._send_reply(context, reply)
```

(4) 编写方法_generate_reply()，功能是根据获取的信息生成对应的回复，此方法特意处理了语音信息，将语音信息转换成文本信息。对应的实现代码如下：

```
def _generate_reply(self, context: Context, reply: Reply = Reply()) -> Reply:
    e_context = PluginManager().emit_event(
        EventContext(
            Event.ON_HANDLE_CONTEXT,
            {"channel": self, "context": context, "reply": reply},
        )
    )
    reply = e_context["reply"]
    if not e_context.is_pass():
        logger.debug("[WX] ready to handle context: type={}, content={}".format
                        (context.type, context.content))
        if context.type == ContextType.TEXT or context.type == ContextType.IMAGE_
                        CREATE:   # 文字和图片消息
            reply = super().build_reply_content(context.content, context)
        elif context.type == ContextType.VOICE:   # 语音消息
            cmsg = context["msg"]
            cmsg.prepare()
            file_path = context.content
            wav_path = os.path.splitext(file_path)[0] + ".wav"
            try:
                any_to_wav(file_path, wav_path)
            except Exception as e:
            # 如果转换失败，则使用原始mp3文件，因为某些api也可以识别mp3格式
                logger.warning("[WX]any to wav error, use raw path. " + str(e))
                wav_path = file_path
            # 语音识别
            reply = super().build_voice_to_text(wav_path)
            # 删除临时文件
            try:
                os.remove(file_path)
                if wav_path != file_path:
                    os.remove(wav_path)
            except Exception as e:
                pass
            if reply.type == ReplyType.TEXT:
                new_context = self._compose_context(ContextType.TEXT,
                            reply.content, **context.kwargs)
                if new_context:
```

```
                    reply = self._generate_reply(new_context)
                else:
                    return
        elif context.type == ContextType.IMAGE:  # 图片消息，当前无默认逻辑
            pass
        else:
            logger.error("[WX] unknown context type: {}".format(context.type))
            return
    return reply
```

(5) 编写方法_decorate_reply()装饰生成的回复信息，展示回复者的微信昵称和聊天内容，如果传递的是图片信息，则会显示这幅图片的网址。对应的实现代码如下：

```
def _decorate_reply(self, context: Context, reply: Reply) -> Reply:
    if reply and reply.type:
        e_context = PluginManager().emit_event(
            EventContext(
                Event.ON_DECORATE_REPLY,
                {"channel": self, "context": context, "reply": reply},
            )
        )
        reply = e_context["reply"]
        desire_rtype = context.get("desire_rtype")
        if not e_context.is_pass() and reply and reply.type:
            if reply.type in self.NOT_SUPPORT_REPLYTYPE:
                logger.error("[WX]reply type not support: " + str(reply.type))
                reply.type = ReplyType.ERROR
                reply.content = "不支持发送的消息类型: " + str(reply.type)

            if reply.type == ReplyType.TEXT:
                reply_text = reply.content
                if desire_rtype == ReplyType.VOICE and ReplyType.VOICE not in
                                    self.NOT_SUPPORT_REPLYTYPE:
                    reply = super().build_text_to_voice(reply.content)
                    return self._decorate_reply(context, reply)
                if context.get("isgroup", False):
                    reply_text = "@" + context["msg"].actual_user_nickname + " " +
                                    reply_text.strip()
                    reply_text = conf().get("group_chat_reply_prefix", "") + reply_text
                else:
                    reply_text = conf().get("single_chat_reply_prefix", "") + reply_text
                reply.content = reply_text
            elif reply.type == ReplyType.ERROR or reply.type == ReplyType.INFO:
                reply.content = "[" + str(reply.type) + "]\n" + reply.content
            elif reply.type == ReplyType.IMAGE_URL or reply.type ==
                                ReplyType.VOICE or reply.type == ReplyType.IMAGE:
                pass
            else:
```

```
            logger.error("[WX] unknown reply type: {}".format(reply.type))
            return
        if desire_rtype and desire_rtype != reply.type and reply.type not in
                [ReplyType.ERROR, ReplyType.INFO]:
            logger.warning("[WX] desire_rtype: {}, but reply type: {}".format
                    (context.get("desire_rtype"), reply.type))
        return reply
```

(6) 编写方法_send_reply()发送回复信息,对应的实现代码如下:

```
def _send_reply(self, context: Context, reply: Reply):
    if reply and reply.type:
        e_context = PluginManager().emit_event(
            EventContext(
                Event.ON_SEND_REPLY,
                {"channel": self, "context": context, "reply": reply},
            )
        )
        reply = e_context["reply"]
        if not e_context.is_pass() and reply and reply.type:
            logger.debug("[WX] ready to send reply: {}, context: {}".format(reply,
                    context))
            self._send(reply, context)
```

(7) 编写消费者方法 consume(),这是一个单线程方法,用于从消息队列中取出消息并进行处理,对应的实现代码如下:

```
def consume(self):
    while True:
        with self.lock:
            session_ids = list(self.sessions.keys())
            for session_id in session_ids:
                context_queue, semaphore = self.sessions[session_id]
                if semaphore.acquire(blocking=False):          #等线程处理完毕才能删除
                    if not context_queue.empty():
                        context = context_queue.get()
                        logger.debug("[WX] consume context: {}".format(context))
                        future: Future = self.handler_pool.submit(self._handle,
                                context)
                        future.add_done_callback(self._thread_pool_callback
                                (session_id, context=context))
                        if session_id not in self.futures:
                            self.futures[session_id] = []
                        self.futures[session_id].append(future)
                    elif semaphore._initial_value == semaphore._value + 1:
                        #除了当前,没有任务再申请到信号量,说明所有任务都处理完毕
                        self.futures[session_id] = [t for t in self.futures[session_id]
                                if not t.done()]
```

```
                    assert len(self.futures[session_id]) == 0, "thread pool error"
                    del self.sessions[session_id]
                else:
                    semaphore.release()
        time.sleep(0.1)
```

(8) 编写方法 cancel_session()，功能是取消 session_id 对应的所有任务，只能取消排队的消息和已提交线程池但未执行的任务，对应的实现代码如下：

```
def cancel_session(self, session_id):
    with self.lock:
        if session_id in self.sessions:
            for future in self.futures[session_id]:
                future.cancel()
            cnt = self.sessions[session_id][0].qsize()
            if cnt > 0:
                logger.info("Cancel {} messages in session {}".format(cnt, session_id))
            self.sessions[session_id][0] = Dequeue()
```

### 6.6.2 微信聊天通道

**1. 构造聊天内容**

编写文件 channel/wechat/wechat_channel.py，功能是将 ChatGPT 和 itchat 进行关联。itchat 是一个开源的微信个人号接口，使用 Python 调用微信从未如此简单。使用不到三十行的代码，就可以完成一个能够处理所有信息的微信聊天机器人。文件 wechat_channel.py 的具体实现流程如下。

(1) 编写方法 qrCallback()，功能是生成可用的微信登录二维码接口，对应的实现代码如下：

```
def qrCallback(uuid, status, qrcode):
    # logger.debug("qrCallback: {} {}".format(uuid,status))
    if status == "0":
        try:
            from PIL import Image

            img = Image.open(io.BytesIO(qrcode))
            _thread = threading.Thread(target=img.show, args=("QRCode",))
            _thread.setDaemon(True)
            _thread.start()
        except Exception as e:
            pass

        import qrcode
```

```
url = f"https://login.weixin.qq.com/l/{uuid}"

qr_api1 = "https://api.isoyu.com/qr/?m=1&e=L&p=20&url={}".format(url)
qr_api2 = "https://api.qrserver.com/v1/create-qr-code/?size=
          400×400&data={}".format(url)
qr_api3 = "https://api.pwmqr.com/qrcode/create/?url={}".format(url)
qr_api4 = "https://my.tv.sohu.com/user/a/wvideo/getQRCode.do?text=
          {}".format(url)
print("You can also scan QRCode in any website below:")
print(qr_api3)
print(qr_api4)
print(qr_api2)
print(qr_api1)

qr = qrcode.QRCode(border=1)
qr.add_data(url)
qr.make(fit=True)
qr.print_ascii(invert=True)
```

接下来创建 handle_* 系列方法，用于在收到聊天消息后构造 Context 内容，然后传入 produce()方法中处理 Context 内容和发送回复。Context 包含了消息的所有信息，包括以下属性。

- type 消息类型，包括 text、voice、image_create。
- content 消息内容，如果是 text 类型，content 就是文本内容；如果是 voice 类型，content 就是语音文件名；如果是 image_create 类型，则 content 表示图片生成命令。
- kwargs 附加参数字典，包含如下所示的 key：
  - session_id：会话 id。
  - isgroup：是不是群聊。
  - receiver：需要回复的对象。
  - msg：ChatMessage 消息对象。
  - origin_ctype：原始消息类型，语音转文字后，私聊时如果匹配前缀失败，会根据初始消息是不是语音来放宽触发规则。
  - desire_rtype：希望回复类型，默认是文本回复，如果设置为 ReplyType.VOICE，则表示是语音回复。

(2) 编写方法 handle_single()实现微信私聊功能，对应的实现代码如下：

```
@time_checker
@_check
def handle_single(self, cmsg: ChatMessage):
    if cmsg.ctype == ContextType.VOICE:
```

```
            if conf().get("speech_recognition") != True:
                return
            logger.debug("[WX]receive voice msg: {}".format(cmsg.content))
        elif cmsg.ctype == ContextType.IMAGE:
            logger.debug("[WX]receive image msg: {}".format(cmsg.content))
        elif cmsg.ctype == ContextType.PATPAT:
            logger.debug("[WX]receive patpat msg: {}".format(cmsg.content))
        elif cmsg.ctype == ContextType.TEXT:
            logger.debug("[WX]receive text msg: {}, cmsg={}".format(json.dumps(cmsg._rawmsg, ensure_ascii=False), cmsg))
        else:
            logger.debug("[WX]receive msg: {}, cmsg={}".format(cmsg.content, cmsg))
        context = self._compose_context(cmsg.ctype, cmsg.content, isgroup=False, msg=cmsg)
        if context:
            self.produce(context)
```

(3) 编写方法 handle_group() 实现微信群聊功能,对应的实现代码如下:

```
    @time_checker
    @_check
    def handle_group(self, cmsg: ChatMessage):
        if cmsg.ctype == ContextType.VOICE:
            if conf().get("speech_recognition") != True:
                return
            logger.debug("[WX]receive voice for group msg: {}".format(cmsg.content))
        elif cmsg.ctype == ContextType.IMAGE:
            logger.debug("[WX]receive image for group msg: {}".format(cmsg.content))
        elif cmsg.ctype in [ContextType.JOIN_GROUP, ContextType.PATPAT]:
            logger.debug("[WX]receive note msg: {}".format(cmsg.content))
        elif cmsg.ctype == ContextType.TEXT:
            # logger.debug("[WX]receive group msg: {}, cmsg={}".format(json.dumps(cmsg._rawmsg, ensure_ascii=False), cmsg))
            pass
        else:
            logger.debug("[WX]receive group msg: {}".format(cmsg.content))
        context = self._compose_context(cmsg.ctype, cmsg.content, isgroup=True, msg=cmsg)
        if context:
            self.produce(context)
```

(4) 编写方法 send() 实现统一的发送聊天新功能,每个 Channel 自行实现,根据 reply 的 type 字段发送不同类型的消息,对应的实现代码如下:

```
    def send(self, reply: Reply, context: Context):
        receiver = context["receiver"]
        if reply.type == ReplyType.TEXT:
            itchat.send(reply.content, toUserName=receiver)
```

```python
            logger.info("[WX] sendMsg={}, receiver={}".format(reply, receiver))
        elif reply.type == ReplyType.ERROR or reply.type == ReplyType.INFO:
            itchat.send(reply.content, toUserName=receiver)
            logger.info("[WX] sendMsg={}, receiver={}".format(reply, receiver))
        elif reply.type == ReplyType.VOICE:
            itchat.send_file(reply.content, toUserName=receiver)
            logger.info("[WX] sendFile={}, receiver={}".format(reply.content, receiver))
        elif reply.type == ReplyType.IMAGE_URL:  # 从网络下载图片
            img_url = reply.content
            pic_res = requests.get(img_url, stream=True)
            image_storage = io.BytesIO()
            for block in pic_res.iter_content(1024):
                image_storage.write(block)
            image_storage.seek(0)
            itchat.send_image(image_storage, toUserName=receiver)
            logger.info("[WX] sendImage url={}, receiver={}".format(img_url, receiver))
        elif reply.type == ReplyType.IMAGE:  # 从文件读取图片
            image_storage = reply.content
            image_storage.seek(0)
            itchat.send_image(image_storage, toUserName=receiver)
            logger.info("[WX] sendImage, receiver={}".format(receiver))
```

### 2. 构造聊天信息

编写文件 channel/wechat/wechat_message.py 构造聊天信息，这里的聊天信息不是在文件 wechat_channel.py 中构造的 Context 内容，而是在聊天过程中产生的相关信息，例如对方的用户昵称、加入群聊提示、拍了拍我提示、发送信息者、接收信息者等信息，这些信息将和具体的聊天内容一起显示在调试界面中。文件 wechat_message.py 的主要实现代码如下：

```python
class WeChatMessage(ChatMessage):
    def __init__(self, itchat_msg, is_group=False):
        super().__init__(itchat_msg)
        self.msg_id = itchat_msg["MsgId"]
        self.create_time = itchat_msg["CreateTime"]
        self.is_group = is_group
        if itchat_msg["Type"] == TEXT:
            self.ctype = ContextType.TEXT
            self.content = itchat_msg["Text"]
        elif itchat_msg["Type"] == VOICE:
            self.ctype = ContextType.VOICE
            self.content = TmpDir().path() + itchat_msg["FileName"]
            # content 直接存临时目录路径
            self.prepare_fn = lambda: itchat_msg.download(self.content)
        elif itchat_msg["Type"] == PICTURE and itchat_msg["MsgType"] == 3:
            self.ctype = ContextType.IMAGE
```

```python
                self.content = TmpDir().path() + itchat_msg["FileName"]
                # content 直接存临时目录路径
                self.prepare_fn = lambda: itchat_msg.download(self.content)
        elif itchat_msg["Type"] == NOTE and itchat_msg["MsgType"] == 10000:
            if is_group and ("加入群聊" in itchat_msg["Content"] or "加入了群聊" in itchat_msg["Content"]):
                self.ctype = ContextType.JOIN_GROUP
                self.content = itchat_msg["Content"]
                # 这里只能得到nickname，actual_user_id还是机器人的id
                if "加入了群聊" in itchat_msg["Content"]:
                    self.actual_user_nickname = re.findall(r"\"(.*?)\"", itchat_msg["Content"])[-1]
                elif "加入群聊" in itchat_msg["Content"]:
                    self.actual_user_nickname = re.findall(r"\"(.*?)\"", itchat_msg["Content"])[0]
            elif "拍了拍我" in itchat_msg["Content"]:
                self.ctype = ContextType.PATPAT
                self.content = itchat_msg["Content"]
                if is_group:
                    self.actual_user_nickname = re.findall(r"\"(.*?)\"", itchat_msg["Content"])[0]
            else:
                raise NotImplementedError("Unsupported note message: " + itchat_msg["Content"])
        else:
            raise NotImplementedError("Unsupported message type: Type:{} MsgType:{}".format(itchat_msg["Type"], itchat_msg["MsgType"]))
        self.from_user_id = itchat_msg["FromUserName"]
        self.to_user_id = itchat_msg["ToUserName"]
        user_id = itchat.instance.storageClass.userName
        nickname = itchat.instance.storageClass.nickName
        # 虽然from_user_id和to_user_id用得比较少，但是为了保持一致性，还是要填充一下
        # 以下很烦琐，一句话总结：能填的都填了
        if self.from_user_id == user_id:
            self.from_user_nickname = nickname
        if self.to_user_id == user_id:
            self.to_user_nickname = nickname
        try:  # 在处理陌生人消息时，User 字段可能不存在
            self.other_user_id = itchat_msg["User"]["UserName"]
            self.other_user_nickname = itchat_msg["User"]["NickName"]
            if self.other_user_id == self.from_user_id:
                self.from_user_nickname = self.other_user_nickname
            if self.other_user_id == self.to_user_id:
                self.to_user_nickname = self.other_user_nickname
        except KeyError as e:  # 处理偶尔没有对方信息的情况
            logger.warn("[WX]get other_user_id failed: " + str(e))
            if self.from_user_id == user_id:
```

```
                self.other_user_id = self.to_user_id
            else:
                self.other_user_id = self.from_user_id
        if self.is_group:
            self.is_at = itchat_msg["IsAt"]
            self.actual_user_id = itchat_msg["ActualUserName"]
            if self.ctype not in [ContextType.JOIN_GROUP, ContextType.PATPAT]:
                self.actual_user_nickname = itchat_msg["ActualNickName"]
```

### 6.6.3 微信公众号通道

鉴于使用个人微信号在服务器上通过 itchat 登录有封号风险，本项目提供了微信公众号 Channel 通道，此通道的聊天机器人功能完全无被封号风险。目前支持订阅号和服务号两种类型的公众号，它们都支持文本交互，语音和图片输入。用户需要注意的是，目前个人主体的微信订阅号无法通过微信认证，回复时间存在限制，图片和声音回复次数每天也存在限制。

在开始部署前，你需要一台拥有公网 IP 的服务器，以提供微信服务器和自己服务器的连接。或者需要进行内网穿透，否则微信服务器无法将消息发送给自己的服务器。此外，还需要在自己的服务器上安装 Python 的 Web 框架 web.py 和 wechatpy。以 Ubuntu 系统为例（在 Ubuntu 22.04 上测试）：

```
pip install web.py
pip install wechatpy
```

接下来，在微信公众平台注册一个公众号，类型选择订阅号，主体为个人即可。然后根据接入指南的说明，在微信公众平台的"设置与开发"|"基本配置"|"服务器配置"中填写服务器地址 URL 和令牌 Token。这里的 URL 是 example.com/wx 的形式，不可以使用 IP，Token 是自己编的一个特定的令牌。消息加/解密方式如果选择了需要加密的模式，需要在配置中填写 wechatmp_aes_key。

在本项目中，相关的服务器验证代码已经写好，不需要再添加任何代码。只需要在本项目根目录的 config.json 中启用如下配置信息即可：

```
# 如果通过了微信认证，将 wechatmp 替换为 wechatmp_service，可极大地优化使用体验
"channel_type": "wechatmp",
"wechatmp_token": "xxxx",              # 微信公众平台的 Token
"wechatmp_port": 8080,                 # 微信公众平台的端口号，需将其端口转发到 80 或 443
"wechatmp_app_id": "xxxx",             # 微信公众平台的 appID
"wechatmp_app_secret": "xxxx",         # 微信公众平台的 appsecret
"wechatmp_aes_key": "",                # 微信公众平台的 EncodingAesKey，加密模式需要
"single_chat_prefix": [""],            # 推荐设置，任意对话都可以触发回复，不添加前缀
"single_chat_reply_prefix": "",        # 推荐设置，回复不设置前缀
```

```
# 推荐设置,在手机微信客户端中,$%^等符号与中文连在一起时会自动显示一段较大的间隔,用户体验不好
"plugin_trigger_prefix": "&",
```

运行 python app.py 启动 Web 服务器,这里会默认监听 8080 端口,但是微信公众号的服务器配置只支持 80/443 端口,有如下两种方法来解决这个问题。

第一种是推荐的方法,使用端口转发命令,将 80 端口转发到 8080 端口:

```
sudo iptables -t nat -A PREROUTING -p tcp --dport 80 -j REDIRECT --to-port 8080
sudo iptables-save > /etc/iptables/rules.v4
```

第二种方法是让 Python 程序直接监听 80 端口,在配置文件中设置 wechatmp_port: 80,在 Linux 上需要使用 sudo python3 app.py 启动程序,这样会导致一系列环境和权限问题,因此不是推荐的方法。

编写文件 channel/wechatmp/wechatmp_channel.py,功能是建立 ChatGPT 和微信公众号进行关联的通道,主要实现代码如下:

```python
@singleton
class WechatMPChannel(ChatChannel):
    def __init__(self, passive_reply=True):
        super().__init__()
        self.passive_reply = passive_reply
        self.NOT_SUPPORT_REPLYTYPE = []
        appid = conf().get("wechatmp_app_id")
        secret = conf().get("wechatmp_app_secret")
        token = conf().get("wechatmp_token")
        aes_key = conf().get("wechatmp_aes_key")
        self.client = WechatMPClient(appid, secret)
        self.crypto = None
        if aes_key:
            self.crypto = WeChatCrypto(token, aes_key, appid)
        if self.passive_reply:
            # Cache the reply to the user's first message
            self.cache_dict = dict()
            # Record whether the current message is being processed
            self.running = set()
            # Count the request from wechat official server by message_id
            self.request_cnt = dict()
            # The permanent media need to be deleted to avoid media number limit
            self.delete_media_loop = asyncio.new_event_loop()
            t = threading.Thread(target=self.start_loop, args=(self.delete_media_loop,))
            t.setDaemon(True)
            t.start()

    def send(self, reply: Reply, context: Context):
        receiver = context["receiver"]
        if self.passive_reply:
```

```python
            if reply.type == ReplyType.TEXT or reply.type == ReplyType.INFO or reply.type == ReplyType.ERROR:
                reply_text = reply.content
                logger.info("[wechatmp] text cached, receiver {}\n{}".format(receiver, reply_text))
                self.cache_dict[receiver] = ("text", reply_text)
            elif reply.type == ReplyType.VOICE:
                try:
                    voice_file_path = reply.content
                    with open(voice_file_path, "rb") as f:
                        # support: <2M, <60s, mp3/wma/wav/amr
                        response = self.client.material.add("voice", f)
                        logger.debug("[wechatmp] upload voice response: {}".format(response))
                        # 根据文件大小估算一个微信自动审核的时间，审核结束前返回将会导致语音无法播放，这个估算有待验证
                        f_size = os.fstat(f.fileno()).st_size
                        time.sleep(1.0 + 2 * f_size / 1024 / 1024)
                        # todo check media_id
                except WeChatClientException as e:
                    logger.error("[wechatmp] upload voice failed: {}".format(e))
                    return
                media_id = response["media_id"]
                logger.info("[wechatmp] voice uploaded, receiver {}, media_id {}".format(receiver, media_id))
                self.cache_dict[receiver] = ("voice", media_id)

            elif reply.type == ReplyType.IMAGE_URL: # 从网络下载图片
                img_url = reply.content
                pic_res = requests.get(img_url, stream=True)
                image_storage = io.BytesIO()
                for block in pic_res.iter_content(1024):
                    image_storage.write(block)
                image_storage.seek(0)
                image_type = imghdr.what(image_storage)
                filename = receiver + "-" + str(context["msg"].msg_id) + "." + image_type
                content_type = "image/" + image_type
                try:
                    response = self.client.material.add("image", (filename, image_storage, content_type))
                    logger.debug("[wechatmp] upload image response: {}".format(response))
                except WeChatClientException as e:
                    logger.error("[wechatmp] upload image failed: {}".format(e))
                    return
                media_id = response["media_id"]
                logger.info("[wechatmp] image uploaded, receiver {}, media_id {}".format(receiver, media_id))
                self.cache_dict[receiver] = ("image", media_id)
            elif reply.type == ReplyType.IMAGE:  # 从文件读取图片
```

```python
                image_storage = reply.content
                image_storage.seek(0)
                image_type = imghdr.what(image_storage)
                filename = receiver + "-" + str(context["msg"].msg_id) + "." + image_type
                content_type = "image/" + image_type
                try:
                    response = self.client.material.add("image", (filename,
image_storage, content_type))
                    logger.debug("[wechatmp] upload image response: {}".format(response))
                except WeChatClientException as e:
                    logger.error("[wechatmp] upload image failed: {}".format(e))
                    return
                media_id = response["media_id"]
                logger.info("[wechatmp] image uploaded, receiver {}, media_id
{}".format(receiver, media_id))
                self.cache_dict[receiver] = ("image", media_id)
        else:
            if reply.type == ReplyType.TEXT or reply.type == ReplyType.INFO or
reply.type == ReplyType.ERROR:
                reply_text = reply.content
                texts = split_string_by_utf8_length(reply_text, MAX_UTF8_LEN)
                if len(texts) > 1:
                    logger.info("[wechatmp] text too long, split into {}
parts".format(len(texts)))
                for text in texts:
                    self.client.message.send_text(receiver, text)
                logger.info("[wechatmp] Do send text to {}: {}".format(receiver,
reply_text))
            elif reply.type == ReplyType.VOICE:
                try:
                    file_path = reply.content
                    file_name = os.path.basename(file_path)
                    file_type = os.path.splitext(file_name)[1]
                    if file_type == ".mp3":
                        file_type = "audio/mpeg"
                    elif file_type == ".amr":
                        file_type = "audio/amr"
                    else:
                        mp3_file = os.path.splitext(file_path)[0] + ".mp3"
                        any_to_mp3(file_path, mp3_file)
                        file_path = mp3_file
                        file_name = os.path.basename(file_path)
                        file_type = "audio/mpeg"
                    logger.info("[wechatmp] file_name: {}, file_type: {}
".format(file_name, file_type))
                    # support: <2M, <60s, AMR\MP3
```

```python
                    response = self.client.media.upload("voice", (file_name,
open(file_path, "rb"), file_type))
                    logger.debug("[wechatmp] upload voice response: {}".format(response))
                except WeChatClientException as e:
                    logger.error("[wechatmp] upload voice failed: {}".format(e))
                    return
                self.client.message.send_voice(receiver, response["media_id"])
                logger.info("[wechatmp] Do send voice to {}".format(receiver))
            elif reply.type == ReplyType.IMAGE_URL:  # 从网络下载图片
                img_url = reply.content
                pic_res = requests.get(img_url, stream=True)
                image_storage = io.BytesIO()
                for block in pic_res.iter_content(1024):
                    image_storage.write(block)
                image_storage.seek(0)
                image_type = imghdr.what(image_storage)
                filename = receiver + "-" + str(context["msg"].msg_id) + "." + image_type
                content_type = "image/" + image_type
                try:
                    response = self.client.media.upload("image", (filename,
image_storage, content_type))
                    logger.debug("[wechatmp] upload image response: {}".format(response))
                except WeChatClientException as e:
                    logger.error("[wechatmp] upload image failed: {}".format(e))
                    return
                self.client.message.send_image(receiver, response["media_id"])
                logger.info("[wechatmp] Do send image to {}".format(receiver))
            elif reply.type == ReplyType.IMAGE:  # 从文件读取图片
                image_storage = reply.content
                image_storage.seek(0)
                image_type = imghdr.what(image_storage)
                filename = receiver + "-" + str(context["msg"].msg_id) + "." + image_type
                content_type = "image/" + image_type
                try:
                    response = self.client.media.upload("image", (filename,
image_storage, content_type))
                    logger.debug("[wechatmp] upload image response: {}".format(response))
                except WeChatClientException as e:
                    logger.error("[wechatmp] upload image failed: {}".format(e))
                    return
                self.client.message.send_image(receiver, response["media_id"])
                logger.info("[wechatmp] Do send image to {}".format(receiver))
        return

    def _success_callback(self, session_id, context, **kwargs):
    # 线程异常结束时的回调函数
```

```
        logger.debug("[wechatmp] Success to generate reply,
msgId={}".format(context["msg"].msg_id))
        if self.passive_reply:
            self.running.remove(session_id)

    def _fail_callback(self, session_id, exception, context, **kwargs):
        # 线程异常结束时的回调函数
        logger.exception("[wechatmp] Fail to generate reply to user, msgId={},
exception={}".format(context["msg"].msg_id, exception))
        if self.passive_reply:
            assert session_id not in self.cache_dict
            self.running.remove(session_id)
```

注意：程序启动并监听端口后，在"服务器配置"中单击"提交"按钮即可验证服务器。随后在微信公众平台启用服务器，关闭手动填写规则的自动回复，即可实现 ChatGPT 的自动回复。之后需要在公众号开发信息下将本机 IP 加入 IP 白名单，否则在启用后发送语音、图片等消息可能会出现如下报错：

```
'errcode': 40164, 'errmsg': 'invalid ip xx.xx.xx.xx not in whitelist rid'
```

## 6.7 对话处理

在构建不同的 Channel 通道后，接下来根据对应的通道实现相应的聊天机器人功能。在本项目中内置了分别基于 OpenAI、ChatGPT 和 Baidu Unit 接口的对话应用功能，接下来将讲解具体的实现过程。

扫码看视频

### 6.7.1 OpenAI 对话

(1) 编写文件 bot/openai/open_ai_bot.py，实现基于 OpenAI 接口的文本对话(包括私聊和群聊)功能，主要实现代码如下：

```
class OpenAIBot(Bot, OpenAIImage):
    def __init__(self):
        super().__init__()
        openai.api_key = conf().get("open_ai_api_key")
        if conf().get("open_ai_api_base"):
            openai.api_base = conf().get("open_ai_api_base")
        proxy = conf().get("proxy")
        if proxy:
            openai.proxy = proxy
```

```python
        self.sessions = SessionManager(OpenAISession, model=conf().get("model") or "text-davinci-003")
        self.args = {
            "model": conf().get("model") or "text-davinci-003",  # 对话模型的名称
            "temperature": conf().get("temperature", 0.9),
            # 值在[0,1]，该值越大表示回复越具有不确定性
            "max_tokens": 1200,  # 回复最大的字符数
            "top_p": 1,
            "frequency_penalty": conf().get("frequency_penalty", 0.0),
            # 值在[-2,2]，该值越大则更倾向于产生不同的内容
            "presence_penalty": conf().get("presence_penalty", 0.0),
            # 值在[-2,2]，该值越大则更倾向于产生不同的内容
            "request_timeout": conf().get("request_timeout", None),
            # 请求超时时间，openai 接口默认设置为 600，对于比较难的问题一般需要较长时间
            "timeout": conf().get("request_timeout", None),
            # 重试超时时间，在这个时间内，将会自动重试
            "stop": ["\n\n\n"],
        }

    def reply(self, query, context=None):
        # acquire reply content
        if context and context.type:
            if context.type == ContextType.TEXT:
                logger.info("[OPEN_AI] query={}".format(query))
                session_id = context["session_id"]
                reply = None
                if query == "#清除记忆":
                    self.sessions.clear_session(session_id)
                    reply = Reply(ReplyType.INFO, "记忆已清除")
                elif query == "#清除所有":
                    self.sessions.clear_all_session()
                    reply = Reply(ReplyType.INFO, "所有人记忆已清除")
                else:
                    session = self.sessions.session_query(query, session_id)
                    result = self.reply_text(session)
                    total_tokens, completion_tokens, reply_content = (
                        result["total_tokens"],
                        result["completion_tokens"],
                        result["content"],
                    )
                    logger.debug(
                        "[OPEN_AI] new_query={}, session_id={}, reply_cont={}, completion_tokens={}".format(str(session), session_id, reply_content, completion_tokens)
                    )

                    if total_tokens == 0:
                        reply = Reply(ReplyType.ERROR, reply_content)
```

```python
            else:
                self.sessions.session_reply(reply_content, session_id, total_tokens)
                reply = Reply(ReplyType.TEXT, reply_content)
        return reply
    elif context.type == ContextType.IMAGE_CREATE:
        ok, retstring = self.create_img(query, 0)
        reply = None
        if ok:
            reply = Reply(ReplyType.IMAGE_URL, retstring)
        else:
            reply = Reply(ReplyType.ERROR, retstring)
        return reply

def reply_text(self, session: OpenAISession, retry_count=0):
    try:
        response = openai.Completion.create(prompt=str(session), **self.args)
        res_content = response.choices[0]["text"].strip().replace("<|endoftext|>", "")
        total_tokens = response["usage"]["total_tokens"]
        completion_tokens = response["usage"]["completion_tokens"]
        logger.info("[OPEN_AI] reply={}".format(res_content))
        return {
            "total_tokens": total_tokens,
            "completion_tokens": completion_tokens,
            "content": res_content,
        }
    except Exception as e:
        need_retry = retry_count < 2
        result = {"completion_tokens": 0, "content": "我现在有点累了，等会再来吧"}
        if isinstance(e, openai.error.RateLimitError):
            logger.warn("[OPEN_AI] RateLimitError: {}".format(e))
            result["content"] = "提问太快啦，请休息一下再问我吧"
            if need_retry:
                time.sleep(20)
        elif isinstance(e, openai.error.Timeout):
            logger.warn("[OPEN_AI] Timeout: {}".format(e))
            result["content"] = "我没有收到你的消息"
            if need_retry:
                time.sleep(5)
        elif isinstance(e, openai.error.APIConnectionError):
            logger.warn("[OPEN_AI] APIConnectionError: {}".format(e))
            need_retry = False
            result["content"] = "我连接不到你的网络"
        else:
            logger.warn("[OPEN_AI] Exception: {}".format(e))
            need_retry = False
            self.sessions.clear_session(session.session_id)
        if need_retry:
```

```
            logger.warn("[OPEN_AI] 第{}次重试".format(retry_count + 1))
            return self.reply_text(session, retry_count + 1)
        else:
            return result
```

(2) 编写文件 bot/openai/open_ai_image.py，实现基于 OpenAI 接口的画图对话功能，具体实现代码如下：

```
class OpenAIImage(object):
    def __init__(self):
        openai.api_key = conf().get("open_ai_api_key")
        if conf().get("rate_limit_dalle"):
            self.tb4dalle = TokenBucket(conf().get("rate_limit_dalle", 50))
    def create_img(self, query, retry_count=0):
        try:
            if conf().get("rate_limit_dalle") and not self.tb4dalle.get_token():
                return False, "请求太快了，请休息一下再问我吧"
            logger.info("[OPEN_AI] image_query={}".format(query))
            response = openai.Image.create(
                prompt=query,    # 图片描述
                n=1,    # 每次生成图片的数量
                size=conf().get("image_create_size", "256x256"),
                # 图片尺寸包括 256×256、512×512、1024×1024
            )
            image_url = response["data"][0]["url"]
            logger.info("[OPEN_AI] image_url={}".format(image_url))
            return True, image_url
        except openai.error.RateLimitError as e:
            logger.warn(e)
            if retry_count < 1:
                time.sleep(5)
                logger.warn("[OPEN_AI] ImgCreate RateLimit exceed, 第{}次重试".format(retry_count + 1))
                return self.create_img(query, retry_count + 1)
            else:
                return False, "提问太快啦，请休息一下再问我吧"
        except Exception as e:
            logger.exception(e)
            return False, str(e)
```

## 6.7.2 ChatGPT 对话

(1) 编写文件 bot/chatgpt/chat_gpt_bot.py，实现基于 ChatGPT 接口的对话(包括私聊和群聊)功能，主要实现代码如下所示。

```python
class ChatGPTBot(Bot, OpenAIImage):
    def __init__(self):
        super().__init__()
        #设置的api_key
        openai.api_key = conf().get("open_ai_api_key")
        if conf().get("open_ai_api_base"):
            openai.api_base = conf().get("open_ai_api_base")
        proxy = conf().get("proxy")
        if proxy:
            openai.proxy = proxy
        if conf().get("rate_limit_chatgpt"):
            self.tb4chatgpt = TokenBucket(conf().get("rate_limit_chatgpt", 20))
        self.sessions = SessionManager(ChatGPTSession, model=conf().get("model") or "gpt-3.5-turbo")
        self.args = {
            "model": conf().get("model") or "gpt-3.5-turbo",  # 对话模型的名称
            "temperature": conf().get("temperature", 0.9),
            # 值在[0,1], 该值越大表示回复越具有不确定性
            # "max_tokens":4096,  # 回复最大的字符数
            "top_p": 1,
            "frequency_penalty": conf().get("frequency_penalty", 0.0),
            # 值在[-2,2], 该值越大则更倾向于产生不同的内容
            "presence_penalty": conf().get("presence_penalty", 0.0),
            # 值在[-2,2], 该值越大则更倾向于产生不同的内容
            "request_timeout": conf().get("request_timeout", None),
            # 请求超时时间, OpenAI接口默认设置为600, 对于难问题一般需要较长时间
            "timeout": conf().get("request_timeout", None),
            # 重试超时时间, 在这个时间内, 将会自动重试
        }

    def reply(self, query, context=None):
        # acquire reply content
        if context.type == ContextType.TEXT:
            logger.info("[CHATGPT] query={}".format(query))
            session_id = context["session_id"]
            reply = None
            clear_memory_commands = conf().get("clear_memory_commands", ["#清除记忆"])
            if query in clear_memory_commands:
                self.sessions.clear_session(session_id)
                reply = Reply(ReplyType.INFO, "记忆已清除")
            elif query == "#清除所有":
                self.sessions.clear_all_session()
                reply = Reply(ReplyType.INFO, "所有人记忆已清除")
            elif query == "#更新配置":
                load_config()
                reply = Reply(ReplyType.INFO, "配置已更新")
            if reply:
```

```python
            return reply
        session = self.sessions.session_query(query, session_id)
        logger.debug("[CHATGPT] session query={}".format(session.messages))
        api_key = context.get("openai_api_key")
        # if context.get('stream'):
        reply_content = self.reply_text(session, api_key)
        logger.debug(
            "[CHATGPT] new_query={}, session_id={}, reply_cont={}, completion_tokens={}".format(
                session.messages,
                session_id,
                reply_content["content"],
                reply_content["completion_tokens"],
            )
        )
        if reply_content["completion_tokens"] == 0 and len(reply_content["content"]) > 0:
            reply = Reply(ReplyType.ERROR, reply_content["content"])
        elif reply_content["completion_tokens"] > 0:
            self.sessions.session_reply(reply_content["content"], session_id, reply_content["total_tokens"])
            reply = Reply(ReplyType.TEXT, reply_content["content"])
        else:
            reply = Reply(ReplyType.ERROR, reply_content["content"])
            logger.debug("[CHATGPT] reply {} used 0 tokens.".format(reply_content))
        return reply
    elif context.type == ContextType.IMAGE_CREATE:
        ok, retstring = self.create_img(query, 0)
        reply = None
        if ok:
            reply = Reply(ReplyType.IMAGE_URL, retstring)
        else:
            reply = Reply(ReplyType.ERROR, retstring)
        return reply
    else:
        reply = Reply(ReplyType.ERROR, "Bot 不支持处理{}类型的消息".format(context.type))
        return reply

def reply_text(self, session: ChatGPTSession, api_key=None, retry_count=0) -> dict:
    """
    调用 openai 的 ChatCompletion 获取回复
    :param session: 一个对话 session
    :param session_id: session id
    :param retry_count: 重试计数
    :return: {}
    """
    try:
```

```python
            if conf().get("rate_limit_chatgpt") and not self.tb4chatgpt.get_token():
                raise openai.error.RateLimitError("RateLimitError: rate limit exceeded")
            response = openai.ChatCompletion.create(api_key=api_key, messages=
                    session.messages, **self.args)
            return {
                "total_tokens": response["usage"]["total_tokens"],
                "completion_tokens": response["usage"]["completion_tokens"],
                "content": response.choices[0]["message"]["content"],
            }
        except Exception as e:
            need_retry = retry_count < 2
            result = {"completion_tokens": 0, "content": "我现在有点累了,等会再来吧"}
            if isinstance(e, openai.error.RateLimitError):
                logger.warn("[CHATGPT] RateLimitError: {}".format(e))
                result["content"] = "提问太快啦,请休息一下再问我吧"
                if need_retry:
                    time.sleep(20)
            elif isinstance(e, openai.error.Timeout):
                logger.warn("[CHATGPT] Timeout: {}".format(e))
                result["content"] = "我没有收到你的消息"
                if need_retry:
                    time.sleep(5)
            elif isinstance(e, openai.error.APIConnectionError):
                logger.warn("[CHATGPT] APIConnectionError: {}".format(e))
                need_retry = False
                result["content"] = "我连接不到你的网络"
            else:
                logger.warn("[CHATGPT] Exception: {}".format(e))
                need_retry = False
                self.sessions.clear_session(session.session_id)
            if need_retry:
                logger.warn("[CHATGPT] 第{}次重试".format(retry_count + 1))
                return self.reply_text(session, api_key, retry_count + 1)
            else:
                return result

class AzureChatGPTBot(ChatGPTBot):
    def __init__(self):
        super().__init__()
        openai.api_type = "azure"
        openai.api_version = "2023-03-15-preview"
        self.args["deployment_id"] = conf().get("azure_deployment_id")
```

(2) 因为OpenAI限制了开发者使用ChatGPT的次数,所以编写文件bot/chatgpt/chat_gpt_session.py实现ChatGPT接口的Session验证功能,验证用户消息是否超过max_tokens,如果超过则显示对应的提示信息。文件chat_gpt_session.py的主要实现代码如下:

```python
class ChatGPTSession(Session):
    def __init__(self, session_id, system_prompt=None, model="gpt-3.5-turbo"):
        super().__init__(session_id, system_prompt)
        self.model = model
        self.reset()

    def discard_exceeding(self, max_tokens, cur_tokens=None):
        precise = True
        try:
            cur_tokens = self.calc_tokens()
        except Exception as e:
            precise = False
            if cur_tokens is None:
                raise e
            logger.debug("Exception when counting tokens precisely for query: {}".format(e))
        while cur_tokens > max_tokens:
            if len(self.messages) > 2:
                self.messages.pop(1)
            elif len(self.messages) == 2 and self.messages[1]["role"] == "assistant":
                self.messages.pop(1)
                if precise:
                    cur_tokens = self.calc_tokens()
                else:
                    cur_tokens = cur_tokens - max_tokens
                break
            elif len(self.messages) == 2 and self.messages[1]["role"] == "user":
                logger.warn("user message exceed max_tokens. total_tokens={}".format(cur_tokens))
                break
            else:
                logger.debug("max_tokens={}, total_tokens={}, len(messages)={}".format(max_tokens, cur_tokens, len(self.messages)))
                break
            if precise:
                cur_tokens = self.calc_tokens()
            else:
                cur_tokens = cur_tokens - max_tokens
        return cur_tokens

    def calc_tokens(self):
        return num_tokens_from_messages(self.messages, self.model)

def num_tokens_from_messages(messages, model):
    """返回消息列表使用的令牌数"""
    import tiktoken
```

```
try:
    encoding = tiktoken.encoding_for_model(model)
except KeyError:
    logger.debug("Warning: model not found. Using cl100k_base encoding.")
    encoding = tiktoken.get_encoding("cl100k_base")
if model == "gpt-3.5-turbo" or model == "gpt-35-turbo":
    return num_tokens_from_messages(messages, model="gpt-3.5-turbo-0301")
elif model == "gpt-4":
    return num_tokens_from_messages(messages, model="gpt-4-0314")
elif model == "gpt-3.5-turbo-0301":
    tokens_per_message = 4  # every message follows <|start|>{role/name}\n{content}<|end|>\n
    tokens_per_name = -1  # if there's a name, the role is omitted
elif model == "gpt-4-0314":
    tokens_per_message = 3
    tokens_per_name = 1
else:
    logger.warn(f"num_tokens_from_messages() is not implemented for model {model}. Returning num tokens assuming gpt-3.5-turbo-0301.")
    return num_tokens_from_messages(messages, model="gpt-3.5-turbo-0301")
num_tokens = 0
for message in messages:
    num_tokens += tokens_per_message
    for key, value in message.items():
        num_tokens += len(encoding.encode(value))
        if key == "name":
            num_tokens += tokens_per_name
num_tokens += 3  # 每个回复都带有<|start|>assistant<|message|>
return num_tokens
```

### 6.7.3 Baidu Unit 对话

编写文件 bot/baidu/baidu_unit_bot.py,实现基于 Baidu Unit 对话接口的对话功能,主要实现代码如下:

```
class BaiduUnitBot(Bot):
    def reply(self, query, context=None):
        token = self.get_token()
        url = "https://aip.baidubce.com/rpc/2.0/unit/service/v3/chat?access_token=" + token
        post_data = (
            '{"version":"3.0","service_id":"S73177","session_id":"","log_id":"7758521","skill_ids":["1221886"],"request":{"terminal_id":"88888","query":"'
            + query
            + '", "hyper_params": {"chat_custom_bot_profile": 1}}}'
```

```
        )
        print(post_data)
        headers = {"content-type": "application/x-www-form-urlencoded"}
        response = requests.post(url, data=post_data.encode(), headers=headers)
        if response:
            reply = Reply(
                ReplyType.TEXT,
                response.json()["result"]["context"]["SYS_PRESUMED_HIST"][1],
            )
            return reply
    def get_token(self):
        access_key = "YOUR_ACCESS_KEY"
        secret_key = "YOUR_SECRET_KEY"
        host = "https://aip.baidubce.com/oauth/2.0/token?grant_type=client_credentials&client_id=" + access_key + "&client_secret=" + secret_key
        response = requests.get(host)
        if response:
            print(response.json())
            return response.json()["access_token"]
```

## 6.8 语音识别

在聊天过程中，本项目机器人可以识别语音消息，可以通过文字或语音进行回复。本项目目前支持 Microsoft Azure、百度、谷歌和 OpenAI 等多种语音模型。

扫码看视频

### 6.8.1 OpenAI 语音识别

编写文件 voice/openai/openai_voice.py，实现基于 OpenAI 接口的语音识别功能，主要实现代码如下：

```
class OpenaiVoice(Voice):
    def __init__(self):
        openai.api_key = conf().get("open_ai_api_key")

    def voiceToText(self, voice_file):
        logger.debug("[Openai] voice file name={}".format(voice_file))
        try:
            file = open(voice_file, "rb")
            result = openai.Audio.transcribe("whisper-1", file)
            text = result["text"]
            reply = Reply(ReplyType.TEXT, text)
```

```
            logger.info("[Openai] voiceToText text={} voice file name={}".format(text,
voice_file))
        except Exception as e:
            reply = Reply(ReplyType.ERROR, str(e))
        finally:
            return reply
```

## 6.8.2 谷歌语音识别

编写文件 voice/google/google_voice.py，实现基于谷歌接口的语音识别功能，主要实现代码如下：

```
class GoogleVoice(Voice):
    recognizer = speech_recognition.Recognizer()
    def __init__(self):
        pass

    def voiceToText(self, voice_file):
        with speech_recognition.AudioFile(voice_file) as source:
            audio = self.recognizer.record(source)
        try:
            text = self.recognizer.recognize_google(audio, language="zh-CN")
            logger.info("[Google] voiceToText text={} voice file
name={}".format(text, voice_file))
            reply = Reply(ReplyType.TEXT, text)
        except speech_recognition.UnknownValueError:
            reply = Reply(ReplyType.ERROR, "抱歉，我听不懂")
        except speech_recognition.RequestError as e:
            reply=Reply(ReplyType.ERROR,"抱歉，无法连接到 Google 语音识别服务；{0}".format(e))
        finally:
            return reply

    def textToVoice(self, text):
        try:
            mp3File = TmpDir().path() + "reply-" + str(int(time.time())) + ".mp3"
            tts = gTTS(text=text, lang="zh")
            tts.save(mp3File)
            logger.info("[Google] textToVoice text={} voice file name={}".format(text,
mp3File))
            reply = Reply(ReplyType.VOICE, mp3File)
        except Exception as e:
            reply = Reply(ReplyType.ERROR, str(e))
        finally:
            return reply
```

## 6.8.3 百度语音识别

编写文件 voice/baidu/baidu_voice.py，实现基于百度接口的语音识别功能。百度语音识别比较特殊，系统中收到的语音文件为 mp3 格式(wx)或者 sil 格式(wxy)，如果要识别需要转换为 pcm 格式，转换后的文件为 16kHz 采样率、单声道、16bit 的 pcm 文件。在发送时又需要将 wx 转换为 mp3 格式，转换后的文件为 16kHz 采样率、单声道、16bit 的 pcm 文件。需要将 wxy 转换为 sil 格式，还要计算声音长度，在发送时需要带上声音长度。文件 baidu_voice.py 的主要实现代码如下：

```python
class BaiduVoice(Voice):
    def __init__(self):
        try:
            curdir = os.path.dirname(__file__)
            config_path = os.path.join(curdir, "config.json")
            bconf = None
            if not os.path.exists(config_path):  # 如果没有配置文件，则创建本地配置文件
                bconf = {"lang": "zh", "ctp": 1, "spd": 5, "pit": 5, "vol": 5, "per": 0}
                with open(config_path, "w") as fw:
                    json.dump(bconf, fw, indent=4)
            else:
                with open(config_path, "r") as fr:
                    bconf = json.load(fr)

            self.app_id = conf().get("baidu_app_id")
            self.api_key = conf().get("baidu_api_key")
            self.secret_key = conf().get("baidu_secret_key")
            self.dev_id = conf().get("baidu_dev_pid")
            self.lang = bconf["lang"]
            self.ctp = bconf["ctp"]
            self.spd = bconf["spd"]
            self.pit = bconf["pit"]
            self.vol = bconf["vol"]
            self.per = bconf["per"]

            self.client = AipSpeech(self.app_id, self.api_key, self.secret_key)
        except Exception as e:
            logger.warn("BaiduVoice init failed: %s, ignore " % e)

    def voiceToText(self, voice_file):
        # 识别本地文件
        logger.debug("[Baidu] voice file name={}".format(voice_file))
        pcm = get_pcm_from_wav(voice_file)
        res = self.client.asr(pcm, "pcm", 16000, {"dev_pid": self.dev_id})
        if res["err_no"] == 0:
            logger.info("百度语音识别到了: {}".format(res["result"]))
```

```
            text = "".join(res["result"])
            reply = Reply(ReplyType.TEXT, text)
        else:
            logger.info("百度语音识别出错了: {}".format(res["err_msg"]))
            if res["err_msg"] == "request pv too much":
                logger.info("  出现这个原因很可能是你的百度语音服务调用量超出限制, 或未开通付费")
            reply = Reply(ReplyType.ERROR, "百度语音识别出错了；{0}".format(res["err_msg"]))
        return reply

    def textToVoice(self, text):
        result = self.client.synthesis(
            text,
            self.lang,
            self.ctp,
            {"spd": self.spd, "pit": self.pit, "vol": self.vol, "per": self.per},
        )
        if not isinstance(result, dict):
            fileName = TmpDir().path() + "reply-" + str(int(time.time())) + ".mp3"
            with open(fileName, "wb") as f:
                f.write(result)
            logger.info("[Baidu] textToVoice text={} voice file name={}".format(text, fileName))
            reply = Reply(ReplyType.VOICE, fileName)
        else:
            logger.error("[Baidu] textToVoice error={}".format(result))
            reply = Reply(ReplyType.ERROR, "抱歉，语音合成失败")
        return reply
```

前面提到过需要将获取的音频进行类型转换，相关转换操作在文件 voice/audio_convert.py 中进行了封装，在使用时直接调用即可。文件 voice/audio_convert.py 的主要实现代码如下：

```
def find_closest_sil_supports(sample_rate):
    """
    找到最接近的支持的采样率
    """
    if sample_rate in sil_supports:
        return sample_rate
    closest = 0
    mindiff = 9999999
    for rate in sil_supports:
        diff = abs(rate - sample_rate)
        if diff < mindiff:
            closest = rate
            mindiff = diff
    return closest

def get_pcm_from_wav(wav_path):
```

```python
    """
    从 wav 文件中读取 pcm
    :param wav_path: wav 文件路径
    :returns: pcm 数据
    """
    wav = wave.open(wav_path, "rb")
    return wav.readframes(wav.getnframes())
def any_to_mp3(any_path, mp3_path):
    """
    把任意格式转成 mp3 文件
    """
    if any_path.endswith(".mp3"):
        shutil.copy2(any_path, mp3_path)
        return
    if any_path.endswith(".sil") or any_path.endswith(".silk") or any_path.endswith(".slk"):
        sil_to_wav(any_path, any_path)
        any_path = mp3_path
    audio = AudioSegment.from_file(any_path)
    audio.export(mp3_path, format="mp3")


def any_to_wav(any_path, wav_path):
    """
    把任意格式转成 wav 文件
    """
    if any_path.endswith(".wav"):
        shutil.copy2(any_path, wav_path)
        return
    if any_path.endswith(".sil") or any_path.endswith(".silk") or any_path.endswith(".slk"):
        return sil_to_wav(any_path, wav_path)
    audio = AudioSegment.from_file(any_path)
    audio.export(wav_path, format="wav")


def any_to_sil(any_path, sil_path):
    """
    把任意格式转成 sil 文件
    """
    if any_path.endswith(".sil") or any_path.endswith(".silk") or any_path.endswith(".slk"):
        shutil.copy2(any_path, sil_path)
        return 10000
    audio = AudioSegment.from_file(any_path)
    rate = find_closest_sil_supports(audio.frame_rate)
    # Convert to PCM_s16
    pcm_s16 = audio.set_sample_width(2)
    pcm_s16 = pcm_s16.set_frame_rate(rate)
```

```
        wav_data = pcm_s16.raw_data
        silk_data = pysilk.encode(wav_data, data_rate=rate, sample_rate=rate)
        with open(sil_path, "wb") as f:
            f.write(silk_data)
        return audio.duration_seconds * 1000

def sil_to_wav(silk_path, wav_path, rate: int = 24000):
    """
    silk 文件转 wav
    """
    wav_data = pysilk.decode_file(silk_path, to_wav=True, sample_rate=rate)
    with open(wav_path, "wb") as f:
        f.write(wav_data)
```

### 6.8.4　Microsoft Azure 语音识别

编写文件 voice/azure/azure_voice.py，实现基于 Microsoft Azure 接口的语音识别功能，主要实现代码如下：

```
class AzureVoice(Voice):
    def __init__(self):
        try:
            curdir = os.path.dirname(__file__)
            config_path = os.path.join(curdir, "config.json")
            config = None
            if not os.path.exists(config_path):  # 如果没有配置文件，创建本地配置文件
                config = {
                    "speech_synthesis_voice_name": "zh-CN-XiaoxiaoNeural",
                    "speech_recognition_language": "zh-CN",
                }
                with open(config_path, "w") as fw:
                    json.dump(config, fw, indent=4)
            else:
                with open(config_path, "r") as fr:
                    config = json.load(fr)
            self.api_key = conf().get("azure_voice_api_key")
            self.api_region = conf().get("azure_voice_region")
            self.speech_config = speechsdk.SpeechConfig(subscription=self.api_key,
region=self.api_region)
            self.speech_config.speech_synthesis_voice_name =
config["speech_synthesis_voice_name"]
            self.speech_config.speech_recognition_language =
config["speech_recognition_language"]
        except Exception as e:
            logger.warn("AzureVoice init failed: %s, ignore " % e)
```

```python
    def voiceToText(self, voice_file):
        audio_config = speechsdk.AudioConfig(filename=voice_file)
        speech_recognizer = speechsdk.SpeechRecognizer
(speech_config=self.speech_config, audio_config=audio_config)
        result = speech_recognizer.recognize_once()
        if result.reason == speechsdk.ResultReason.RecognizedSpeech:
            logger.info("[Azure] voiceToText voice file name={} text={}".format(voice_file, result.text))
            reply = Reply(ReplyType.TEXT, result.text)
        else:
            logger.error("[Azure] voiceToText error, result={}, canceldetails={}".format(result, result.cancellation_details))
            reply = Reply(ReplyType.ERROR, "抱歉，语音识别失败")
        return reply

    def textToVoice(self, text):
        fileName = TmpDir().path() + "reply-" + str(int(time.time())) + ".wav"
        audio_config = speechsdk.AudioConfig(filename=fileName)
        speech_synthesizer = speechsdk.SpeechSynthesizer
(speech_config=self.speech_config, audio_config=audio_config)
        result = speech_synthesizer.speak_text(text)
        if result.reason == speechsdk.ResultReason.SynthesizingAudioCompleted:
            logger.info("[Azure] textToVoice text={} voice file name={}".format(text, fileName))
            reply = Reply(ReplyType.VOICE, fileName)
        else:
            logger.error("[Azure] textToVoice error, result={}, canceldetails={}".format(result, result.cancellation_details))
            reply = Reply(ReplyType.ERROR, "抱歉，语音合成失败")
        return reply
```

## 6.9 调试运行

通过 python app.py 命令运行程序，会弹出二维码供用户用微信扫描登录，如图 6-13 所示。

随便选一个微信好友，发送以"@bob"开头的聊天内容即可启动 ChatGPT 机器人，笔者是跟文件传输助手进行对话的，如图 6-14 所示。最简单的方法是设置一个昵称为"@bob"的好友进行对话。发送以"@画"开头的聊天内容，激活图片生成功能，如图 6-15 所示。

图 6-13　生成微信登录二维码

图 6-14　聊天内容

图 6-15　生成图片

# 第 7 章

## 移动机器人智能物体识别系统

近年来,随着深度学习与云计算的跳跃式发展,带动了物体识别技术产生质的飞跃。高分辨率图像和检测的实时性要求越来越高。在本章的内容中,将详细讲解使用人工智能技术为移动机器人开发一个物体识别检测系统的过程,包括项目的架构分析、创建模型和具体实现等内容。具体流程由 TensorFlow Lite+TensorFlow+Android+iOS 实现。

## 7.1 背景介绍

随着机电一体化技术的快速发展,作为其典型代表的机器人的智能化越来越受到人们关注。在复杂环境中工作的机器人通过使用视觉技术,能够对周围环境中的物体进行精确识别是机器人智能化的重要标志。与传统机器人不同,具有"视觉"且能够识别物体的机器人可以对外部世界进行感知(即获取图像),分析所得信息,并做出合理的决策。这种技术恰恰满足了人们对机器人智能化的需求,对机器人的工作和未来机器人智能化的发展具有重要意义。

扫码看视频

机器人视觉的核心技术在于物体识别,物体识别通俗来说即是运用计算机技术使机器人具有和人类一样的,能够对任意环境下观察到的物体进行检测、分割和识别的能力。物体识别的作用主要有:对汽车或车牌的识别,并附以其他处理(速度计算等),也可以对交通进行智能监控;工厂中的智能机器人可以识别零件种类,对零件进行相应操作(搬运、组装等);家用机器人对各种物体的识别可以帮助人类做更多的工作,而不是像传统机器人那样只能做一些简单的重复性的事情,这会使机器人更加智能化,发挥更大的作用。在各种各样的物体识别中,人脸识别是最典型的识别之一,更准确地讲应该是人脸检测,两者的区别在于,"识别"(recognition)是从图像中找到能与特定人脸相匹配的部分;而"检测"(detection)只是识别的一部分,即在图像中检测出人脸并标记位置。目前,人脸检测已经满足了"物体识别"的要求,它完全可以代表其他物体(如汽车、杯子等)的识别,人脸检测已经被应用于很多领域。比如,家用机器人可以从复杂环境中判断主人的位置,数码相机可以通过人脸识别来对人脸进行准确对焦等。

## 7.2 物体识别

大千世界的物体种类繁多,人们主要通过视觉系统对形形色色的物体进行分类和辨别,统称为物体识别。通过模拟人类视觉系统的视觉信息获取和处理功能以便于计算机具有人类识别物体的能力,出现了计算机视觉和模式识别等研究领域。物体识别(object recognition)是当前国内外计算机视觉与模式识别领域的一个活跃的研究方向,在很多方面有了很大的进步。比如,对人类的视觉系统有了更进一步的认识,数学工具更高级,计算效率越来越高,涌现越来越多的具有挑战性的数据库收集等,这些进步使得物体识别更加引起人们的关注。

扫码看视频

## 7.2.1 物体识别介绍

物体识别是机器智能的基本功能之一,它是任何一个以图像或视频作为输入的实际应用系统中的核心问题和关键技术。这类系统的性能和应用前景都依赖其中物体的知识表示和分类识别所能达到的水平。物体识别技术无论是在军事还是在民用中都有着广泛的需求和应用,如智能视频监控、视觉导航、人机交互、计算机取证、各类身份识别和认证系统、数字图书馆,以及互联网中的在海量图像库和视频中的基于内容的检索、编码与压缩,等等。

基于图像的物体识别的过程通常表现为:首先建立待识别物体图像的一种知识表示模型,在一定量的训练样本中学习得到一组满足预定要求的模型参数;同时根据物体图像的表示模型,建立一套从实际图像中进行推理的识别算法,通过在实际图像中测试可获得系统的泛化能力对其进行性能评估。由此可见,物体识别技术的提高,无论在军事还是民用方面都有非常重要的意义。

所谓一般物体识别,通俗来说,即是使计算机具有和人类一样的,对于在任意环境下观察到的物体进行检测、分割和识别的能力,它作为计算机视觉领域的一个特定并极为重要的任务,要求在给予一定量的训练样本的前提下,计算机能够学习有关指定物体类别的知识,并在观察到从属于旧类别的新物体时,给出识别的结果。

一般物体识别与特定物体识别(specific object recognition)的主要区别在于,特定物体识别通过构造高度特化的特征提取及机器学习方法,使用海量的训练样本进行训练,仅仅处理某种物体或是某类物体,典型例子如汽车检测与人脸检测。一般物体识别面对的问题则要困难得多,概括地说,它必须使用物体类间通用的一般特征,而不能为某个特定类别定义特征;它必须能处理多种分类及增量学习,在此前提下无法使用给定类别的海量样本进行训练。

研究一般物体识别,无论对于理论还是实践都有极其重大的意义。计算机视觉的核心在于识别,而一般物体识别又是识别中最为复杂核心的问题,对于神经科学而言,破解了一般物体识别就相当于将神秘莫测的人脑工作机制拉开了帷幕的一角,由此展开进一步的深入研究,意义不言而喻。在实践中,一般物体识别的研究则能给人类生活的方方面面,尤其是交通、国防、教育带来极为重大的影响,甚至能改变人们的生活方式,对整个社会有着深远的意义。

## 7.2.2 图像特征的提取方法

图像特征提取就是提取出一幅图像中不同于其他图像的根本属性,以区别不同的图像,如灰度、亮度、纹理和形状等特征都是与图像的视觉外观相对应的;还有一些则缺少自然的对应性,如颜色直方图、灰度直方图和空间频谱图等。基于图像特征进行物体识别实际上是根据提取到图像的特征来判断图像中物体属于什么类别。形状、纹理和颜色等特征是最常用的视觉特征,也是现阶段基于图像的物体识别技术中采用的主要特征。下面分别介绍一下图像的形状、纹理和颜色特征的提取方法。

1) 图像形状特征提取

形状特征是用来描述和区分物体形状的特征。在图像处理和计算机视觉中,形状特征可以帮助我们识别和分析物体的轮廓、边界和整体形状。常用的图像形状特征提取方法有两种:基于轮廓的方法和基于区域的方法。这两种方法的不同之处在于:

- 对于基于轮廓的方法来说,图像的轮廓特征主要针对物体的外边界,描述形状的轮廓特征的主要方法有样条、链码和多边形逼近等。
- 在基于区域的方法中,图像的区域特征则关系到整个形状区域,描述形状区域特征的主要方法有区域的面积、凹凸面积、形状的主轴方向、纵横比、形状的不变矩等。

2) 图像纹理特征提取

图像的纹理是与物体表面结构和材质有关的图像的内在特征,反映出来的是图像的全局特征。图像的纹理可以描述为:一个邻域内像素的灰度级发生变化的空间分布规律,包括表面组织结构、与周围环境关系等许多重要的图像信息。典型的图像纹理特征提取方法有统计方法(灰度共生矩阵纹理特征分析方法就是典型的统计方法之一)、几何法(建立在基本的纹理元素理论基础上的一种纹理特征分析方法)、模型法(将图像的构造模型的参数作为纹理特征)、信号处理法(主要以小波变换为主)。

3) 图像颜色特征提取

图像的颜色特征描述了图像或图像区域的物体的表面性质,反映出的是图像的全局特征。一般来说,图像的颜色特征是基于像素点的特征,只要是属于图像或图像区域内的像素点都将会有贡献。典型的图像颜色特征提取方法有颜色直方图、颜色集和颜色矩,具体说明如下。

- 颜色直方图:颜色直方图是最常用的表达颜色特征的方法,它的优点是能简单描述图像中不同色彩在整幅图像中所占的比例,特别适用于描述一些不需要考虑物体空间位置的图像和难以自动分割的图像。而颜色直方图的缺点是它无法描述图像中的某一具体的物体,无法区分局部颜色信息。
- 颜色集:颜色集可以看成是颜色直方图的一种近似表达。具体方法是:首先将图像从 RGB 颜色空间转换到视觉均衡的颜色空间;然后将视觉均衡的颜色空间量化;最后,采用色彩分割技术自动地将图像分为几个区域,用量化的颜色空间中的某个颜色分量来表示每个区域的索引,这样就可以用一个二进制的颜色索引集来表示一幅图像了。

❑ 颜色矩：颜色矩是一种描述图像颜色分布和色彩特征的数学特征。它可以用来对图像进行特征提取和分析，常用于图像识别、图像分类、图像检索等领域。颜色矩可以用来表示图像的颜色分布情况，包括灰度矩、彩色矩等不同类型的矩。通过计算颜色矩，可以对图像的颜色特征进行定量分析和比较，从而实现图像处理和识别算法的应用。

## 7.3 系统介绍

对于给定的图片或者视频流，机器人的物体检测系统可以识别出已知的物体和该物体在图片中的位置。物体检测模块被训练用于检测多种物体的存在以及它们的位置，例如，模型可使用包含多种水果的图片和水果(如苹果、香蕉、草莓)所分别代表的 label 进行训练，返回的数据指明了图像中对象所出现的位置。随后，当我们为模型提供图片时，模型将会返回一个列表，其中包含检测到的对象、对象矩形框的坐标和代表检测可信度的分数。本项目的具体结构如图 7-1 所示。

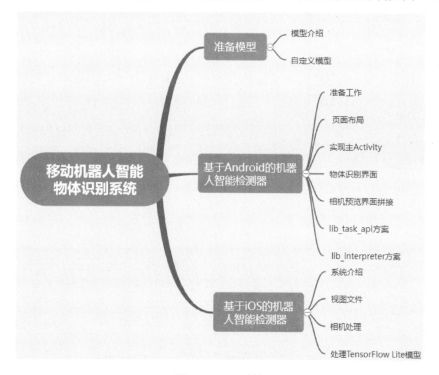

图 7-1 项目结构

## 7.4 准备模型

本项目使用的是 TensorFlow 官方提供的现成模型，用户可以登录 TensorFlow 官方网站下载模型文件 detect.tflite。

### 7.4.1 模型介绍

扫码看视频

在本项目中，在文件 download_model.gradle 中设置了使用的初始模型和标签文件。文件 download_model.gradle 的具体实现代码如下：

```
task downloadModelFile(type: Download) {
  src 'https://tfhub.dev/tensorflow/lite-model/ssd_mobilenet_v1/1/metadata/2?lite-format=tflite'
  dest project.ext.ASSET_DIR + '/detect.tflite'
  overwrite false
}
```

这个物体检测模型 detect.tflite 最多能够在一张图中识别和定位 10 个物体，目前支持 80 种物体的识别。

1) 输入

模型使用单个图片作为输入，理想的图片尺寸大小是 300×300 像素，每个像素点有 3 个通道(红、蓝和绿)。这将反馈给模块一个 27000 字节(300×300×3)的扁平化缓存。由于该模块经过标准化处理，每一个字节代表了 0～255 的一个值。

2) 输出

该模型输出 4 个数组，分别对应索引的 0～3。前 3 个数组描述 10 个被检测到的物体，每个数组的最后一个元素匹配每个对象，检测到的物体数量总是 10。各个索引的具体说明如表 7-1 所示。

表 7-1 索引说明

| 索引 | 名称 | 描述 |
| --- | --- | --- |
| 0 | 坐标 | [10][4] 多维数组，每一个元素是从 0～1 的浮点数，内部数组表示了矩形边框的 [top, left, bottom, right] |
| 1 | 类型 | 10 个整型元素组成的数组(输出为浮点型值)，每一个元素代表标签文件中的索引 |
| 2 | 分数 | 10 个整型元素组成的数组，元素值为 0～1 的浮点数，代表检测到的类型 |
| 3 | 检测到的物体和数量 | 长度为 1 的数组，元素为检测到的总数 |

## 7.4.2 自定义模型

开发者可以使用转移学习等技术来重新训练模型,从而能够辨识初始设置之外的物品种类,例如,可以重新训练模型来辨识各种蔬菜,哪怕原始训练数据中只有一种蔬菜。为达成此目标,为每一个需要训练的标签准备一系列训练图片。

接下来将介绍在 Oxford-IIIT Pet 数据集上训练新对象检测模型的过程,该模型将能够检测猫和狗的位置并识别每种动物的品种。假设我们在 Ubuntu 16.04 系统上运行,在开始之前需要设置开发环境,具体如下。

- 设置 Google Cloud 项目,配置计费并启用必要的 Cloud API;
- 设置 Google Cloud SDK;
- 安装 TensorFlow。

1) 安装 TensorFlow 对象检测 API

假设已经安装了 TensorFlow,那么可以使用以下命令安装对象检测 API 和其他依赖项:

```
git clone https://github.com/tensorflow/models
cd models/research
sudo apt-get install protobuf-compiler python-pil python-lxml
protoc object_detection/protos/*.proto --python_out=.
export PYTHONPATH=$PYTHONPATH:'pwd':'pwd'/slim
```

通过运行以下命令来测试安装:

```
python object_detection/builders/model_builder_test.py
```

2) 下载 Oxford-IIIT Pet Dataset

开始下载 Oxford-IIIT Pet Dataset 数据集,然后转换为 TFRecords 并上传到 GCS。Tensorflow 对象检测 API 使用 TFRecord 格式进行训练和验证数据集。使用以下命令下载 Oxford-IIIT Pet 数据集并转换为 TFRecords:

```
wget http://www.robots.ox.ac.uk/~vgg/data/pets/data/images.tar.gz
wget http://www.robots.ox.ac.uk/~vgg/data/pets/data/annotations.tar.gz
tar -xvf annotations.tar.gz
tar -xvf images.tar.gz
python object_detection/dataset_tools/create_pet_tf_record.py \
   --label_map_path=object_detection/data/pet_label_map.pbtxt \
   --data_dir='pwd' \
   --output_dir='pwd'
```

接下来应该会看到两个新生成的文件 pet_train.record 和 pet_val.record。要在 GCS(Google Cloud Storage,谷歌提供的一个云存储服务)上使用数据集,需要使用以下命令将其上传到我们的 Cloud Storage。请注意,我们同样上传了一个"标签地图"(包含在 git 存储库中),

它将我们的模型预测的数字索引与类别名称对应起来(例如，4 -> basset hound、5 -> beagle)。

```
gsutil cp pet_train_with_masks.record ${YOUR_GCS_BUCKET}/data/pet_train.record
gsutil cp pet_val_with_masks.record ${YOUR_GCS_BUCKET}/data/pet_val.record
gsutil cp object_detection/data/pet_label_map.pbtxt \
    ${YOUR_GCS_BUCKET}/data/pet_label_map.pbtxt
```

3) 上传用于迁移学习的预训练 COCO 模型

从头开始训练一个物体检测器模型可能需要几天时间，为了加快训练速度，将使用提供的模型中的参数初始化宠物模型，该模型已经在 COCO 数据集上进行了预训练。这个基于 ResNet101 的 Faster R-CNN 模型的权重将成为我们新模型的起点(微调检查点)，并将训练时间从几天缩短到几个小时。要从此模型初始化，需要下载它并将其放入 Cloud Storage。

```
wget https://storage.googleapis.com/download.tensorflow.org/models/
object_detection/faster_rcnn_resnet101_coco_11_06_2017.tar.gz
tar -xvf faster_rcnn_resnet101_coco_11_06_2017.tar.gz
gsutil cp faster_rcnn_resnet101_coco_11_06_2017/model.ckpt.*
${YOUR_GCS_BUCKET}/data/
```

4) 配置管道

使用 TensorFlow 对象检测 API 中的协议缓冲区配置，可以在 object_detection/samples/configs/中找到本项目的配置文件。这些配置文件可用于调整模型和训练参数(例如学习率、dropout 和正则化参数)。我们需要修改提供的配置文件，以了解上传数据集的位置并微调检查点。需要更改 PATH_TO_BE_CONFIGURED 字符串，以便它们只向上传到 Cloud Storage 存储分区的数据集文件和微调检查点。之后，还需要将配置文件本身上传到 Cloud Storage。

```
sed -i "s|PATH_TO_BE_CONFIGURED|"${YOUR_GCS_BUCKET}"/data|g"
object_detection/samples/configs/faster_rcnn_resnet101_pets.config
gsutil cp object_detection/samples/configs/faster_rcnn_resnet101_pets.config \
    ${YOUR_GCS_BUCKET}/data/faster_rcnn_resnet101_pets.config
```

5) 运行训练和评估

在 GCS 上运行之前，必须先打包 TensorFlow Object Detection API 和 TF Slim。

```
python setup.py sdist
(cd slim && python setup.py sdist)
```

仔细检查是否已将数据集上传到 Cloud Storage 存储分区，可以使用 Cloud Storage 浏览器检查存储分区。目录结构应如下所示：

```
+ ${YOUR_GCS_BUCKET}/
  + data/
    - faster_rcnn_resnet101_pets.config
```

```
- model.ckpt.index
- model.ckpt.meta
- model.ckpt.data-00000-of-00001
- pet_label_map.pbtxt
- pet_train.record
- pet_val.record
```

代码打包后,准备开始训练和评估工作:

```
gcloud ml-engine jobs submit training 'whoami'_object_detection_'date +%s' \
--job-dir=${YOUR_GCS_BUCKET}/train \
--packages dist/object_detection-0.1.tar.gz,slim/dist/slim-0.1.tar.gz \
--module-name object_detection.train \
```

此时可以在机器学习引擎仪表板上看到您的作业并检查日志以确保作业正在进行中。请注意,此训练作业使用具有五个工作 GPU 和三个参数服务器的分布式异步梯度下降。

6) 导出 TensorFlow 图

现在已经训练了一个宠物检测器,为了在训练后对一些示例图像运行检测,建议使用 Jupyter notebook 演示。但是,在此之前,需要将经过训练的模型导出到 TensorFlow 图形原型,并将学习到的权重作为常量进行处理。首先,需要确定要导出的候选检查点,可以使用 Google Cloud Storage Browser 搜索存储分区。检查点应存储在"${YOUR_GCS_BUCKET}/train"目录下。检查点通常由以下三个文件组成:

❑ model.ckpt-${CHECKPOINT_NUMBER}.data-00000-of-00001;
❑ model.ckpt-${CHECKPOINT_NUMBER}.index;
❑ model.ckpt-${CHECKPOINT_NUMBER}.meta。

确定要导出的候选检查点(通常是最新的)后,从 tensorflow/models 目录运行以下命令:

```
# Please define CEHCKPOINT_NUMBER based on the checkpoint you'd like to export
export CHECKPOINT_NUMBER=${CHECKPOINT_NUMBER}

# From tensorflow/models
gsutil cp ${YOUR_GCS_BUCKET}/train/model.ckpt-${CHECKPOINT_NUMBER}.* .
python object_detection/export_inference_graph \
    --input_type image_tensor \
    --pipeline_config_path
object_detection/samples/configs/faster_rcnn_resnet101_pets.config \
    --checkpoint_path model.ckpt-${CHECKPOINT_NUMBER} \
    --inference_graph_path output_inference_graph.pb
```

导出的图形将存储在名为 output_inference_graph.pb 的文件中。

## 7.5 基于 Android 的机器人智能检测器

在准备好 TensorFlow Lite 模型后,接下来将使用这个模型开发一个基于 Android 系统的物体检测识别器系统。本项目提供了两种情感分析解决方案。

扫码看视频

❑ lib_task_api:直接使用现成的 Task 库集成模型 API 进行 Tnference 推断识别;

❑ lib_interpreter:使用 TensorFlow Lite Interpreter Java API 创建自定义推断管道。
在本项目的内部 app 文件 build.gradle 中,设置了使用上述方案的方法。

### 7.5.1 准备工作

(1) 使用 Android Studio 导入本项目源码工程 object_detection,如图 7-2 所示。

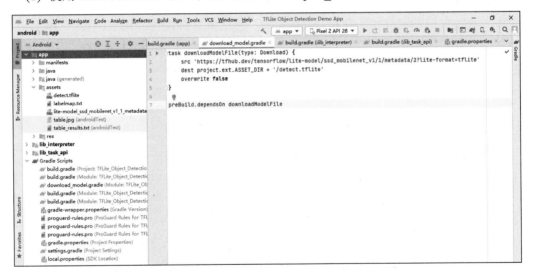

图 7-2 导入工程

(2) 更新 build.gradle。

打开 app 模块中的文件 build.gradle,分别设置 Android 的编译版本和运行版本,设置需要使用的库文件,添加对 TensorFlow Lite 模型库的引用。对应的代码如下:

```
apply plugin: 'com.android.application'
apply plugin: 'de.undercouch.download'

android {
```

```groovy
    compileSdkVersion 30
    defaultConfig {
        applicationId "org.tensorflow.lite.examples.detection"
        minSdkVersion 21
        targetSdkVersion 30
        versionCode 1
        versionName "1.0"

        testInstrumentationRunner "androidx.test.runner.AndroidJUnitRunner"
    }
    buildTypes {
        release {
            minifyEnabled false
            proguardFiles getDefaultProguardFile('proguard-android.txt'),
'proguard-rules.pro'
        }
    }
    aaptOptions {
        noCompress "tflite"
    }
    compileOptions {
        sourceCompatibility = '1.8'
        targetCompatibility = '1.8'
    }
    lintOptions {
        abortOnError false
    }
    flavorDimensions "tfliteInference"
    productFlavors {
        // TFLite 推断是使用 TFLiteJava 解释器构建的
        interpreter {
            dimension "tfliteInference"
        }
        // 默认：TFLite 推断是使用 TFLite 任务库(高级 API)构建的
        taskApi {
            getIsDefault().set(true)
            dimension "tfliteInference"
        }
    }
}

//导入下载模型任务
project.ext.ASSET_DIR = projectDir.toString() + '/src/main/assets'
project.ext.TMP_DIR   = project.buildDir.toString() + '/downloads'

//下载默认模型；如果希望使用自己的模型，请将它们放在 assets 目录中，并注释掉这一行
apply from:'download_model.gradle'
```

```
dependencies {
    implementation fileTree(dir: 'libs', include: ['*.jar','*.aar'])
    interpreterImplementation project(":lib_interpreter")
    taskApiImplementation project(":lib_task_api")
    implementation 'androidx.appcompat:appcompat:1.0.0'
    implementation 'androidx.coordinatorlayout:coordinatorlayout:1.0.0'
    implementation 'com.google.android.material:material:1.0.0'
    androidTestImplementation 'androidx.test.ext:junit:1.1.1'
    androidTestImplementation 'com.google.truth:truth:1.0.1'
    androidTestImplementation 'androidx.test:runner:1.2.0'
    androidTestImplementation 'androidx.test:rules:1.1.0'
}
```

## 7.5.2 页面布局

(1) 本项目主界面的页面布局文件是 tfe_od_activity_camera.xml，功能是在 Android 屏幕上方分别显示相机预览窗口，在屏幕下方显示悬浮式的系统配置参数。

(2) 在上面的页面布局文件 tfe_od_activity_camera.xml 中，通过调用文件 tfe_od_layout_bottom_sheet.xml 显示在主界面屏幕下方显示的悬浮式配置面板。

## 7.5.3 实现主 Activity

本项目的主 Activity 功能是由文件 CameraActivity.java 实现的，功能是调用前面的布局文件 tfe_od_activity_camera.xml，在 Android 屏幕上方显示相机预览窗口，在屏幕下方显示悬浮式的系统配置参数。文件 CameraActivity.java 的具体实现流程如下。

(1) 设置摄像头预览界面的公共属性，对应代码如下：

```
public abstract class CameraActivity extends AppCompatActivity
    implements OnImageAvailableListener,
        Camera.PreviewCallback,
        CompoundButton.OnCheckedChangeListener,
        View.OnClickListener {
  private static final Logger LOGGER = new Logger();

  private static final int PERMISSIONS_REQUEST = 1;

  private static final String PERMISSION_CAMERA = Manifest.permission.CAMERA;
  protected int previewWidth = 0;
  protected int previewHeight = 0;
  private boolean debug = false;
  private Handler handler;
  private HandlerThread handlerThread;
```

```
private boolean useCamera2API;
private boolean isProcessingFrame = false;
private byte[][] yuvBytes = new byte[3][];
private int[] rgbBytes = null;
```

(2) 在初始化函数 onCreate() 中加载布局文件 tfe_od_activity_camera.xml,对应代码如下:

```
@Override
protected void onCreate(final Bundle savedInstanceState) {
  LOGGER.d("onCreate " + this);
  super.onCreate(null);
  getWindow().addFlags(WindowManager.LayoutParams.FLAG_KEEP_SCREEN_ON);

  setContentView(R.layout.tfe_od_activity_camera);
  Toolbar toolbar = findViewById(R.id.toolbar);
  setSupportActionBar(toolbar);
  getSupportActionBar().setDisplayShowTitleEnabled(false);

  if (hasPermission()) {
    setFragment();
  } else {
    requestPermission();
  }
}
```

(3) 获取悬浮面板中的配置参数,系统将根据这些配置参数加载显示预览界面。对应的代码如下:

```
threadsTextView = findViewById(R.id.threads);
plusImageView = findViewById(R.id.plus);
minusImageView = findViewById(R.id.minus);
apiSwitchCompat = findViewById(R.id.api_info_switch);
bottomSheetLayout = findViewById(R.id.bottom_sheet_layout);
gestureLayout = findViewById(R.id.gesture_layout);
sheetBehavior = BottomSheetBehavior.from(bottomSheetLayout);
bottomSheetArrowImageView = findViewById(R.id.bottom_sheet_arrow);
```

(4) 获取视图树观察者对象,设置底页回调处理事件。对应的代码如下:

```
ViewTreeObserver vto = gestureLayout.getViewTreeObserver();
vto.addOnGlobalLayoutListener(
    new ViewTreeObserver.OnGlobalLayoutListener() {
      @Override
      public void onGlobalLayout() {
        if (Build.VERSION.SDK_INT < Build.VERSION_CODES.JELLY_BEAN) {
          gestureLayout.getViewTreeObserver().removeGlobalOnLayoutListener(this);
        } else {
          gestureLayout.getViewTreeObserver().removeOnGlobalLayoutListener(this);
        }
```

```java
            //int width = bottomSheetLayout.getMeasuredWidth();
            int height = gestureLayout.getMeasuredHeight();

            sheetBehavior.setPeekHeight(height);
        }
    });
    sheetBehavior.setHideable(false);

    sheetBehavior.setBottomSheetCallback(
        new BottomSheetBehavior.BottomSheetCallback() {
            @Override
            public void onStateChanged(@NonNull View bottomSheet, int newState) {
                switch (newState) {
                    case BottomSheetBehavior.STATE_HIDDEN:
                        break;
                    case BottomSheetBehavior.STATE_EXPANDED:
                    {
                        bottomSheetArrowImageView.setImageResource(R.drawable.icn_chevron_down);
                    }
                        break;
                    case BottomSheetBehavior.STATE_COLLAPSED:
                    {
                        bottomSheetArrowImageView.setImageResource(R.drawable.icn_chevron_up);
                    }
                        break;
                    case BottomSheetBehavior.STATE_DRAGGING:
                        break;
                    case BottomSheetBehavior.STATE_SETTLING:
                        bottomSheetArrowImageView.setImageResource(R.drawable.icn_chevron_up);
                        break;
                }
            }

            @Override
            public void onSlide(@NonNull View bottomSheet, float slideOffset) {}
        });
    frameValueTextView = findViewById(R.id.frame_info);
    cropValueTextView = findViewById(R.id.crop_info);
    inferenceTimeTextView = findViewById(R.id.inference_info);

    apiSwitchCompat.setOnCheckedChangeListener(this);
    plusImageView.setOnClickListener(this);
    minusImageView.setOnClickListener(this);
}
```

（5）创建 android.hardware.Camera API 的回调，打开手机中的相机预览界面，使用函数 ImageUtils.convertYUV420SPToARGB8888()将相机 data 转换成 rgbBytes。对应的代码如下：

```java
@Override
public void onPreviewFrame(final byte[] bytes, final Camera camera) {
  if (isProcessingFrame) {
    LOGGER.w("Dropping frame!");
    return;
  }
  try {
    //已知分辨率,初始化存储位图一次
    if (rgbBytes == null) {
      Camera.Size previewSize = camera.getParameters().getPreviewSize();
      previewHeight = previewSize.height;
      previewWidth = previewSize.width;
      rgbBytes = new int[previewWidth * previewHeight];
      onPreviewSizeChosen(new Size(previewSize.width, previewSize.height), 90);
    }
  } catch (final Exception e) {
    LOGGER.e(e, "Exception!");
    return;
  }
  isProcessingFrame = true;
  yuvBytes[0] = bytes;
  yRowStride = previewWidth;
  imageConverter =
      new Runnable() {
        @Override
        public void run() {
          ImageUtils.convertYUV420SPToARGB8888(bytes, previewWidth, previewHeight, rgbBytes);
        }
      };

  postInferenceCallback =
      new Runnable() {
        @Override
        public void run() {
          camera.addCallbackBuffer(bytes);
          isProcessingFrame = false;
        }
      };
  processImage();
}
```

(6) 编写函数 onImageAvailable(),实现 Camera2 API 的回调,对应代码如下:

```java
@Override
public void onImageAvailable(final ImageReader reader) {
  //需要等待,直到从 onPreviewSizeChosen 得到一些尺寸
  if (previewWidth == 0 || previewHeight == 0) {
```

```java
      return;
    }
    if (rgbBytes == null) {
      rgbBytes = new int[previewWidth * previewHeight];
    }
    try {
      final Image image = reader.acquireLatestImage();

      if (image == null) {
        return;
      }

      if (isProcessingFrame) {
        image.close();
        return;
      }
      isProcessingFrame = true;
      Trace.beginSection("imageAvailable");
      final Plane[] planes = image.getPlanes();
      fillBytes(planes, yuvBytes);
      yRowStride = planes[0].getRowStride();
      final int uvRowStride = planes[1].getRowStride();
      final int uvPixelStride = planes[1].getPixelStride();

      imageConverter =
          new Runnable() {
            @Override
            public void run() {
              ImageUtils.convertYUV420ToARGB8888(
                  yuvBytes[0],
                  yuvBytes[1],
                  yuvBytes[2],
                  previewWidth,
                  previewHeight,
                  yRowStride,
                  uvRowStride,
                  uvPixelStride,
                  rgbBytes);
            }
          };
      postInferenceCallback =
          new Runnable() {
            @Override
            public void run() {
              image.close();
              isProcessingFrame = false;
            }
```

```
    };
    processImage();
} catch (final Exception e) {
    LOGGER.e(e, "Exception!");
    Trace.endSection();
    return;
}
Trace.endSection();
}
```

(7) 编写函数 onImageAvailable()，功能是判断当前手机设备是否支持所需的硬件级别或更高级别，如果是，则返回 true。对应的代码如下：

```
private boolean isHardwareLevelSupported(
    CameraCharacteristics characteristics, int requiredLevel) {
  int deviceLevel =
characteristics.get(CameraCharacteristics.INFO_SUPPORTED_HARDWARE_LEVEL);
  if (deviceLevel == CameraCharacteristics.INFO_SUPPORTED_HARDWARE_LEVEL_LEGACY) {
    return requiredLevel == deviceLevel;
  }
  //使用数字排序
  return requiredLevel <= deviceLevel;
}
```

(8) 启用当前设备中的摄像头功能，对应代码如下：

```
private String chooseCamera() {
  final CameraManager manager = (CameraManager) getSystemService(Context.CAMERA_SERVICE);
  try {
    for (final String cameraId : manager.getCameraIdList()) {
      final CameraCharacteristics characteristics =
manager.getCameraCharacteristics(cameraId);

      //不使用前向摄像头
      final Integer facing = characteristics.get(CameraCharacteristics.LENS_FACING);
      if (facing != null && facing == CameraCharacteristics.LENS_FACING_FRONT) {
        continue;
      }

      final StreamConfigurationMap map =
          characteristics.get(CameraCharacteristics.SCALER_STREAM_CONFIGURATION_MAP);
      if (map == null) {
        continue;
      }

      //对于没有完全支持的内部摄像头，请返回 Camera API，这将有助于解决使用 Camera2 API
      //导致预览失真或损坏的遗留问题
```

```
            useCamera2API =
                (facing == CameraCharacteristics.LENS_FACING_EXTERNAL)
                    || isHardwareLevelSupported(
                        characteristics, CameraCharacteristics.INFO_SUPPORTED_HARDWARE_
LEVEL_FULL);
            LOGGER.i("Camera API lv2?: %s", useCamera2API);
            return cameraId;
        }
    } catch (CameraAccessException e) {
      LOGGER.e(e, "Not allowed to access camera");
    }
    return null;
  }
```

### 7.5.4 物体识别界面

本实例的物体识别界面 Activity 是由文件 DetectorActivity.java 实现的，功能是调用 lib_task_api 或 lib_interpreter 方案实现物体识别。文件 DetectorActivity.java 的具体实现流程如下。

（1）在设置了 Camera 捕获图片的一些参数后，例如图片预览大小 previewSize，摄像头方向 sensorOrientation 等，最重要的是回调之前传入到 fragment 中的 cameraConnectionCallback 的 onPreviewSizeChosen()函数，这是预览图片的宽、高确定后执行的回调函数。对应的代码如下：

```
  public void onPreviewSizeChosen(final Size size, final int rotation) {
    final float textSizePx =
        TypedValue.applyDimension(
            TypedValue.COMPLEX_UNIT_DIP, TEXT_SIZE_DIP,
getResources().getDisplayMetrics());
    borderedText = new BorderedText(textSizePx);
    borderedText.setTypeface(Typeface.MONOSPACE);

    tracker = new MultiBoxTracker(this);
    int cropSize = TF_OD_API_INPUT_SIZE;
    try {
      detector =
          TFLiteObjectDetectionAPIModel.create(
              this,
              TF_OD_API_MODEL_FILE,
              TF_OD_API_LABELS_FILE,
              TF_OD_API_INPUT_SIZE,
              TF_OD_API_IS_QUANTIZED);
      cropSize = TF_OD_API_INPUT_SIZE;
    } catch (final IOException e) {
```

```java
        e.printStackTrace();
      LOGGER.e(e, "Exception initializing Detector!");
      Toast toast =
          Toast.makeText(
              getApplicationContext(), "Detector could not be initialized",
Toast.LENGTH_SHORT);
      toast.show();
      finish();
    }
    previewWidth = size.getWidth();
    previewHeight = size.getHeight();

    sensorOrientation = rotation - getScreenOrientation();
    LOGGER.i("Camera orientation relative to screen canvas: %d", sensorOrientation);

    LOGGER.i("Initializing at size %dx%d", previewWidth, previewHeight);
    rgbFrameBitmap = Bitmap.createBitmap(previewWidth, previewHeight, Config.ARGB_8888);
    croppedBitmap = Bitmap.createBitmap(cropSize, cropSize, Config.ARGB_8888);

    frameToCropTransform =
        ImageUtils.getTransformationMatrix(
            previewWidth, previewHeight,
            cropSize, cropSize,
            sensorOrientation, MAINTAIN_ASPECT);

    cropToFrameTransform = new Matrix();
    frameToCropTransform.invert(cropToFrameTransform);

    trackingOverlay = (OverlayView) findViewById(R.id.tracking_overlay);
    trackingOverlay.addCallback(
        new DrawCallback() {
          @Override
          public void drawCallback(final Canvas canvas) {
            tracker.draw(canvas);
            if (isDebug()) {
              tracker.drawDebug(canvas);
            }
          }
        });
    tracker.setFrameConfiguration(previewWidth, previewHeight,
sensorOrientation);
  }
```

(2) 处理摄像头中的图像,将流式 YUV420_888 图像转换为可理解的图像,会自动启动一个处理图像的线程,这意味着可以随意使用而不会崩溃。如果图像处理无法跟上相机的进给速度,则会丢弃相框。对应的代码如下:

```java
protected void processImage() {
  ++timestamp;
  final long currTimestamp = timestamp;
  trackingOverlay.postInvalidate();

  //不需要互斥,因为此方法不可重入
  if (computingDetection) {
    readyForNextImage();
    return;
  }
  computingDetection = true;
  LOGGER.i("Preparing image " + currTimestamp + " for detection in bg thread.");

  rgbFrameBitmap.setPixels(getRgbBytes(), 0, previewWidth, 0, 0, previewWidth, previewHeight);

  readyForNextImage();

  final Canvas canvas = new Canvas(croppedBitmap);
  canvas.drawBitmap(rgbFrameBitmap, frameToCropTransform, null);
  //用于检查实际TF输入
  if (SAVE_PREVIEW_BITMAP) {
    ImageUtils.saveBitmap(croppedBitmap);
  }

  runInBackground(
      new Runnable() {
        @Override
        public void run() {
          LOGGER.i("Running detection on image " + currTimestamp);
          final long startTime = SystemClock.uptimeMillis();
          final List<Detector.Recognition> results =
detector.recognizeImage(croppedBitmap);
          lastProcessingTimeMs = SystemClock.uptimeMillis() - startTime;

          cropCopyBitmap = Bitmap.createBitmap(croppedBitmap);
          final Canvas canvas = new Canvas(cropCopyBitmap);
          final Paint paint = new Paint();
          paint.setColor(Color.RED);
          paint.setStyle(Style.STROKE);
          paint.setStrokeWidth(2.0f);

          float minimumConfidence = MINIMUM_CONFIDENCE_TF_OD_API;
          switch (MODE) {
            case TF_OD_API:
              minimumConfidence = MINIMUM_CONFIDENCE_TF_OD_API;
              break;
```

```java
    }

    final List<Detector.Recognition> mappedRecognitions =
        new ArrayList<Detector.Recognition>();

    for (final Detector.Recognition result : results) {
      final RectF location = result.getLocation();
      if (location != null && result.getConfidence() >= minimumConfidence) {
        canvas.drawRect(location, paint);

        cropToFrameTransform.mapRect(location);

        result.setLocation(location);
        mappedRecognitions.add(result);
      }
    }

    tracker.trackResults(mappedRecognitions, currTimestamp);
    trackingOverlay.postInvalidate();

    computingDetection = false;

    runOnUiThread(
        new Runnable() {
          @Override
          public void run() {
            showFrameInfo(previewWidth + "x" + previewHeight);
            showCropInfo(cropCopyBitmap.getWidth() + "x" + cropCopyBitmap.getHeight());
            showInference(lastProcessingTimeMs + "ms");
          }
        });
  }
});
}
```

## 7.5.5 相机预览界面拼接

编写文件 CameraConnectionFragment.java，功能是在摄像头中识别物体后会用文字标注识别结果，并将识别结果和摄像头预览界面拼接在一起，构成一幅完整的图形。文件 CameraConnectionFragment.java 的具体实现流程如下所示。

(1) 设置常量属性，例如设置相机的预览大小为 320，这将被设置为能够容纳所需大小正方形的最小逐像素帧大小。对应的代码如下：

```
private static final int MINIMUM_PREVIEW_SIZE = 320;

/**从屏幕旋转到JPEG方向的转换 */
private static final SparseIntArray ORIENTATIONS = new SparseIntArray();

private static final String FRAGMENT_DIALOG = "dialog";

static {
  ORIENTATIONS.append(Surface.ROTATION_0, 90);
  ORIENTATIONS.append(Surface.ROTATION_90, 0);
  ORIENTATIONS.append(Surface.ROTATION_180, 270);
  ORIENTATIONS.append(Surface.ROTATION_270, 180);
}

/**一个{@link Semaphore}用于在关闭摄像头之前阻止应用程序退出 */
private final Semaphore cameraOpenCloseLock = new Semaphore(1);
/** 用于接收可用帧的{@link OnImageAvailableListener} */
private final OnImageAvailableListener imageListener;
/**TensorFlow所需的输入大小(正方形位图的宽度和高度)，以像素为单位 */
private final Size inputSize;
/**设置布局标识符 */
private final int layout;
```

(2) 使用 TextureView.SurfaceTextureListener 处理 TextureView 上的多个生命周期事件，对应的代码如下：

```
private final TextureView.SurfaceTextureListener surfaceTextureListener =
    new TextureView.SurfaceTextureListener() {
      @Override
      public void onSurfaceTextureAvailable(
          final SurfaceTexture texture, final int width, final int height) {
        openCamera(width, height);
      }

      @Override
      public void onSurfaceTextureSizeChanged(
          final SurfaceTexture texture, final int width, final int height) {
        configureTransform(width, height);
      }

      @Override
      public boolean onSurfaceTextureDestroyed(final SurfaceTexture texture) {
        return true;
      }

      @Override
      public void onSurfaceTextureUpdated(final SurfaceTexture texture) {}
    };
```

```
private CameraConnectionFragment(
    final ConnectionCallback connectionCallback,
    final OnImageAvailableListener imageListener,
    final int layout,
    final Size inputSize) {
  this.cameraConnectionCallback = connectionCallback;
  this.imageListener = imageListener;
  this.layout = layout;
  this.inputSize = inputSize;
}
```

(3) 编写函数 chooseOptimalSize()，设置相机的参数，会根据设置的参数返回最佳大小的预览界面。如果没有足够大的界面，则返回任意值，其中设置的宽度和高度至少与两者最小值相同。或者如果有可能，可以选择完全匹配的值。各个参数的具体说明如下。

- choices：相机可以支持的所有预览尺寸的列表；
- width：所需的最小宽度；
- height：所需的最小高度。

函数 chooseOptimalSize()的具体实现代码如下：

```
protected static Size chooseOptimalSize(final Size[] choices, final int width,
final int height) {
  final int minSize = Math.max(Math.min(width, height), MINIMUM_PREVIEW_SIZE);
  final Size desiredSize = new Size(width, height);

  //收集至少与预览界面一样大的支持分辨率
  boolean exactSizeFound = false;
  final List<Size> bigEnough = new ArrayList<Size>();
  final List<Size> tooSmall = new ArrayList<Size>();
  for (final Size option : choices) {
    if (option.equals(desiredSize)) {
      //设置大小，但不要返回，以便仍记录剩余的大小
      exactSizeFound = true;
    }

    if (option.getHeight() >= minSize && option.getWidth() >= minSize) {
      bigEnough.add(option);
    } else {
      tooSmall.add(option);
    }
  }

  LOGGER.i("Desired size: " + desiredSize + ", min size: " + minSize + "x" + minSize);
  LOGGER.i("Valid preview sizes: [" + TextUtils.join(", ", bigEnough) + "]");
  LOGGER.i("Rejected preview sizes: [" + TextUtils.join(", ", tooSmall) + "]");
```

```
    if (exactSizeFound) {
      LOGGER.i("Exact size match found.");
      return desiredSize;
    }

    //挑选最小的
    if (bigEnough.size() > 0) {
      final Size chosenSize = Collections.min(bigEnough, new CompareSizesByArea());
      LOGGER.i("Chosen size: " + chosenSize.getWidth() + "x" +
chosenSize.getHeight());
      return chosenSize;
    } else {
      LOGGER.e("Couldn't find any suitable preview size");
      return choices[0];
    }
  }
```

(4) 编写函数 showToast()，功能是显示 UI 线程上的要显示的提醒消息，对应的代码如下：

```
  private void showToast(final String text) {
    final Activity activity = getActivity();
    if (activity != null) {
      activity.runOnUiThread(
          new Runnable() {
            @Override
            public void run() {
              Toast.makeText(activity, text, Toast.LENGTH_SHORT).show();
            }
          });
    }
  }
```

(5) 编写函数 setUpCameraOutputs()，功能是设置与相机相关的成员变量，对应的代码如下：

```
  private void setUpCameraOutputs() {
    final Activity activity = getActivity();
    final CameraManager manager = (CameraManager)
activity.getSystemService(Context.CAMERA_SERVICE);
    try {
      final CameraCharacteristics characteristics =
manager.getCameraCharacteristics(cameraId);

      final StreamConfigurationMap map =
          characteristics.get(CameraCharacteristics.SCALER_STREAM_CONFIGURATION_MAP);
```

```
      sensorOrientation =
characteristics.get(CameraCharacteristics.SENSOR_ORIENTATION);

      //如果尝试使用过大的预览尺寸，可能会超出相机总线的带宽限制，这会导致虽然预览效果很好，但
      //实际存储的捕获数据可能会损坏
      previewSize =
          chooseOptimalSize(
              map.getOutputSizes(SurfaceTexture.class),
              inputSize.getWidth(),
              inputSize.getHeight());

      //将TextureView的纵横比与拾取的预览大小相匹配
      final int orientation = getResources().getConfiguration().orientation;
      if (orientation == Configuration.ORIENTATION_LANDSCAPE) {
        textureView.setAspectRatio(previewSize.getWidth(), previewSize.getHeight());
      } else {
        textureView.setAspectRatio(previewSize.getHeight(),
previewSize.getWidth());
      }
    } catch (final CameraAccessException e) {
      LOGGER.e(e, "Exception!");
    } catch (final NullPointerException e) {
      //当使用Camera2 API，但此代码运行的设备不支持时，会引发NPE
      ErrorDialog.newInstance(getString(R.string.tfe_od_camera_error))
          .show(getChildFragmentManager(), FRAGMENT_DIALOG);
      throw new IllegalStateException(getString(R.string.tfe_od_camera_error));
    }

    cameraConnectionCallback.onPreviewSizeChosen(previewSize,
sensorOrientation);
  }
```

（6）编写函数 openCamera()，功能是打开由 CameraConnectionFragmen 指定的相机，对应代码如下：

```
  private void openCamera(final int width, final int height) {
    setUpCameraOutputs();
    configureTransform(width, height);
    final Activity activity = getActivity();
    final CameraManager manager = (CameraManager)
activity.getSystemService(Context.CAMERA_SERVICE);
    try {
      if (!cameraOpenCloseLock.tryAcquire(2500, TimeUnit.MILLISECONDS)) {
        throw new RuntimeException("Time out waiting to lock camera opening.");
      }
      manager.openCamera(cameraId, stateCallback, backgroundHandler);
```

```
    } catch (final CameraAccessException e) {
        LOGGER.e(e, "Exception!");
    } catch (final InterruptedException e) {
        throw new RuntimeException("Interrupted while trying to lock camera opening.", e);
    }
}
```

(7) 编写函数 closeCamera()，功能是关闭当前的 CameraDevice 相机，对应的代码如下：

```
private void closeCamera() {
    try {
        cameraOpenCloseLock.acquire();
        if (null != captureSession) {
            captureSession.close();
            captureSession = null;
        }
        if (null != cameraDevice) {
            cameraDevice.close();
            cameraDevice = null;
        }
        if (null != previewReader) {
            previewReader.close();
            previewReader = null;
        }
    } catch (final InterruptedException e) {
        throw new RuntimeException("Interrupted while trying to lock camera closing.", e);
    } finally {
        cameraOpenCloseLock.release();
    }
}
```

(8) 分别启动前台线程和后台线程，对应的代码如下：

```
/**启动后台线程及其{@link Handler} */
private void startBackgroundThread() {
    backgroundThread = new HandlerThread("ImageListener");
    backgroundThread.start();
    backgroundHandler = new Handler(backgroundThread.getLooper());
}

/**停止后台线程及其{@link Handler} */
private void stopBackgroundThread() {
    backgroundThread.quitSafely();
    try {
        backgroundThread.join();
        backgroundThread = null;
        backgroundHandler = null;
    } catch (final InterruptedException e) {
```

```
        LOGGER.e(e, "Exception!");
    }
}
```

(9) 为相机预览界面创建新的 CameraCaptureSession 缓存，对应的代码如下：

```
private void createCameraPreviewSession() {
  try {
    final SurfaceTexture texture = textureView.getSurfaceTexture();
    assert texture != null;

    //将默认缓冲区的大小配置为所需的相机预览大小
    texture.setDefaultBufferSize(previewSize.getWidth(),
previewSize.getHeight());

    //这是我们需要开始预览的输出界面
    final Surface surface = new Surface(texture);

    //用输出界面设置了CaptureRequest.Builder
    previewRequestBuilder = cameraDevice.createCaptureRequest
(CameraDevice.TEMPLATE_PREVIEW);
    previewRequestBuilder.addTarget(surface);

    LOGGER.i("Opening camera preview: " + previewSize.getWidth() + "x" +
previewSize.getHeight());

    //为预览帧创建读取器
    previewReader =
        ImageReader.newInstance(
            previewSize.getWidth(), previewSize.getHeight(),
ImageFormat.YUV_420_888, 2);

    previewReader.setOnImageAvailableListener(imageListener,
backgroundHandler);
    previewRequestBuilder.addTarget(previewReader.getSurface());

    //为相机预览创建一个CameraCaptureSession
    cameraDevice.createCaptureSession(
        Arrays.asList(surface, previewReader.getSurface()),
        new CameraCaptureSession.StateCallback() {

          @Override
          public void onConfigured(final CameraCaptureSession cameraCaptureSession) {
            //相机已经关闭
            if (null == cameraDevice) {
              return;
            }
```

```
            //当会话准备就绪时开始显示预览
            captureSession = cameraCaptureSession;
            try {
                //自动对焦，连续用于相机预览
                previewRequestBuilder.set(
                    CaptureRequest.CONTROL_AF_MODE,
                    CaptureRequest.CONTROL_AF_MODE_CONTINUOUS_PICTURE);
                // 在必要时自动启用闪存
                previewRequestBuilder.set(
                    CaptureRequest.CONTROL_AE_MODE,
CaptureRequest.CONTROL_AE_MODE_ON_AUTO_FLASH);

                //最后，开始显示相机预览
                previewRequest = previewRequestBuilder.build();
                captureSession.setRepeatingRequest(
                    previewRequest, captureCallback, backgroundHandler);
            } catch (final CameraAccessException e) {
                LOGGER.e(e, "Exception!");
            }
        }

        @Override
        public void onConfigureFailed(final CameraCaptureSession cameraCaptureSession) {
            showToast("Failed");
        }
    },
    null);
} catch (final CameraAccessException e) {
    LOGGER.e(e, "Exception!");
}
}
```

(10) 编写函数 configureTransform()，功能是将必要的 Matrix 转换配置为 mTextureView。在 setUpCameraOutputs 中确定相机预览画面的大小，并且在固定 mTextureView 的大小后需要调用此方法。其中参数 viewWidth 表示 mTextureView 的宽度，参数 viewHeight 表示 mTextureView 的高度。对应的代码如下：

```
private void configureTransform(final int viewWidth, final int viewHeight) {
    final Activity activity = getActivity();
    if (null == textureView || null == previewSize || null == activity) {
        return;
    }
    final int rotation = activity.getWindowManager().getDefaultDisplay().getRotation();
    final Matrix matrix = new Matrix();
    final RectF viewRect = new RectF(0, 0, viewWidth, viewHeight);
    final RectF bufferRect = new RectF(0, 0, previewSize.getHeight(),
previewSize.getWidth());
```

```
    final float centerX = viewRect.centerX();
    final float centerY = viewRect.centerY();
    if (Surface.ROTATION_90 == rotation || Surface.ROTATION_270 == rotation) {
      bufferRect.offset(centerX - bufferRect.centerX(), centerY - bufferRect.centerY());
      matrix.setRectToRect(viewRect, bufferRect, Matrix.ScaleToFit.FILL);
      final float scale =
          Math.max(
              (float) viewHeight / previewSize.getHeight(),
              (float) viewWidth / previewSize.getWidth());
      matrix.postScale(scale, scale, centerX, centerY);
      matrix.postRotate(90 * (rotation - 2), centerX, centerY);
    } else if (Surface.ROTATION_180 == rotation) {
      matrix.postRotate(180, centerX, centerY);
    }
    textureView.setTransform(matrix);
}
```

## 7.5.6 lib_task_api 方案

本项目默认使用 TensorFlow Lite 任务库中的开箱即用 API 实现物体检测和识别功能，通过文件 TFLiteObjectDetectionAPIModel.java 调用 TensorFlow 对象检测 API 训练的检测模型包装器。对应的代码如下：

```
/**
使用 TensorFlow 对象检测 API 训练的检测模型包装器
*/
public class TFLiteObjectDetectionAPIModel implements Detector {
  private static final String TAG = "TFLiteObjectDetectionAPIModelWithTaskApi";

  /** 只返回这么多结果 */
  private static final int NUM_DETECTIONS = 10;

  private final MappedByteBuffer modelBuffer;

  /**使用 TensorFlow Lite 运行模型推断的驱动程序类的实例 */
  private ObjectDetector objectDetector;

  /**用于配置 ObjectDetector 选项的生成器 */
  private final ObjectDetectorOptions.Builder optionsBuilder;

  /**
  * 初始化对图像进行分类的 TensorFlow 会话
  * {@code-labelFilename}、{@code-inputSize}和{@code-isQuantized}不是必需的，而是为了与使用 TFLite 解释器 Java API 的实现保持一致。
```

```java
 * *@param modelFilename 模型文件路径
 * *@param labelFilename 标签文件路径
 * *@param inputSize 图像输入的大小
 * *@param isQuantized 布尔值,表示模型是否量化
 */
public static Detector create(
    final Context context,
    final String modelFilename,
    final String labelFilename,
    final int inputSize,
    final boolean isQuantized)
    throws IOException {
  return new TFLiteObjectDetectionAPIModel(context, modelFilename);
}

private TFLiteObjectDetectionAPIModel(Context context, String modelFilename) throws IOException {
    modelBuffer = FileUtil.loadMappedFile(context, modelFilename);
    optionsBuilder = ObjectDetectorOptions.builder().setMaxResults(NUM_DETECTIONS);
    objectDetector = ObjectDetector.createFromBufferAndOptions(modelBuffer, optionsBuilder.build());
}

@Override
public List<Recognition> recognizeImage(final Bitmap bitmap) {
    //记录此方法,以便使用systrace进行分析
    Trace.beginSection("recognizeImage");
    List<Detection> results = objectDetector.detect(TensorImage.fromBitmap(bitmap));

    // 将{@link Detection}对象列表转换为{@link Recognition}对象列表,以匹配其他推理方法的
    // 接口,例如,使用TFLite Java API
    final ArrayList<Recognition> recognitions = new ArrayList<>();
    int cnt = 0;
    for (Detection detection : results) {
      recognitions.add(
          new Recognition(
              "" + cnt++,
              detection.getCategories().get(0).getLabel(),
              detection.getCategories().get(0).getScore(),
              detection.getBoundingBox()));
    }
    Trace.endSection(); // "recognizeImage"
    return recognitions;
}
@Override
```

```java
public void enableStatLogging(final boolean logStats) {}

@Override
public String getStatString() {
  return "";
}
@Override
public void close() {
  if (objectDetector != null) {
    objectDetector.close();
  }
}

@Override
public void setNumThreads(int numThreads) {
  if (objectDetector != null) {
    optionsBuilder.setNumThreads(numThreads);
    recreateDetector();
  }
}

@Override
public void setUseNNAPI(boolean isChecked) {
  throw new UnsupportedOperationException(
      "在此任务中不允许操作硬件加速器,只允许使用CPU!");
}

private void recreateDetector() {
  objectDetector.close();
  objectDetector = ObjectDetector.createFromBufferAndOptions(modelBuffer,
optionsBuilder.build());
}
}
```

## 7.5.7 lib_interpreter 方案

本项目还可以使用 lib_interpreter 方案实现物体检测和识别功能,本方案使用 TensorFlow Lite 中的 Interpreter Java API 创建自定义识别函数。本功能主要由文件 TFLiteObjectDetectionAPIModel.java 实现,对应的代码如下:

```java
/**内存映射资源中的模型文件 */
private static MappedByteBuffer loadModelFile(AssetManager assets, String modelFilename)
    throws IOException {
  AssetFileDescriptor fileDescriptor = assets.openFd(modelFilename);
```

```java
    FileInputStream inputStream = new FileInputStream(fileDescriptor.getFileDescriptor());
    FileChannel fileChannel = inputStream.getChannel();
    long startOffset = fileDescriptor.getStartOffset();
    long declaredLength = fileDescriptor.getDeclaredLength();
    return fileChannel.map(FileChannel.MapMode.READ_ONLY, startOffset, declaredLength);
}
/**
 * 初始化用于对图像进行分类的本机 TensorFlow 会话。
 * *@param modelFilename 模型文件路径
 * *@param labelFilename 标签文件路径
 * *@param inputSize 图像输入的大小
 * *@param isQuantized 布尔值,表示模型是否量化
 */
public static Detector create(
    final Context context,
    final String modelFilename,
    final String labelFilename,
    final int inputSize,
    final boolean isQuantized)
    throws IOException {
  final TFLiteObjectDetectionAPIModel d = new TFLiteObjectDetectionAPIModel();

  MappedByteBuffer modelFile = loadModelFile(context.getAssets(), modelFilename);
  MetadataExtractor metadata = new MetadataExtractor(modelFile);
  try (BufferedReader br =
      new BufferedReader(
          new InputStreamReader(
              metadata.getAssociatedFile(labelFilename), Charset.defaultCharset()))) {
    String line;
    while ((line = br.readLine()) != null) {
      Log.w(TAG, line);
      d.labels.add(line);
    }
  }

  d.inputSize = inputSize;

  try {
    Interpreter.Options options = new Interpreter.Options();
    options.setNumThreads(NUM_THREADS);
    options.setUseXNNPACK(true);
    d.tfLite = new Interpreter(modelFile, options);
    d.tfLiteModel = modelFile;
    d.tfLiteOptions = options;
  } catch (Exception e) {
    throw new RuntimeException(e);
```

```
    }
    d.isModelQuantized = isQuantized;
    //预先分配缓冲区
    int numBytesPerChannel;
    if (isQuantized) {
      numBytesPerChannel = 1;  //量化
    } else {
      numBytesPerChannel = 4;  //浮点数
    }
    d.imgData = ByteBuffer.allocateDirect(1 * d.inputSize * d.inputSize * 3 *
numBytesPerChannel);
    d.imgData.order(ByteOrder.nativeOrder());
    d.intValues = new int[d.inputSize * d.inputSize];

    d.outputLocations = new float[1][NUM_DETECTIONS][4];
    d.outputClasses = new float[1][NUM_DETECTIONS];
    d.outputScores = new float[1][NUM_DETECTIONS];
    d.numDetections = new float[1];
    return d;
  }

  @Override
  public List<Recognition> recognizeImage(final Bitmap bitmap) {
    //记录此方法,以便使用 systrace 进行分析
    Trace.beginSection("recognizeImage");

    Trace.beginSection("preprocessBitmap");
    //根据提供的参数,将图像数据从 0~255 整数型预处理为标准化浮点
    bitmap.getPixels(intValues, 0, bitmap.getWidth(), 0, 0, bitmap.getWidth(),
bitmap.getHeight());

    imgData.rewind();
    for (int i = 0; i < inputSize; ++i) {
      for (int j = 0; j < inputSize; ++j) {
        int pixelValue = intValues[i * inputSize + j];
        if (isModelQuantized) {
          //量化模型
          imgData.put((byte) ((pixelValue >> 16) & 0xFF));
          imgData.put((byte) ((pixelValue >> 8) & 0xFF));
          imgData.put((byte) (pixelValue & 0xFF));
        } else { // Float model
          imgData.putFloat((((pixelValue >> 16) & 0xFF) - IMAGE_MEAN) / IMAGE_STD);
          imgData.putFloat((((pixelValue >> 8) & 0xFF) - IMAGE_MEAN) / IMAGE_STD);
          imgData.putFloat(((pixelValue & 0xFF) - IMAGE_MEAN) / IMAGE_STD);
        }
      }
    }
```

```
                }
                Trace.endSection(); //预处理位图

                //将输入数据复制到TensorFlow中
                Trace.beginSection("feed");
                outputLocations = new float[1][NUM_DETECTIONS][4];
                outputClasses = new float[1][NUM_DETECTIONS];
                outputScores = new float[1][NUM_DETECTIONS];
                numDetections = new float[1];

                Object[] inputArray = {imgData};
                Map<Integer, Object> outputMap = new HashMap<>();
                outputMap.put(0, outputLocations);
                outputMap.put(1, outputClasses);
                outputMap.put(2, outputScores);
                outputMap.put(3, numDetections);
                Trace.endSection();

                //运行推断调用
                Trace.beginSection("run");
                tfLite.runForMultipleInputsOutputs(inputArray, outputMap);
                Trace.endSection();

                //显示最佳检测结果
                //将其缩放回输入大小后,需要使用输出中的检测数,而不是顶部声明的NUM_DETECTONS变量
                //因为在某些模型上,它们并不总是输出相同的检测总数
                //例如,模型的NUM_DETECTIONS=20,但有时它只输出16个预测
                //如果不使用输出的numDetections,将获得无意义的数据
                int numDetectionsOutput =
                    min(
                        NUM_DETECTIONS,
                        (int) numDetections[0]); //从浮点转换为整数,使用最小值以确保安全

                final ArrayList<Recognition> recognitions = new
ArrayList<>(numDetectionsOutput);
                for (int i = 0; i < numDetectionsOutput; ++i) {
                    final RectF detection =
                        new RectF(
                            outputLocations[0][i][1] * inputSize,
                            outputLocations[0][i][0] * inputSize,
                            outputLocations[0][i][3] * inputSize,
                            outputLocations[0][i][2] * inputSize);

                    recognitions.add(
                        new Recognition(
                            "" + i, labels.get((int) outputClasses[0][i]), outputScores[0][i],
detection));
```

```
    }
    Trace.endSection(); // "recognizeImage"
    return recognitions;
}
```

上述两种方案的识别文件都是 Detector.java，功能是调用各自方案下面的文件 TFLiteObjectDetectionAPIModel.java 实现具体识别功能，对应的代码如下：

```
/**与不同识别引擎交互的通用接口 */
public interface Detector {
  List<Recognition> recognizeImage(Bitmap bitmap);
  void enableStatLogging(final boolean debug);
  String getStatString();
  void close();
  void setNumThreads(int numThreads);
  void setUseNNAPI(boolean isChecked);
  /**检测器返回一个不变的结果，描述识别的内容 */
  public class Recognition {
    /**
     * 已识别内容的唯一标识符。特定于类，而不是对象的实例
     */
    private final String id;
    /** 用于识别的显示名称 */
    private final String title;
    /**
     * 识别度相对于其他可能性的可排序分数，分数越高越好
     */
    private final Float confidence;
    /**源图像中用于识别对象位置的可选位置 */
    private RectF location;
    public Recognition(
        final String id, final String title, final Float confidence, final RectF location) {
      this.id = id;
      this.title = title;
      this.confidence = confidence;
      this.location = location;
    }
    public String getId() {
      return id;
    }
    public String getTitle() {
      return title;
    }
    public Float getConfidence() {
      return confidence;
    }
```

```java
public RectF getLocation() {
  return new RectF(location);
}
public void setLocation(RectF location) {
  this.location = location;
}
@Override
public String toString() {
  String resultString = "";
  if (id != null) {
    resultString += "[" + id + "] ";
  }
  if (title != null) {
    resultString += title + " ";
  }
  if (confidence != null) {
    resultString += String.format("(%.1f%%) ", confidence * 100.0f);
  }
  if (location != null) {
    resultString += location + " ";
  }
  return resultString.trim();
}
```

到此为止,整个项目工程全部开发完毕。

## 7.6 基于 iOS 的机器人智能检测器

在前面讲解了基于 Android 系统为机器人开发物体检测识别器的过程,接下来,将详细讲解基于 iOS 系统使用 TensorFlow Lite 模型开发物体检测识别器的过程。

扫码看视频

### 7.6.1 系统介绍

使用 Xcode 导入本项目的 iOS 源码,如图 7-3 所示。
在 Model 目录下保存了需要使用的 TensorFlow Lite 模型文件,如图 7-4 所示。
通过故事板 Main.storyboard 文件设计 iOS 应用程序的 UI,如图 7-5 所示。

# 第 7 章 移动机器人智能物体识别系统

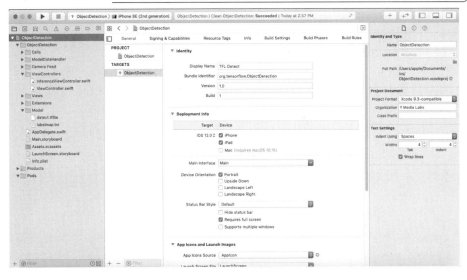

图 7-3 使用 Xcode 导入 iOS 源码

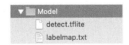

图 7-4 TensorFlow Lite 模型文件

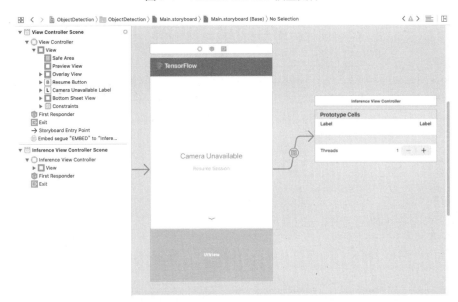

图 7-5 故事板 Main.storyboard 文件

## 7.6.2 视图文件

在 Xcode 工程的 ViewController 目录下保存了本项目视图文件,视图文件和故事板 Main.storyboard 文件相互结合,构建 iOS 应用程序的 UI。

(1) 编写主视图控制器文件 ViewController.swift,具体实现流程如下。

① 分别设置整个系统需要的公用 UI 参数,包括连接 Storyboards 故事板中的组件参数、常量参数、实例变量和实现视图管理功能的控制器。对应的代码如下:

```
import UIKit

class ViewController: UIViewController {

 //MARK: 连接Storyboards故事板中的组件参数
 @IBOutlet weak var previewView: PreviewView!
 @IBOutlet weak var overlayView: OverlayView!
 @IBOutlet weak var resumeButton: UIButton!
 @IBOutlet weak var cameraUnavailableLabel: UILabel!

 @IBOutlet weak var bottomSheetStateImageView: UIImageView!
 @IBOutlet weak var bottomSheetView: UIView!
 @IBOutlet weak var bottomSheetViewBottomSpace: NSLayoutConstraint!
 //MARK: 常量
 private let displayFont = UIFont.systemFont(ofSize: 14.0, weight: .medium)
 private let edgeOffset: CGFloat = 2.0
 private let labelOffset: CGFloat = 10.0
 private let animationDuration = 0.5
 private let collapseTransitionThreshold: CGFloat = -30.0
 private let expandTransitionThreshold: CGFloat = 30.0
 private let delayBetweenInferencesMs: Double = 200
 //MARK: 实例变量
 private var initialBottomSpace: CGFloat = 0.0
 //随时保存结果
 private var result: Result?
 private var previousInferenceTimeMs: TimeInterval =
Date.distantPast.timeIntervalSince1970 * 1000
 //MARK: 管理功能的控制器
 private lazy var cameraFeedManager = CameraFeedManager(previewView: previewView)
 private var modelDataHandler: ModelDataHandler? =
   ModelDataHandler(modelFileInfo: MobileNetSSD.modelInfo, labelsFileInfo:
MobileNetSSD.labelsInfo)
 private var inferenceViewController: InferenceViewController?

 //视图处理方法
```

```
override func viewDidLoad() {
  super.viewDidLoad()

  guard modelDataHandler != nil else {
    fatalError("Failed to load model")
  }
  cameraFeedManager.delegate = self
  overlayView.clearsContextBeforeDrawing = true

  addPanGesture()
}
override func didReceiveMemoryWarning() {
  super.didReceiveMemoryWarning()      //处理所有可以重新创建的资源
}
```

② 编写函数 onClickResumeButton(),实现单击 Button 按钮后的处理程序,对应的代码如下:

```
@IBAction func onClickResumeButton(_ sender: Any) {
  cameraFeedManager.resumeInterruptedSession { (complete) in
    if complete {
      self.resumeButton.isHidden = true
      self.cameraUnavailableLabel.isHidden = true
    }
    else {
      self.presentUnableToResumeSessionAlert()
    }
  }
}
```

③ 编写函数 prepare(),实现故事板 Segue 处理器,对应的代码如下:

```
override func prepare(for segue: UIStoryboardSegue, sender: Any?) {
  super.prepare(for: segue, sender: sender)
  if segue.identifier == "EMBED" {
    guard let tempModelDataHandler = modelDataHandler else {
      return
    }
    inferenceViewController = segue.destination as? InferenceViewController
    inferenceViewController?.wantedInputHeight = tempModelDataHandler.inputHeight
    inferenceViewController?.wantedInputWidth = tempModelDataHandler.inputWidth
    inferenceViewController?.threadCountLimit = tempModelDataHandler.threadCountLimit
    inferenceViewController?.currentThreadCount = tempModelDataHandler.threadCount
    inferenceViewController?.delegate = self
    guard let tempResult = result else {
      return
    }
    inferenceViewController?.inferenceTime = tempResult.inferenceTime
```

```
    }
  }
}
```

④ 通过 extension 扩展实现推断视图控制器,对应的代码如下:

```
extension ViewController: InferenceViewControllerDelegate {
  func didChangeThreadCount(to count: Int) {
    if modelDataHandler?.threadCount == count { return }
    modelDataHandler = ModelDataHandler(
      modelFileInfo: MobileNetSSD.modelInfo,
      labelsFileInfo: MobileNetSSD.labelsInfo,
      threadCount: count
    )
  }
}
```

⑤ 通过 extension 扩展实现相机管理器的委托方法,对应的代码如下:

```
extension ViewController: CameraFeedManagerDelegate {

  func didOutput(pixelBuffer: CVPixelBuffer) {
    runModel(onPixelBuffer: pixelBuffer)
  }
```

⑥ 编写自定义函数分别实现会话处理,包括实现会话处理提示框、会话中断时更新 UI、会话中断结束后更新 UI,对应的代码如下:

```
  //MARK: 会话处理提示框
  func sessionRunTimeErrorOccurred() {
    //通过更新 UI 并提供一个按钮(如果可以手动恢复会话)来处理会话运行时错误
    self.resumeButton.isHidden = false
  }
  func sessionWasInterrupted(canResumeManually resumeManually: Bool) {
    //会话中断时更新 UI
    if resumeManually {
      self.resumeButton.isHidden = false
    }
    else {
      self.cameraUnavailableLabel.isHidden = false
    }
  }
  func sessionInterruptionEnded() {
    //会话中断结束后更新 UI
    if !self.cameraUnavailableLabel.isHidden {
      self.cameraUnavailableLabel.isHidden = true
    }
    if !self.resumeButton.isHidden {
```

```
        self.resumeButton.isHidden = true
    }
}
```

⑦ 如果发生错误，则调用方法 presentVideoConfigurationErrorAlert()，弹出提示框，对应的代码如下：

```
func presentVideoConfigurationErrorAlert() {
    let alertController = UIAlertController(title: "Configuration Failed", message: "Configuration of camera has failed.", preferredStyle: .alert)
    let okAction = UIAlertAction(title: "OK", style: .cancel, handler: nil)
    alertController.addAction(okAction)
    present(alertController, animated: true, completion: nil)
}
```

⑧ 编写函数 presentCameraPermissionsDeniedAlert()，如果当前没有获得相机权限，则弹出提示框。对应的代码如下：

```
func presentCameraPermissionsDeniedAlert() {
    let alertController = UIAlertController(title: "Camera Permissions Denied", message: "Camera permissions have been denied for this app. You can change this by going to Settings", preferredStyle: .alert)
    let cancelAction = UIAlertAction(title: "Cancel", style: .cancel, handler: nil)
    let settingsAction = UIAlertAction(title: "Settings", style: .default) { (action) in
      UIApplication.shared.open(URL(string: UIApplication.openSettingsURLString)!, options: [:], completionHandler: nil)
    }
    alertController.addAction(cancelAction)
    alertController.addAction(settingsAction)
    present(alertController, animated: true, completion: nil)
}
```

⑨ 编写方法 runModel()，功能是通过 TensorFlow 运行实时相机像素缓冲区，对应的代码如下：

```
@objc  func runModel(onPixelBuffer pixelBuffer: CVPixelBuffer) {

    //通过 TensorFlow 运行 pixelBuffer 相机以获得实时结果
    let currentTimeMs = Date().timeIntervalSince1970 * 1000
    guard  (currentTimeMs - previousInferenceTimeMs) >= delayBetweenInferencesMs else {
      return
    }
    previousInferenceTimeMs = currentTimeMs
    result = self.modelDataHandler?.runModel(onFrame: pixelBuffer)
    guard let displayResult = result else {
      return
```

```
      }
      let width = CVPixelBufferGetWidth(pixelBuffer)
      let height = CVPixelBufferGetHeight(pixelBuffer)
      DispatchQueue.main.async {
        //通过传递给推断视图控制器来显示结果
        self.inferenceViewController?.resolution = CGSize(width: width, height: height)
        var inferenceTime: Double = 0
        if let resultInferenceTime = self.result?.inferenceTime {
          inferenceTime = resultInferenceTime
        }
        self.inferenceViewController?.inferenceTime = inferenceTime
        self.inferenceViewController?.tableView.reloadData()
        //绘制边界框并显示类名和置信度分数
        self.drawAfterPerformingCalculations(onInferences: displayResult.inferences, withImageSize: CGSize(width: CGFloat(width), height: CGFloat(height)))
      }
    }
```

⑩ 编写方法 drawAfterPerformingCalculations()获取识别结果，将边界框矩形转换为当前视图，绘制边界框、类名和推断的置信度分数。对应的代码如下：

```
  func drawAfterPerformingCalculations(onInferences inferences: [Inference], withImageSize imageSize:CGSize) {
    self.overlayView.objectOverlays = []
    self.overlayView.setNeedsDisplay()
    guard !inferences.isEmpty else {
      return
    }
    var objectOverlays: [ObjectOverlay] = []
    for inference in inferences {
      //将边界框矩形转换为当前视图
      var convertedRect = inference.rect.applying(CGAffineTransform(scaleX: self.overlayView.bounds.size.width / imageSize.width, y: self.overlayView.bounds.size.height / imageSize.height))
      if convertedRect.origin.x < 0 {
        convertedRect.origin.x = self.edgeOffset
      }
      if convertedRect.origin.y < 0 {
        convertedRect.origin.y = self.edgeOffset
      }
      if convertedRect.maxY > self.overlayView.bounds.maxY {
        convertedRect.size.height = self.overlayView.bounds.maxY - convertedRect.origin.y - self.edgeOffset
      }
      if convertedRect.maxX > self.overlayView.bounds.maxX {
        convertedRect.size.width = self.overlayView.bounds.maxX - convertedRect.origin.x - self.edgeOffset
      }
```

```
        let confidenceValue = Int(inference.confidence * 100.0)
        let string = "\(inference.className)  (\(confidenceValue)%)"
        let size = string.size(usingFont: self.displayFont)
        let objectOverlay = ObjectOverlay(name: string, borderRect: convertedRect,
nameStringSize: size, color: inference.displayColor, font: self.displayFont)
        objectOverlays.append(objectOverlay)
    }
    //将绘图交给覆盖视图
    self.draw(objectOverlays: objectOverlays)
}
```

⑪ 编写方法 draw()，功能是使用检测到的边界框和类名更新覆盖视图，对应的代码如下：

```
func draw(objectOverlays: [ObjectOverlay]) {
    self.overlayView.objectOverlays = objectOverlays
    self.overlayView.setNeedsDisplay()
}
```

⑫ 编写方法 addPanGesture()，添加平移手势处理功能，以使底部选项具有交互性，对应的代码如下：

```
private func addPanGesture() {
    let panGesture = UIPanGestureRecognizer(target: self, action:
#selector(ViewController.didPan(panGesture:)))
    bottomSheetView.addGestureRecognizer(panGesture)
}
```

⑬ 编写方法 changeBottomViewState()，功能是更改底部选项应处于展开还是折叠状态，对应的代码如下：

```
private func changeBottomViewState() {
    guard let inferenceVC = inferenceViewController else {
        return
    }
    if bottomSheetViewBottomSpace.constant == inferenceVC.collapsedHeight -
bottomSheetView.bounds.size.height {
        bottomSheetViewBottomSpace.constant = 0.0
    }
    else {
        bottomSheetViewBottomSpace.constant = inferenceVC.collapsedHeight -
bottomSheetView.bounds.size.height
    }
    setImageBasedOnBottomViewState()
}
```

⑭ 编写方法 setImageBasedOnBottomViewState()，功能是根据底部选项图标是展开还

是折叠设置显示图像。对应的代码如下：

```
private func setImageBasedOnBottomViewState() {
  if bottomSheetViewBottomSpace.constant == 0.0 {
    bottomSheetStateImageView.image = UIImage(named: "down_icon")
  }
  else {
    bottomSheetStateImageView.image = UIImage(named: "up_icon")
  }
}
```

⑮ 编写方法 didPan()，响应用户在底部选项表上的平移操作，对应的代码如下：

```
@objc func didPan(panGesture: UIPanGestureRecognizer) {
  //根据用户与底部选项表的交互，打开或关闭底部工作表
  let translation = panGesture.translation(in: view)
  switch panGesture.state {
  case .began:
    initialBottomSpace = bottomSheetViewBottomSpace.constant
    translateBottomSheet(withVerticalTranslation: translation.y)
  case .changed:
    translateBottomSheet(withVerticalTranslation: translation.y)
  case .cancelled:
    setBottomSheetLayout(withBottomSpace: initialBottomSpace)
  case .ended:
    translateBottomSheetAtEndOfPan(withVerticalTranslation: translation.y)
    setImageBasedOnBottomViewState()
    initialBottomSpace = 0.0
  default:
    break
  }
}
```

⑯ 编写方法 translateBottomSheet()，在平移手势状态不断变化时设置底部选项平移，对应的代码如下：

```
private func translateBottomSheet(withVerticalTranslation verticalTranslation: CGFloat) {
  let bottomSpace = initialBottomSpace - verticalTranslation
  guard bottomSpace <= 0.0 && bottomSpace >= inferenceViewController!
      .collapsedHeight - bottomSheetView.bounds.size.height else {
    return
  }
  setBottomSheetLayout(withBottomSpace: bottomSpace)
}
```

⑰ 编写方法 translateBottomSheetAtEndOfPan()，功能是将底部选项状态更改为在平移结束时完全展开或闭合，对应的代码如下：

```
private func translateBottomSheetAtEndOfPan(withVerticalTranslation
verticalTranslation: CGFloat) {
  //将底部选项状态更改为在平移结束时完全打开或关闭
  let bottomSpace = bottomSpaceAtEndOfPan(withVerticalTranslation: verticalTranslation)
  setBottomSheetLayout(withBottomSpace: bottomSpace)
}
```

⑱ 编写方法 bottomSpaceAtEndOfpan()，功能是返回要保留的底部图纸视图的最终状态（完全折叠或展开），对应的代码如下：

```
private func bottomSpaceAtEndOfPan(withVerticalTranslation verticalTranslation:
CGFloat) -> CGFloat {
  //计算在平移手势结束时是完全展开还是折叠底部选项
  var bottomSpace = initialBottomSpace - verticalTranslation
  var height: CGFloat = 0.0
  if initialBottomSpace == 0.0 {
    height = bottomSheetView.bounds.size.height
  }
  else {
    height = inferenceViewController!.collapsedHeight
  }
  let currentHeight = bottomSheetView.bounds.size.height + bottomSpace
  if currentHeight - height <= collapseTransitionThreshold {
    bottomSpace = inferenceViewController!.collapsedHeight -
bottomSheetView.bounds.size.height
  }
  else if currentHeight - height >= expandTransitionThreshold {
    bottomSpace = 0.0
  }
  else {
    bottomSpace = initialBottomSpace
  }
  return bottomSpace
}
```

⑲ 编写方法 setBottomSheetLayout()，布局底部选项的底部空间相对于此控制器管理的视图的更改，对应的代码如下：

```
func setBottomSheetLayout(withBottomSpace bottomSpace: CGFloat) {
  view.setNeedsLayout()
  bottomSheetViewBottomSpace.constant = bottomSpace
  view.setNeedsLayout()
}
```

(2) 编写推断视图控制器文件 InferenceViewController.swift，具体实现流程如下。
① 创建继承于主视图类 UIViewController 的子类 InferenceViewController，在视图界面

中显示识别信息。对应的代码如下:

```swift
import UIKit
protocol InferenceViewControllerDelegate {
  /**
   当用户更改步进器值以更新用于推断的线程数时,将调用此方法
   */
  func didChangeThreadCount(to count: Int)
}
class InferenceViewController: UIViewController {
  //MARK: 要显示的信息
  private enum InferenceSections: Int, CaseIterable {
    case InferenceInfo
  }
  private enum InferenceInfo: Int, CaseIterable {
    case Resolution
    case Crop
    case InferenceTime
    func displayString() -> String {
      var toReturn = ""
      switch self {
      case .Resolution:
        toReturn = "Resolution"
      case .Crop:
        toReturn = "Crop"
      case .InferenceTime:
        toReturn = "Inference Time"
      }
      return toReturn
    }
  }
  //MARK: 故事板的Outlets输出
  @IBOutlet weak var tableView: UITableView!
  @IBOutlet weak var threadStepper: UIStepper!
  @IBOutlet weak var stepperValueLabel: UILabel!
  //MARK: 常量
  private let normalCellHeight: CGFloat = 27.0
  private let separatorCellHeight: CGFloat = 42.0
  private let bottomSpacing: CGFloat = 21.0
  private let minThreadCount = 1
  private let bottomSheetButtonDisplayHeight: CGFloat = 60.0
  private let infoTextColor = UIColor.black
  private let lightTextInfoColor = UIColor(displayP3Red: 117.0/255.0, green: 117.0/255.0, blue: 117.0/255.0, alpha: 1.0)
  private let infoFont = UIFont.systemFont(ofSize: 14.0, weight: .regular)
  private let highlightedFont = UIFont.systemFont(ofSize: 14.0, weight: .medium)
```

```swift
//MARK:实例变量
var inferenceTime: Double = 0
var wantedInputWidth: Int = 0
var wantedInputHeight: Int = 0
var resolution: CGSize = CGSize.zero
var threadCountLimit: Int = 0
var currentThreadCount: Int = 0
//MARK: 委托
var delegate: InferenceViewControllerDelegate?
//MARK: 计算属性
var collapsedHeight: CGFloat {
  return bottomSheetButtonDisplayHeight
}
override func viewDidLoad() {
  super.viewDidLoad()
  //设置步进器
  threadStepper.isUserInteractionEnabled = true
  threadStepper.maximumValue = Double(threadCountLimit)
  threadStepper.minimumValue = Double(minThreadCount)
  threadStepper.value = Double(currentThreadCount)
}
```

② 将线程数的更改委托给 View Controller 并更改显示效果，对应的代码如下：

```swift
@IBAction func onClickThreadStepper(_ sender: Any) {
  delegate?.didChangeThreadCount(to: Int(threadStepper.value))
  currentThreadCount = Int(threadStepper.value)
  stepperValueLabel.text = "\(currentThreadCount)"
  }
}
//MARK: UITableView 数据源
extension InferenceViewController: UITableViewDelegate, UITableViewDataSource {
  func numberOfSections(in tableView: UITableView) -> Int {
    return InferenceSections.allCases.count
  }
  func tableView(_ tableView: UITableView, numberOfRowsInSection section: Int) -> Int {
    guard let inferenceSection = InferenceSections(rawValue: section) else {
      return 0
    }
    var rowCount = 0
    switch inferenceSection {
    case .InferenceInfo:
      rowCount = InferenceInfo.allCases.count
    }
    return rowCount
  }
```

```
func tableView(_ tableView: UITableView, heightForRowAt indexPath: IndexPath) -> CGFloat {
    var height: CGFloat = 0.0
    guard let inferenceSection = InferenceSections(rawValue: indexPath.section) else {
        return height
    }
    switch inferenceSection {
    case .InferenceInfo:
        if indexPath.row == InferenceInfo.allCases.count - 1 {
            height = separatorCellHeight + bottomSpacing
        }
        else {
            height = normalCellHeight
        }
    }
    return height
}
```

③ 设置底部工作表中信息的显示格式，将格式化显示与推断相关的附加信息，对应的代码如下：

```
func displayStringsForInferenceInfo(atRow row: Int) -> (String, String) {
    var fieldName: String = ""
    var info: String = ""
    guard let inferenceInfo = InferenceInfo(rawValue: row) else {
        return (fieldName, info)
    }
    fieldName = inferenceInfo.displayString()
    switch inferenceInfo {
    case .Resolution:
        info = "\(Int(resolution.width))x\(Int(resolution.height))"
    case .Crop:
        info = "\(wantedInputWidth)x\(wantedInputHeight)"
    case .InferenceTime:
        info = String(format: "%.2fms", inferenceTime)
    }
    return (fieldName, info)
}
```

(3) 在View目录下编写文件CurvedView.swift，功能是创建一个CurvedView视图，它的左上角和右上角是圆形的，具体实现代码如下：

```
import UIKit
class CurvedView: UIView {
    let cornerRadius: CGFloat = 24.0
```

```
override func layoutSubviews() {
  super.layoutSubviews()
  setMask()
}
/**在视图上设置遮罩以使其拐角圆化
 */
func setMask() {
  let maskPath = UIBezierPath(roundedRect:self.bounds,
                    byRoundingCorners: [.topLeft, .topRight],
                    cornerRadii: CGSize(width: cornerRadius, height: cornerRadius))
  let shape = CAShapeLayer()
  shape.path = maskPath.cgPath
  self.layer.mask = shape
}
}
```

(4) 在 View 目录下编写文件 OverlayView.swift，功能是创建一个覆盖视图，这样可以在 UI 中显示识别结果的文字内容，具体实现代码如下：

```
import UIKit
/**
结构体 ObjectOverlay 保存在检测到的对象上用于绘制覆盖层的显示参数
 */
struct ObjectOverlay {
  let name: String
  let borderRect: CGRect
  let nameStringSize: CGSize
  let color: UIColor
  let font: UIFont
}
/**
此 UIView 在检测到的对象上绘制覆盖
 */
class OverlayView: UIView {
  var objectOverlays: [ObjectOverlay] = []
  private let cornerRadius: CGFloat = 10.0
  private let stringBgAlpha: CGFloat = 0.7
  private let lineWidth: CGFloat = 3
  private let stringFontColor = UIColor.white
  private let stringHorizontalSpacing: CGFloat = 17.0
  private let stringVerticalSpacing: CGFloat = 7.0
  override func draw(_ rect: CGRect) {
    //绘制代码
    for objectOverlay in objectOverlays {
      drawBorders(of: objectOverlay)
      drawBackground(of: objectOverlay)
```

```
    drawName(of: objectOverlay)
  }
}
/**
  此方法绘制检测到的对象的边界
 */
func drawBorders(of objectOverlay: ObjectOverlay) {
  let path = UIBezierPath(rect: objectOverlay.borderRect)
  path.lineWidth = lineWidth
  objectOverlay.color.setStroke()
  path.stroke()
}
/**
  此方法绘制字符串的背景
 */
func drawBackground(of objectOverlay: ObjectOverlay) {
  let stringBgRect = CGRect(x: objectOverlay.borderRect.origin.x, y: objectOverlay.borderRect.origin.y , width: 2 * stringHorizontalSpacing + objectOverlay.nameStringSize.width, height: 2 * stringVerticalSpacing + objectOverlay.nameStringSize.height
  )
  let stringBgPath = UIBezierPath(rect: stringBgRect)
  objectOverlay.color.withAlphaComponent(stringBgAlpha).setFill()
  stringBgPath.fill()
}
/**
  此方法绘制对象覆盖的名称
 */
func drawName(of objectOverlay: ObjectOverlay) {
  //绘制字符串
  let stringRect = CGRect(x: objectOverlay.borderRect.origin.x + stringHorizontalSpacing, y: objectOverlay.borderRect.origin.y + stringVerticalSpacing, width: objectOverlay.nameStringSize.width, height: objectOverlay.nameStringSize.height)

  let attributedString = NSAttributedString(string: objectOverlay.name, attributes: [NSAttributedString.Key.foregroundColor : stringFontColor, NSAttributedString.Key.font : objectOverlay.font])
  attributedString.draw(in: stringRect)
 }
}
```

### 7.6.3　相机处理

在 Xcode 工程的 Camera Feed 目录下保存实现相机功能的程序文件,会要求使用相机权限采集图像，然后会识别结果。

(1) 编写文件 PreviewView.swift，功能是显示相机采集到的画面的预览结果，具体实现代码如下：

```swift
import UIKit
import AVFoundation
/**
 相机帧将显示在此视图上
 */
class PreviewView: UIView {
  var previewLayer: AVCaptureVideoPreviewLayer {
    guard let layer = layer as? AVCaptureVideoPreviewLayer else {
      fatalError("Layer expected is of type VideoPreviewLayer")
    }
    return layer
  }
  var session: AVCaptureSession? {
    get {
      return previewLayer.session
    }
    set {
      previewLayer.session = newValue
    }
  }
  override class var layerClass: AnyClass {
    return AVCaptureVideoPreviewLayer.self
  }
}
```

(2) 编写文件 CameraFeedManager.swift 实现相机采集处理功能，具体实现流程如下。

① 创建枚举保存相机初始化的状态，对应的代码如下：

```swift
enum CameraConfiguration {
  case success
  case failed
  case permissionDenied
}
```

② 创建类 CameraFeedManager，用于管理所有与相机相关的功能，对应的代码如下：

```swift
class CameraFeedManager: NSObject {
  private let session: AVCaptureSession = AVCaptureSession()
  private let previewView: PreviewView
  private let sessionQueue = DispatchQueue(label: "sessionQueue")
  private var cameraConfiguration: CameraConfiguration = .failed
  private lazy var videoDataOutput = AVCaptureVideoDataOutput()
  private var isSessionRunning = false
```

```
//MARK：相机馈送管理器代理
weak var delegate: CameraFeedManagerDelegate?
//MARK：初始化
init(previewView: PreviewView) {
  self.previewView = previewView
  super.init()
  //初始化会话
  session.sessionPreset = .high
  self.previewView.session = session
  self.previewView.previewLayer.connection?.videoOrientation = .portrait
  self.previewView.previewLayer.videoGravity = .resizeAspectFill
  self.attemptToConfigureSession()
}
```

③ 编写方法 checkCameraConfigurationAndStartSession()，功能是根据相机配置是否成功启动 AVCaptureSession，对应的代码如下：

```
func checkCameraConfigurationAndStartSession() {
  sessionQueue.async {
    switch self.cameraConfiguration {
    case .success:
      self.addObservers()
      self.startSession()
    case .failed:
      DispatchQueue.main.async {
        self.delegate?.presentVideoConfigurationErrorAlert()
      }
    case .permissionDenied:
      DispatchQueue.main.async {
        self.delegate?.presentCameraPermissionsDeniedAlert()
      }
    }
  }
}
```

④ 编写方法 stopSession()，停止运行 AVCaptureSession，对应的代码如下：

```
func stopSession() {
  self.removeObservers()
  sessionQueue.async {
    if self.session.isRunning {
      self.session.stopRunning()
      self.isSessionRunning = self.session.isRunning
    }
  }
}
```

⑤ 编写方法 resumeInterruptedSession()，恢复中断的 AVCaptureSession，对应的代码

如下：

```
func resumeInterruptedSession(withCompletion completion: @escaping (Bool) -> ()) {
  sessionQueue.async {
    self.startSession()
    DispatchQueue.main.async {
      completion(self.isSessionRunning)
    }
  }
}
```

⑥ 编写方法 startSession()，启动 AVCaptureSession，对应的代码如下：

```
private func startSession() {
  self.session.startRunning()
  self.isSessionRunning = self.session.isRunning
}
```

⑦ 编写方法 attemptToConfigureSession()，请求相机的权限，处理请求会话配置并存储配置结果，对应的代码如下：

```
private func attemptToConfigureSession() {
  switch AVCaptureDevice.authorizationStatus(for: .video) {
  case .authorized:
    self.cameraConfiguration = .success
  case .notDetermined:
    self.sessionQueue.suspend()
    self.requestCameraAccess(completion: { (granted) in
      self.sessionQueue.resume()
    })
  case .denied:
    self.cameraConfiguration = .permissionDenied
  default:
    break
  }
  self.sessionQueue.async {
    self.configureSession()
  }
}
```

⑧ 编写方法 requestCameraAccess()，请求获取相机权限，对应的代码如下：

```
private func requestCameraAccess(completion: @escaping (Bool) -> ()) {
  AVCaptureDevice.requestAccess(for: .video) { (granted) in
    if !granted {
      self.cameraConfiguration = .permissionDenied
    }
    else {
```

```
        self.cameraConfiguration = .success
    }
    completion(granted)
  }
}
```

⑨ 编写方法 configureSession()，处理配置 AVCaptureSession 的所有步骤，对应的代码如下：

```
private func configureSession() {
  guard cameraConfiguration == .success else {
    return
  }
  session.beginConfiguration()
  //尝试添加 AVCaptureDeviceInput
  guard addVideoDeviceInput() == true else {
    self.session.commitConfiguration()
    self.cameraConfiguration = .failed
    return
  }
  //尝试添加 AVCaptureVideoDataOutput
  guard addVideoDataOutput() else {
    self.session.commitConfiguration()
    self.cameraConfiguration = .failed
    return
  }
  session.commitConfiguration()
  self.cameraConfiguration = .success
}
```

⑩ 编写方法 addVideoDeviceInput()，功能是尝试将 AVCaptureDeviceInput 添加到当前 AVCaptureSession，对应的代码如下：

```
private func addVideoDeviceInput() -> Bool {
  /**尝试获取默认的后置摄像头
   */
  guard let camera = AVCaptureDevice.default(.builtInWideAngleCamera,
for: .video, position: .back) else {
    fatalError("Cannot find camera")
  }
  do {
    let videoDeviceInput = try AVCaptureDeviceInput(device: camera)
    if session.canAddInput(videoDeviceInput) {
      session.addInput(videoDeviceInput)
      return true
    }
    else {
      return false
```

```
    }
  }
  catch {
    fatalError("Cannot create video device input")
  }
}
```

⑪ 编写方法 addVideoDataOutput()，将 AVCaptureVideoDataOutput 添加到当前 AVCaptureSession，对应的代码如下：

```
private func addVideoDataOutput() -> Bool {
  let sampleBufferQueue = DispatchQueue(label: "sampleBufferQueue")
  videoDataOutput.setSampleBufferDelegate(self, queue: sampleBufferQueue)
  videoDataOutput.alwaysDiscardsLateVideoFrames = true
  videoDataOutput.videoSettings = [ String(kCVPixelBufferPixelFormatTypeKey) : kCMPixelFormat_32BGRA]
  if session.canAddOutput(videoDataOutput) {
    session.addOutput(videoDataOutput)
    videoDataOutput.connection(with: .video)?.videoOrientation = .portrait
    return true
  }
  return false
}
```

⑫ 编写用于通知 Observers 观察处理器的方法 addObservers()，对应的代码如下：

```
private func addObservers() {
  NotificationCenter.default.addObserver(self, selector: #selector(CameraFeedManager.sessionRuntimeErrorOccurred(notification:)), name: NSNotification.Name.AVCaptureSessionRuntimeError, object: session)
  NotificationCenter.default.addObserver(self, selector: #selector(CameraFeedManager.sessionWasInterrupted(notification:)), name: NSNotification.Name.AVCaptureSessionWasInterrupted, object: session)
  NotificationCenter.default.addObserver(self, selector: #selector(CameraFeedManager.sessionInterruptionEnded), name: NSNotification.Name.AVCaptureSessionInterruptionEnded, object: session)
}
private func removeObservers() {
  NotificationCenter.default.removeObserver(self, name: NSNotification.Name.AVCaptureSessionRuntimeError, object: session)
  NotificationCenter.default.removeObserver(self, name: NSNotification.Name.AVCaptureSessionWasInterrupted, object: session)
  NotificationCenter.default.removeObserver(self, name: NSNotification.Name.AVCaptureSessionInterruptionEnded, object: session)
}
//MARK: 通知 Observers
@objc func sessionWasInterrupted(notification: Notification) {
```

```
        if let userInfoValue = notification.userInfo?[AVCaptureSessionInterruptionReasonKey]
as AnyObject?,
            let reasonIntegerValue = userInfoValue.integerValue,
            let reason = AVCaptureSession.InterruptionReason(rawValue: reasonIntegerValue) {
            print("Capture session was interrupted with reason \(reason)")
            var canResumeManually = false
            if reason == .videoDeviceInUseByAnotherClient {
                canResumeManually = true
            } else if reason == .videoDeviceNotAvailableWithMultipleForegroundApps {
                canResumeManually = false
            }
            self.delegate?.sessionWasInterrupted(canResumeManually: canResumeManually)
        }
    }

    @objc func sessionInterruptionEnded(notification: Notification) {
        self.delegate?.sessionInterruptionEnded()
    }

    @objc func sessionRuntimeErrorOccurred(notification: Notification) {
        guard let error = notification.userInfo?[AVCaptureSessionErrorKey] as? AVError else {
            return
        }
        print("Capture session runtime error: \(error)")
        if error.code == .mediaServicesWereReset {
            sessionQueue.async {
                if self.isSessionRunning {
                    self.startSession()
                } else {
                    DispatchQueue.main.async {
                        self.delegate?.sessionRunTimeErrorOccurred()
                    }
                }
            }
        } else {
            self.delegate?.sessionRunTimeErrorOccurred()
        }
    }
}
```

⑬ 创建扩展 CameraFeedManager，用于将 AVCapture 视频数据输出样本缓冲区的处理委托给相应的处理方法，对应的代码如下：

```
extension CameraFeedManager: AVCaptureVideoDataOutputSampleBufferDelegate {
    func captureOutput(_ output: AVCaptureOutput, didOutput sampleBuffer:
CMSampleBuffer, from connection: AVCaptureConnection) {
```

```
  //将CMSampleBuffer 转换为CVPixelBuffer
  let pixelBuffer: CVPixelBuffer? = CMSampleBufferGetImageBuffer(sampleBuffer)

  guard let imagePixelBuffer = pixelBuffer else {
    return
  }
  //将像素缓冲区委托给ViewController
  delegate?.didOutput(pixelBuffer: imagePixelBuffer)
  }
}
```

## 7.6.4 处理 TensorFlow Lite 模型

在 Xcode 工程的 ModelDataHandler 目录下编写文件 ModelDataHandler.swift，用于使用 TensorFlow Lite 模型实现物体检测识别功能，具体实现流程如下。

(1) 定义结构体 Result，存储通过 Interpreter 成功实现物体识别的结果，对应的代码如下：

```
struct Result {
  let inferenceTime: Double
  let inferences: [Inference]
}
```

(2) 使用 Inference 存储一个格式化的推断，对应的代码如下：

```
struct Inference {
  let confidence: Float
  let className: String
  let rect: CGRect
  let displayColor: UIColor
}
///有关模型文件或标签文件的信息
typealias FileInfo = (name: String, extension: String)
```

(3) 通过枚举 MobileNetSSD 存储有关 MobileNet SSD 型号的信息，对应的代码如下：

```
enum MobileNetSSD {
  static let modelInfo: FileInfo = (name: "detect", extension: "tflite")
  static let labelsInfo: FileInfo = (name: "labelmap", extension: "txt")
}
```

(4) 定义类 ModelDataHandler，处理所有的预处理数据，并通过调用 Interpreter 在给定帧上运行推断。然后格式化获得的推断结果，并返回成功推断中的前 N 个结果。对应的代码如下：

```
class ModelDataHandler: NSObject {
  //MARK：内部属性
```

```
///TensorFlow Lite 解释器使用的当前线程计数
let threadCount: Int
let threadCountLimit = 10
let threshold: Float = 0.5
//MARK: 模型参数
let batchSize = 1
let inputChannels = 3
let inputWidth = 300
let inputHeight = 300
//浮动模型的图像平均值和标准差应与模型训练中使用的参数一致
let imageMean: Float = 127.5
let imageStd:  Float = 127.5

//MARK: 私有属性
private var labels: [String] = []

///TensorFlow Lite 的 Interpreter 对象,用于对给定模型执行推理
private var interpreter: Interpreter

private let bgraPixel = (channels: 4, alphaComponent: 3, lastBgrComponent: 2)
private let rgbPixelChannels = 3
private let colorStrideValue = 10
private let colors = [
  UIColor.red,
  UIColor(displayP3Red: 90.0/255.0, green: 200.0/255.0, blue: 250.0/255.0, alpha: 1.0),
  UIColor.green,
  UIColor.orange,
  UIColor.blue,
  UIColor.purple,
  UIColor.magenta,
  UIColor.yellow,
  UIColor.cyan,
  UIColor.brown
]
```

(5) 编写方法 init?(),实现初始化操作,设置 ModelDataHandler 的可失败初始值设定项,如果从应用程序的主捆绑包成功加载模型和标签文件,则会创建一个新实例,默认的 threadCount 值为 1,对应的代码如下:

```
init?(modelFileInfo: FileInfo, labelsFileInfo: FileInfo, threadCount: Int = 1) {
  let modelFilename = modelFileInfo.name
  //构造模型文件的路径
  guard let modelPath = Bundle.main.path(
    forResource: modelFilename,
    ofType: modelFileInfo.extension
  ) else {
    print("Failed to load the model file with name: \(modelFilename).")
```

```swift
    return nil
  }
  //指定该选项的 Interpreter 选项
  self.threadCount = threadCount
  var options = Interpreter.Options()
  options.threadCount = threadCount
  do {
    //创建 Interpreter
    interpreter = try Interpreter(modelPath: modelPath, options: options)
    //为模型输入 Tensor 的分配内存
    try interpreter.allocateTensors()
  } catch let error {
    print("Failed to create the interpreter with error: \(error.localizedDescription)")
    return nil
  }
  super.init()
  //加载标签文件中列出的类
  loadLabels(fileInfo: labelsFileInfo)
}
```

(6) 编写方法 runModel()，处理所有的预处理数据，并通过 Interpreter 调用在指定的帧上运行推断。然后格式化处理推断结果，并返回成功推断中的前 N 个结果。对应的代码如下：

```swift
func runModel(onFrame pixelBuffer: CVPixelBuffer) -> Result? {
  let imageWidth = CVPixelBufferGetWidth(pixelBuffer)
  let imageHeight = CVPixelBufferGetHeight(pixelBuffer)
  let sourcePixelFormat = CVPixelBufferGetPixelFormatType(pixelBuffer)
  assert(sourcePixelFormat == kCVPixelFormatType_32ARGB ||
         sourcePixelFormat == kCVPixelFormatType_32BGRA ||
         sourcePixelFormat == kCVPixelFormatType_32RGBA)
  let imageChannels = 4
  assert(imageChannels >= inputChannels)

  //将图像裁剪到中心最大的正方形，并将其缩小到模型尺寸
  let scaledSize = CGSize(width: inputWidth, height: inputHeight)
  guard let scaledPixelBuffer = pixelBuffer.resized(to: scaledSize) else {
    return nil
  }
  let interval: TimeInterval
  let outputBoundingBox: Tensor
  let outputClasses: Tensor
  let outputScores: Tensor
  let outputCount: Tensor
  do {
    let inputTensor = try interpreter.input(at: 0)
    //从图像缓冲区中删除 alpha 组件以获取 RGB 数据
    guard let rgbData = rgbDataFromBuffer(
```

```
    scaledPixelBuffer,
    byteCount: batchSize * inputWidth * inputHeight * inputChannels,
    isModelQuantized: inputTensor.dataType == .uInt8
  ) else {
    print("Failed to convert the image buffer to RGB data.")
    return nil
  }
  //将 RGB 数据复制到输入张量
  try interpreter.copy(rgbData, toInputAt: 0)
  //调用 Interpreter
  let startDate = Date()
  try interpreter.invoke()
  interval = Date().timeIntervalSince(startDate) * 1000
  outputBoundingBox = try interpreter.output(at: 0)
  outputClasses = try interpreter.output(at: 1)
  outputScores = try interpreter.output(at: 2)
  outputCount = try interpreter.output(at: 3)
} catch let error {
  print("Failed to invoke the interpreter with error: \(error.localizedDescription)")
  return nil
}
//格式化结果
let resultArray = formatResults(
  boundingBox: [Float](unsafeData: outputBoundingBox.data) ?? [],
  outputClasses: [Float](unsafeData: outputClasses.data) ?? [],
  outputScores: [Float](unsafeData: outputScores.data) ?? [],
  outputCount: Int(([Float](unsafeData: outputCount.data) ?? [0])[0]),
  width: CGFloat(imageWidth),
  height: CGFloat(imageHeight)
)
//返回推断时间和推断结果
let result = Result(inferenceTime: interval, inferences: resultArray)
return result
}
```

(7) 编写方法 formatResults()，筛选出置信度"得分<阈值"的所有结果，并返回按降序排列的前 N 个结果，对应的代码如下：

```
func formatResults(boundingBox: [Float], outputClasses: [Float], outputScores:
[Float], outputCount: Int, width: CGFloat, height: CGFloat) -> [Inference]{
  var resultsArray: [Inference] = []
  if (outputCount == 0) {
    return resultsArray
  }
  for i in 0...outputCount - 1 {

    let score = outputScores[i]
```

```swift
  //筛选 confidence < threshold 的结果
  guard score >= threshold else {
    continue
  }
  //从标签列表中获取检测到的类别名称
  let outputClassIndex = Int(outputClasses[i])
  let outputClass = labels[outputClassIndex + 1]
  var rect: CGRect = CGRect.zero
  //将检测到的边界框转换为CGRect
  rect.origin.y = CGFloat(boundingBox[4*i])
  rect.origin.x = CGFloat(boundingBox[4*i+1])
  rect.size.height = CGFloat(boundingBox[4*i+2]) - rect.origin.y
  rect.size.width = CGFloat(boundingBox[4*i+3]) - rect.origin.x
  //检测到的角点用于模型尺寸,所以根据实际的图像尺寸来缩放 rect
  let newRect = rect.applying(CGAffineTransform(scaleX: width, y: height))
  //获取为类指定的颜色
  let colorToAssign = colorForClass(withIndex: outputClassIndex + 1)
  let inference = Inference(confidence: score,
               className: outputClass,
               rect: newRect,
               displayColor: colorToAssign)
  resultsArray.append(inference)
}
//排序结果按可信度的降序排列
resultsArray.sort { (first, second) -> Bool in
  return first.confidence > second.confidence
}
return resultsArray
}
```

(8) 编写方法 loadLabels(),加载标签,并将其存储在 labels 属性中,对应的代码如下:

```swift
private func loadLabels(fileInfo: FileInfo) {
  let filename = fileInfo.name
  let fileExtension = fileInfo.extension
  guard let fileURL = Bundle.main.url(forResource: filename, withExtension: fileExtension) else {
    fatalError("Labels file not found in bundle. Please add a labels file with name " +
          "\(filename).\(fileExtension) and try again.")
  }
  do {
    let contents = try String(contentsOf: fileURL, encoding: .utf8)
    labels = contents.components(separatedBy: .newlines)
  } catch {
    fatalError("Labels file named \(filename).\(fileExtension) cannot be read. Please add a " + "valid labels file and try again.")
  }
}
```

(9) 编写方法 rgbDataFromBuffer()，返回具有指定值的给定图像缓冲区的 RGB 数据表示形式，各个参数的说明如下。

- buffer：用于转换为 RGB 数据的 BGRA 像素缓冲区。
- byteCount：使用模型的训练内容为 batchSize*imageWidth*imageHeight*ComponentScont。
- isModelQuantized：模型是否量化(即固定点值，而非浮点值)。

返回值是图像缓冲区的 RGB 数据表示形式，如果无法创建缓冲区，则返回 nil。方法 rgbDataFromBuffer() 的实现代码如下：

```swift
private func rgbDataFromBuffer(
  _ buffer: CVPixelBuffer,
  byteCount: Int,
  isModelQuantized: Bool
) -> Data? {
  CVPixelBufferLockBaseAddress(buffer, .readOnly)
  defer {
    CVPixelBufferUnlockBaseAddress(buffer, .readOnly)
  }
  guard let sourceData = CVPixelBufferGetBaseAddress(buffer) else {
    return nil
  }
  let width = CVPixelBufferGetWidth(buffer)
  let height = CVPixelBufferGetHeight(buffer)
  let sourceBytesPerRow = CVPixelBufferGetBytesPerRow(buffer)
  let destinationChannelCount = 3
  let destinationBytesPerRow = destinationChannelCount * width
  var sourceBuffer = vImage_Buffer(data: sourceData,
                        height: vImagePixelCount(height),
                        width: vImagePixelCount(width),
                        rowBytes: sourceBytesPerRow)
  guard let destinationData = malloc(height * destinationBytesPerRow) else {
    print("Error: out of memory")
    return nil
  }
  defer {
    free(destinationData)
  }
  var destinationBuffer = vImage_Buffer(data: destinationData,
                            height: vImagePixelCount(height),
                            width: vImagePixelCount(width),
                            rowBytes: destinationBytesPerRow)
  if (CVPixelBufferGetPixelFormatType(buffer) == kCVPixelFormatType_32BGRA){
    vImageConvert_BGRA8888toRGB888(&sourceBuffer, &destinationBuffer, UInt32(kvImageNoFlags))
  } else if (CVPixelBufferGetPixelFormatType(buffer) == kCVPixelFormatType_32ARGB) {
```

```
    vImageConvert_ARGB8888toRGB888(&sourceBuffer, &destinationBuffer,
UInt32(kvImageNoFlags))
  }
  let byteData = Data(bytes: destinationBuffer.data, count:
destinationBuffer.rowBytes * height)
  if isModelQuantized {
    return byteData
  }
  //未量化,转换为浮点数
  let bytes = Array<UInt8>(unsafeData: byteData)!
  var floats = [Float]()
  for i in 0..<bytes.count {
    floats.append((Float(bytes[i]) - imageMean) / imageStd)
  }
  return Data(copyingBufferOf: floats)
}
```

(10) 编写方法 colorForClass(),为特定类指定颜色,对应的代码如下:

```
  private func colorForClass(withIndex index: Int) -> UIColor {
    //有一组颜色,会根据每个对象的索引为每个对象指定基础颜色的变化
    let baseColor = colors[index % colors.count]
    var colorToAssign = baseColor
    let percentage = CGFloat((colorStrideValue / 2 - index / colors.count) *
colorStrideValue)
    if let modifiedColor = baseColor.getModified(byPercentage: percentage) {
      colorToAssign = modifiedColor
    }
    return colorToAssign
  }
}
```

(11) 创建扩展 Data,用于通过给定数组的缓冲区指针生成新的缓冲区,对应的代码如下:

```
extension Data {
  init<T>(copyingBufferOf array: [T]) {
    self = array.withUnsafeBufferPointer(Data.init)
  }
}
```

(12) 创建扩展 Array,功能是根据指定不安全数据的字节创建新的数组,对应的代码如下:

```
extension Array {
  init?(unsafeData: Data) {
    guard unsafeData.count % MemoryLayout<Element>.stride == 0 else { return nil }
    #if swift(>=5.0)
    self = unsafeData.withUnsafeBytes { .init($0.bindMemory(to: Element.self)) }
    #else
```

```
    self = unsafeData.withUnsafeBytes {
      .init(UnsafeBufferPointer<Element>(
        start: $0,
        count: unsafeData.count / MemoryLayout<Element>.stride
      ))
    }
    #endif // swift(>=5.0)
  }
}
```

另外，在 Xcode 工程的 Cells 目录下编写文件 InfoCell.swift，功能是以单元格形式显示识别结果列表，具体实现代码如下：

```
import UIKit
class InfoCell: UITableViewCell {
  @IBOutlet weak var fieldNameLabel: UILabel!
  @IBOutlet weak var infoLabel: UILabel!
}
```

## 7.7 调试运行

无论是在 Android 机器人设备，还是在 iOS 机器人设备中，运行后都可以实时显示自带相机中物体的识别结果，执行效果如图 7-6 所示。

扫码看视频

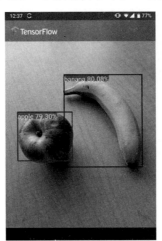

图 7-6　执行效果

# 第8章

# AI 考勤管理系统

现在已经进入了一个人工智能(AI)飞速发展的时代,在商业办公领域,考勤打卡应用已经实现了无纸化处理。在本章的内容中,将详细介绍开发一个 AI 考勤打卡系统的过程,以及使用 Face-recognition、Matplotlib、Django、Scikit-learn、Dlib 等技术开发一个大型人工智能项目的方法。

## 8.1 背景介绍

随着企业人事管理的日趋复杂和企业人员的增多，企业的考勤管理变得越来越复杂，拥有一个比较完善的考勤管理系统变得非常重要。在这个科学技术高速发展的时代，智能化的考勤系统在很多场所被广泛使用，例如企业、学校、事业单位等办公场所。而且智能化考勤的方式也很多，尤其是基于生物识别方式的考勤系统更是成为研究的热点。在各种各样的通过生物识别技术来进行考勤管理的方式中，人脸识别这种生物识别方式脱颖而出，其原因在于它在使用与操作上更加方便快捷，并且就系统本身而言更安全准确，广泛被用户接受。

扫码看视频

## 8.2 系统介绍

考勤管理系统是使用计算机管理方式代替以前手工处理的工作，应用计算机技术和通信技术建立一个高效率的、无差错的考勤管理系统，能够有效地帮助企业实现"公正考勤，高效薪资"，使企业的管理水平登上一个新的台阶。企业职工考勤管理系统可用于各种机构的职工考勤管理、查询、更新与维护，使用方便，易用性强，图形界面清晰明了。解决目前员工出勤管理问题，实现员工出勤信息和缺勤信息对企业领导透明，使管理人员及时把握员工的情况，及时与员工沟通，提高生产效率。

扫码看视频

在现实应用中，企业建设 AI 考勤管理系统具有以下几个方面的意义。

1) 提升企业员工管理工作质量

企业以往采用的人工考勤方式主要依靠人力资源部门进行考勤，在工作过程中存在着比较严重的主观性以及其他人为因素，员工考勤管理效率低下，考勤结果可靠性差，无法起到对员工进行约束的作用。而通过本系统的开发，企业的员工考勤系统实现完全信息化与自动化，不仅大大简化了学校的员工考勤流程，同时对于员工的考勤管理工作质量的提高也有着重要的意义。

2) 极大节约了企业日常管理的成本

随着信息化进程的日益推进，企业建立了完善的内部网，如何有效利用企业现有的网络资源，提高企业的日常管理工作，不仅可以为企业节省大量的管理费用开支，同时也可以提高企业的管理水平。

3) 提高了企业的办公质量，促进企业良好文化精神的形成

在采用考勤系统之后，由于人脸具有唯一性的特点，可以从本质上杜绝员工代打卡或

代签到的现象。另外，对于员工迟到、早退、旷工等情况也可以做到准确记录与统计，同时还可以有效节约企业的资源。因此，本系统的实施可以在很大程度上提高企业的办公质量，并对企业文化的改善有着重要的帮助作用。

## 8.3 系统需求分析

需求分析是介于系统分析和软件设计阶段之间的桥梁，好的需求分析是项目成功的基石。一方面，需求分析以系统规格说明和项目规划作为分析活动的基本出发点，并从软件角度对它们进行检查与调整；另一方面，需求分析又是软件设计、实现、测试直至维护的主要基础。良好的分析活动有助于避免或尽早剔除错误，从而提高软件生产率，降低开发成本，改进软件质量。

扫码看视频

### 8.3.1 可行性分析

考勤管理是企业管理中非常重要的一环，作为公司主管考勤的人员能够通过考勤管理系统清楚地看到公司员工的签到时间、签离时间，以及是否迟到、早退等诸多信息，还能够通过所有员工出勤记录的比较来发现企业管理和员工作业方面的诸多问题，更是员工工资及福利待遇方面重要的参考依据。

### 8.3.2 系统操作流程分析

(1) 职工用户登录系统，上下班时进行签到考勤，系统验证通过后该员工签到成功。

(2) 管理用户登录系统，输入用户名和密码，系统进行验证，验证通过进入程序主界面，在主界面对普通用户的信息进行录入，使用摄像头采集员工的人脸信息，然后通过机器学习技术创建学习模型。

### 8.3.3 系统模块设计

1) 登录验证模块

通过登录表单登录系统，整个系统分为管理员用户和普通用户。

2) 考勤打卡

普通用户登录系统后，可以分别实现在线上班打卡签到和下班打卡签退功能。

3) 添加新用户信息

管理员用户可以在后台添加新的员工信息，分别添加新员工的用户名和密码。

4)采集照片和人脸信息

管理员用户可以在后台采集员工的照片,输入用户名,然后通过摄像头采集员工的人脸信息。

5)训练照片模型

使用机器学习技术训练采集到的员工照片,供员工打卡签到使用。

6)考勤统计管理

使用可视化工具绘制员工的考勤数据,使用折线图统计最近两周每天到场的员工人数。

本项目的功能模块如图 8-1 所示。

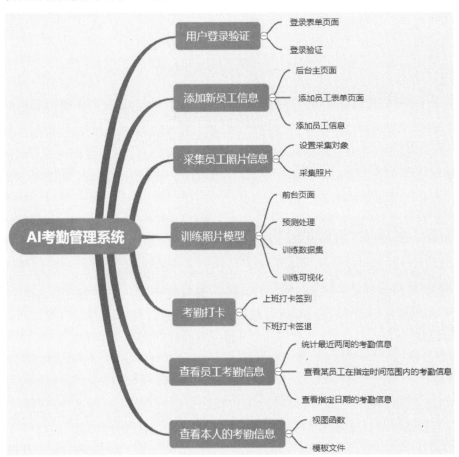

图 8-1 功能模块

## 8.4 系统配置

本系统是使用库 Django 实现的 Web 项目，在创建 Django Web 后会自动生成配置文件，开发者需要根据项目的要求设置这些配置文件。

### 8.4.1 Django 配置文件

扫码看视频

文件 settings.py 是 Django 项目的配置文件，主要用于设置整个 Django 项目所用到的程序文件和配置信息。在本项目中，需要设置 SQLite3 数据库的名字 db.sqlite3，并分别设置系统主页、登录页面和登录成功页面的 URL。文件 settings.py 的主要实现代码如下：

```
DATABASES = {
    'default': {
        'ENGINE': 'django.db.backends.sqlite3',
        'NAME': os.path.join(BASE_DIR, 'db.sqlite3'),
    }
}

STATIC_URL = '/static/'
CRISPY_TEMPLATE_PACK = 'bootstrap4'
LOGIN_URL='login'
LOGOUT_REDIRECT_URL = 'home'

LOGIN_REDIRECT_URL='dashboard'
```

### 8.4.2 路径导航文件

在 Django Web 项目中会自动创建路径导航文件 urls.py，在里面设置整个 Web 中所有页面对应的视图模块。本实例文件 urls.py 的主要实现代码如下：

```
urlpatterns = [
    path('admin/', admin.site.urls),
    path('', recog_views.home, name='home'),

    path('dashboard/', recog_views.dashboard, name='dashboard'),
    path('train/', recog_views.train, name='train'),
    path('add_photos/', recog_views.add_photos, name='add-photos'),

    path('login/',auth_views.LoginView.as_view(template_name='users/login.html'),
        name='login'),
```

```python
    path('logout/',auth_views.LogoutView.as_view(template_name='recognition/
        home.html'),name='logout'),
    path('register/', users_views.register, name='register'),
    path('mark_your_attendance', recog_views.mark_your_attendance,
        name='mark-your-attendance'),
    path('mark_your_attendance_out', recog_views.mark_your_attendance_out,
        name='mark-your-attendance-out'),
    path('view_attendance_home', recog_views.view_attendance_home,
        name='view-attendance-home'),
    path('view_attendance_date', recog_views.view_attendance_date,
        name='view-attendance-date'),
    path('view_attendance_employee', recog_views.view_attendance_employee,
        name='view-attendance-employee'),
    path('view_my_attendance', recog_views.view_my_attendance_employee_login,
        name='view-my-attendance-employee-login'),
    path('not_authorised', recog_views.not_authorised, name='not-authorised')
]
```

### 8.4.3 设计数据模型

在 Django Web 项目中，使用模型文件 models.py 设计项目中需要的数据库结构。因为本项目使用 django.contrib.auth 模块实现登录验证功能，该模块提供了完整的用户数据库表方案，所以在文件 models.py 中无须为会员用户设计数据库结构。模型文件 models.py 的主要实现代码如下：

```python
from django.db import models
from django.contrib.auth.models import User

import datetime

class Present(models.Model):
    user=models.ForeignKey(User,on_delete=models.CASCADE)
    date = models.DateField(default=datetime.date.today)
    present=models.BooleanField(default=False)

class Time(models.Model):
    user=models.ForeignKey(User,on_delete=models.CASCADE)
    date = models.DateField(default=datetime.date.today)
    time=models.DateTimeField(null=True,blank=True)
    out=models.BooleanField(default=False)
```

通过上述代码设计了两个数据库表。

- Present：保存当前的打卡信息；
- Time：保存打卡的时间信息。

## 8.5 用户登录验证

为了提高开发效率，本项目使用库 Django 中的 django.contrib.auth 模块实现用户注册和登录验证功能。这样做的好处是减少代码编写量，节省开发时间。在本节的内容中，主要讲解用户登录验证模块的实现过程。

扫码看视频

### 8.5.1 登录表单页面

从文件 urls.py 中的如下代码可知，用户登录页面对应的模板文件是 login.html，此文件提供了用户登录表单，调用 django.contrib.auth 模块验证表单中的数据是否合法。

```
path('login/',auth_views.LoginView.as_view(template_name='users/login.html'),
name='login'),
```

文件 login.html 的主要实现代码如下：

```
{% load static %}
{% load crispy_forms_tags %}
######此处省略部分代码
  <a href="{% url 'home' %}"><h5 class="text-left">系统主页</h5></a>
</div>

<div class="col-lg-4" style="background: rgba(0,0,0,0.6);margin-top:300px ;
padding-top:1em;padding-bottom:3em;color:#fff;border-radius:10px;-webkit-box-
shadow: 2px 2px 15px 0px rgba(0, 3, 0, 0.7);
-moz-box-shadow:    2px 2px 15px 0px rgba(0, 3, 0, 0.7);
box-shadow:         2px 2px 15px 0px rgba(0, 3, 0, 0.7); margin-left:auto;
margin-right: auto; ">
  <form method="POST" >
    {% csrf_token %}
    <fieldset class="form-group">
      <legend class="border-bottom mb-4"> 登录系统 </legend>
      {{form| crispy}}
    </fieldset>
    <div class="form-group">
      <button class="btn btn-outline-info" type="submit">登录</button>
    </div>
  </form>
</div>
```

用户登录验证表单页面的执行效果如图 8-2 所示。

图 8-2　用户登录验证表单页面

## 8.5.2　登录验证

前面曾经说过，开发者无须编写自定义验证方法，只需调用 django.contrib.auth 模块验证登录表单中的数据是否合法即可。在 LoginView.as_view 中提供了内置的验证方法，其中登录验证方法 login() 的主要实现代码如下：

```python
def login(self, request, extra_context=None):
    if request.method == 'GET' and self.has_permission(request):
        index_path = reverse('admin:index', current_app=self.name)
        return HttpResponseRedirect(index_path)

    from django.contrib.auth.views import LoginView
    from django.contrib.admin.forms import AdminAuthenticationForm
    context = {
        **self.each_context(request),
        'title': _('Log in'),
        'app_path': request.get_full_path(),
        'username': request.user.get_username(),
    }
    if (REDIRECT_FIELD_NAME not in request.GET and
            REDIRECT_FIELD_NAME not in request.POST):
        context[REDIRECT_FIELD_NAME] = reverse('admin:index', current_app=self.name)
    context.update(extra_context or {})

    defaults = {
        'extra_context': context,
        'authentication_form': self.login_form or AdminAuthenticationForm,
        'template_name': self.login_template or 'admin/login.html',
    }
    request.current_app = self.name
    return LoginView.as_view(**defaults)(request)
```

## 8.6 添加新员工信息

管理员用户成功登录系统后会来到后台主页 http://127.0.0.1:8000/dashboard/，页面效果如图 8-3 所示。

扫码看视频

图 8-3 后台主页

在管理员登录系统后，就可以向系统中添加新员工的信息了。在本节的内容中，将向读者讲解添加新员工信息的实现过程。

### 8.6.1 后台主页面

本项目后台主页面的模板文件是 recognition/templates/recognition/admin_dashboard.html，主要实现代码如下：

```
<script type='text/javascript'
src='https://ajax.googleapis.com/ajax/libs/jquery/2.1.1/jquery.min.js'>
$(document).ready(function(){
  $("#train").click(function(){
    alert("开始训练,这可能需要 5 分钟,请您稍等...");
  });
});
</script>

</head>
<body>
    <div class="col-sm-12">
      <a href="{% url 'logout' %}"><h5 class="text-right"><i class="fa fa-user-circle"
            aria-hidden="true"></i> 退出登录</h5></a>
      <h1 class="text-center section-title" style="margin-bottom:2em">管理员您好! </h1>
```

293

```
        </div>
        <div class="row">
            <div class="col-md-3">
            <a href="{%url 'register' %}"><img src="{% static 'recognition/img/
                register.png' %}" class="img-responsive" style="width:300px;
                height:300px ;" /></a>

            </div>
            <div class="col-md-3">
            <a href="{%url 'add-photos'%}"><img src="{% static 'recognition/img/
                addphotos.png' %}" class="img-responsive" style="width:300px;
                height:300px ;" /></a>
            </div>
            <div class="col-md-3">
            <a href="{%url 'train' %}" id="train" ><img src="{% static 'recognition/img/
                train.jpeg' %}" class="img-responsive" style="width:300px;
                height:300px ;" /></a>
            </div>
            <div class="col-md-3">
            <a href="{%url 'view-attendance-home' %}"><img src="{% static 'recognition/
                img/reports.png' %}" class="img-responsive" style="width:300px;
                height:300px"/></a>
            </div>
        </div>
        <div class="col">
            <div class="col md-3">
                <h4>添加新员工的信息</h4>
        </div>
            <div class="col md-3">
                <h4>添加员工的照片信息</h4>
        </div>
            <div class="col md-3">
                <h4>使用深度学习训练数据</h4>
        </div>
            <div class="col md-3">
                <h4>查看员工考勤信息 </h4>
            </div>

        </div>
{% if messages %}
    {% for message in messages%}
    <div class="alert alert-{{message.tags}}" > {{message}}
    </div>
    {%endfor %}
  {%endif%}
</div>
```

## 8.6.2 添加员工表单页面

根据文件 urls.py 中的如下代码可知,新用户注册页面对应的功能模块是 users_views.register,对应的模板文件是 register.html。

```
path('register/', users_views.register, name='register'),
```

模板文件 register.html 用于提供添加员工信息的表单,主要实现代码如下:

```
<div class="col-lg-12" style="background: rgba(0,0,0,0.6);max-height: 20px ;
padding-top:1em;padding-bottom:3em;color:#fff;border-radius:10px;-webkit-box-
shadow: 2px 2px 15px 0px rgba(0, 3, 0, 0.7);
-moz-box-shadow:    2px 2px 15px 0px rgba(0, 3, 0, 0.7);
box-shadow:         2px 2px 15px 0px rgba(0, 3, 0, 0.7); margin-left:auto;
margin-right: auto; ">

 <a href="{% url 'dashboard' %}"><h5 class="text-left">Back</h5></a>
</div>

<div class="col-lg-4" style="background: rgba(0,0,0,0.6);margin-top:150px ;
padding-top:1em;padding-bottom:3em;color:#fff;border-radius:10px;-webkit-box-
shadow: 2px 2px 15px 0px rgba(0, 3, 0, 0.7);
-moz-box-shadow:    2px 2px 15px 0px rgba(0, 3, 0, 0.7);
box-shadow:         2px 2px 15px 0px rgba(0, 3, 0, 0.7); margin-left:auto;
margin-right: auto; ">

<form method="POST" >
   {% csrf_token %}
   <fieldset class="form-group">
      <legend class="border-bottom mb-4">添加新员工信息</legend>
      {{form| crispy}}
   </fieldset>
   <div class="form-group">
      <button class="btn btn-outline-info" type="submit"> 添加</button>
   </div>
 </form>
</div>
```

添加员工信息表单页面的执行效果如图 8-4 所示。

图 8-4 添加新用户表单页面

### 8.6.3 添加员工信息

在文件 views.py 中，函数 register()用于获取注册表单中的员工信息，然后将这些信息添加到系统数据库中。文件 views.py 的主要实现代码如下：

```
@login_required
def register(request):
    if request.user.username!='admin':
        return redirect('not-authorised')
    if request.method=='POST':
        form=UserCreationForm(request.POST)
        if form.is_valid():
            form.save()  ###添加用户数据
            messages.success(request, f'Employee registered successfully!')
            return redirect('dashboard')

    else:
        form=UserCreationForm()
    return render(request,'users/register.html', {'form' : form})
```

## 8.7 采集员工照片信息

添加新的员工信息后，接下来需要采集员工的照片信息。在本节的内容中，将详细讲解采集员工照片信息的过程。

扫码看视频

## 8.7.1 设置采集对象

管理员可以在后台采集员工的照片,单击"添加员工的照片信息"上面的"+"按钮,会在打开的页面中看到如图8-5所示的表单,在表单中输入被采集员工的用户名。

图 8-5 输入被采集的员工用户名

根据文件 urls.py 中的如下代码可知,输入被采集对象用户名页面对应的视图模块是 recog_views.add_photos。

```
path('add_photos/', recog_views.add_photos, name='add-photos'),
```

在文件 views.py 中,视图函数 add_photos()的功能是获取在表单中输入的用户名,验证该用户名是否在数据库中存在,如果存在,就继续完成照片采集工作。函数 add_photos()的具体实现代码如下:

```
@login_required
def add_photos(request):
    if request.user.username!='admin':
        return redirect('not-authorised')
    if request.method=='POST':
        form=usernameForm(request.POST)
        data = request.POST.copy()
        username=data.get('username')
        if username_present(username):
            create_dataset(username)
            messages.success(request, f'Dataset Created')
            return redirect('add-photos')
        else:
            messages.warning(request, f'No such username found. Please register employee first.')
            return redirect('dashboard')
    else:

        form=usernameForm()
        return render(request,'recognition/add_photos.html', {'form' : form})
```

文件add_photos.html中提供了输入的被采集对象的用户名表单，主要实现代码如下：

```html
<form method="POST" >
    {% csrf_token %}
    <fieldset class="form-group">
      <legend class="border-bottom mb-4"> Enter Username </legend>
      {{form| crispy}}
    </fieldset>

    <div class="form-group">
      <button class="btn btn-outline-info" type="submit"> Submit</button>
    </div>
  </form>
</div>
<div class="col-lg-12" style="padding-top: 100px;">
 {% if messages %}
    {% for message in messages%}
    <div class="alert alert-{{message.tags}}" > {{message}}
    </div>
    {%endfor %}


  {%endif%}
 </div>
```

## 8.7.2 采集照片

在采集表单中输入用户名并单击"添加"按钮后，打开当前电脑中的摄像头采集照片，然后采集照片中的人脸，并将这些人脸数据创建为Dataset文件。在文件views.py中，视图函数create_dataset()的功能是将采集的照片创建为Dataset文件，具体实现代码如下：

```python
def create_dataset(username):
    id = username
    if(os.path.exists('face_recognition_data/training_dataset/{}/'.format(id))==False):
        os.makedirs('face_recognition_data/training_dataset/{}/'.format(id))
    directory='face_recognition_data/training_dataset/{}/'.format(id)

    #检测人脸
    print("[INFO] Loading the facial detector")
    detector = dlib.get_frontal_face_detector()
    predictor = dlib.shape_predictor('face_recognition_data/shape_predictor_68_face_landmarks.dat')   #向形状预测器添加路径#######稍后更改为相对路径
    fa = FaceAligner(predictor , desiredFaceWidth = 96)
    #从摄像头捕获图像并处理和检测人脸
    #初始化视频流
    print("[INFO] Initializing Video stream")
```

```python
vs = VideoStream(src=0).start()
#time.sleep(2.0)  ####CHECK######

#识别码,将把 id 放在这里,并将 id 与一张脸一起存储,以便稍后可以识别它是谁的脸
#设置初始值是 0
sampleNum = 0
#一张一张地捕捉人脸,检测出人脸并显示在窗口上
while(True):
    #拍摄图像,使用 vs.read 读取每一帧
    frame = vs.read()
    #调整每张图像的大小
    frame = imutils.resize(frame ,width = 800)
    #返回的 img 是一张彩色图像,但是为了使分类器工作,需要一个灰度图像来转换
    gray_frame = cv2.cvtColor(frame, cv2.COLOR_BGR2GRAY)
    #存储人脸,检测当前帧中的所有图像,并返回图像中人脸的坐标和其他一些参数,以获得准确的结果
    faces = detector(gray_frame,0)
    #在上面的 faces 变量中,可以有多个人脸,因此必须得到每个人脸,并在其周围绘制一个矩形
    for face in faces:
        print("inside for loop")
        (x,y,w,h) = face_utils.rect_to_bb(face)

        face_aligned = fa.align(frame,gray_frame,face)
        #每当程序捕捉到人脸时,都会把它写成一个文件夹
        #在捕获人脸之前,需要告诉脚本它是为谁的人脸创建的,需要一个标识符,这里称为 id
        #抓到一张人脸后需要把它写进一个文件
        sampleNum = sampleNum+1
        #保存图像数据集,但只保存面部,裁剪掉其余的部分
        if face is None:
            print("face is none")
            continue

        cv2.imwrite(directory+'/'+str(sampleNum)+'.jpg', face_aligned)
        face_aligned = imutils.resize(face_aligned ,width = 400)
        #cv2.imshow("Image Captured",face_aligned)
        #@params 矩形的初始点是(x,y),终点是 x 的宽度和 y 的高度
        #@params 矩形的颜色
        #@params 矩形的厚度
        cv2.rectangle(frame,(x,y),(x+w,y+h),(0,255,0),1)
        # 在继续下一个循环之前,设置 50 毫秒的暂停等待键
        cv2.waitKey(50)

    #在另一个窗口中显示图像,创建一个窗口 Face,图像为 img
    cv2.imshow("Add Images",frame)
    #在关闭它之前,需要给出一个 wait 命令,否则 OpenCV 将无法工作,通过以下代码设置延迟 1 毫秒
    cv2.waitKey(1)
```

```
        #跳出循环
        if(sampleNum>300):
            break

    #停止视频流
    vs.stop()
    #销毁所有窗口
    cv2.destroyAllWindows()
```

## 8.8 训练照片模型

在采集完员工的照片信息后,会在 training_dataset 目录下存储员工的照片,接下来需要使用 Scikit-learn 将这些照片训练为深度学习模型,为员工的考勤打卡提供人脸识别和检测功能。

扫码看视频

单击后台主页中的"使用深度学习训练数据"按钮,使用机器学习技术 Scikit-learn 训练 Dataset 数据集文件。根据文件 urls.py 中的如下代码可知,本项目通过 recog_views.train 模块训练 Dataset 数据集文件。

```
path('train/', recog_views.train, name='train'),
```

### 8.8.1 前台页面

前台页面的模板文件是 recognition/templates/recognition/train.html,主要实现代码如下:

```
<div class="col-lg-12" style="padding-top: 100px;">
  {% if messages %}
    {% for message in messages%}
    <div class="alert alert-{{message.tags}}" > {{message}}
    </div>
    {%endfor %}
  {%endif%}
</div>
```

### 8.8.2 预测处理

在视图文件 views.py 中,函数 predict() 的功能是实现预测处理,具体实现代码如下:

```
def predict(face_aligned,svc,threshold=0.7):
    face_encodings=np.zeros((1,128))
    try:
```

```
            x_face_locations=face_recognition.face_locations(face_aligned)
            faces_encodings=face_recognition.face_encodings(face_aligned,
known_face_locations=x_face_locations)
            if(len(faces_encodings)==0):
                return ([-1],[0])
        except:
            return ([-1],[0])

        prob=svc.predict_proba(faces_encodings)
        result=np.where(prob[0]==np.amax(prob[0]))
        if(prob[0][result[0]]<=threshold):
            return ([-1],prob[0][result[0]])

        return (result[0],prob[0][result[0]])
```

### 8.8.3 训练数据集

在视图文件 views.py 中，函数 train() 的功能是训练数据集文件，具体实现代码如下：

```
@login_required
def train(request):
    if request.user.username!='admin':
        return redirect('not-authorised')
    training_dir='face_recognition_data/training_dataset'

    count=0
    for person_name in os.listdir(training_dir):
        curr_directory=os.path.join(training_dir,person_name)
        if not os.path.isdir(curr_directory):
            continue
        for imagefile in image_files_in_folder(curr_directory):
            count+=1

    X=[]
    y=[]
    i=0

    for person_name in os.listdir(training_dir):
        print(str(person_name))
        curr_directory=os.path.join(training_dir,person_name)
        if not os.path.isdir(curr_directory):
            continue
        for imagefile in image_files_in_folder(curr_directory):
            print(str(imagefile))
            image=cv2.imread(imagefile)
            try:
```

```
                X.append((face_recognition.face_encodings(image)[0]).tolist())
                y.append(person_name)
                i+=1
            except:
                print("removed")
                os.remove(imagefile)
    targets=np.array(y)
    encoder = LabelEncoder()
    encoder.fit(y)
    y=encoder.transform(y)
    X1=np.array(X)
    print("shape: "+ str(X1.shape))
    np.save('face_recognition_data/classes.npy', encoder.classes_)
    svc = SVC(kernel='linear',probability=True)
    svc.fit(X1,y)
    svc_save_path="face_recognition_data/svc.sav"
    with open(svc_save_path, 'wb') as f:
        pickle.dump(svc,f)
    vizualize_Data(X1,targets)
    messages.success(request, f'Training Complete.')
    return render(request,"recognition/train.html")
```

## 8.8.4 训练可视化

训练完毕后会可视化展示训练结果,在视图文件 views.py 中,函数 vizualize_Data()的功能是将训练结果可视化展示出来,并保存到本地。函数 vizualize_Data()的具体实现代码如下:

```
def vizualize_Data(embedded, targets,):
    X_embedded = TSNE(n_components=2).fit_transform(embedded)

    for i, t in enumerate(set(targets)):
        idx = targets == t
        plt.scatter(X_embedded[idx, 0], X_embedded[idx, 1], label=t)

    plt.legend(bbox_to_anchor=(1, 1));
    rcParams.update({'figure.autolayout': True})
    plt.tight_layout()
    plt.savefig('./recognition/static/recognition/img/training_visualisation.png')
    plt.close()
```

执行效果如图 8-6 所示,说明本项目在目前只是采集了 3 名员工的照片信息。

第 8 章　AI 考勤管理系统

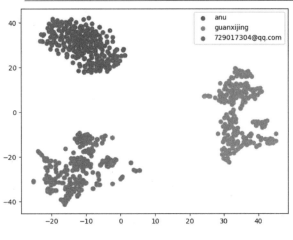

图 8-6　可视化训练结果

## 8.9　考勤打卡

员工登录系统主页后，可以分别实现在线上班打卡签到和下班打卡签退功能。在本节的内容中，将详细讲解实现考勤打卡功能的过程。

扫码看视频

### 8.9.1　上班打卡签到

在系统主页中单击"打卡 - 上班打卡"上面的图标链接，来到上班打卡页面，根据文件 urls.py 中的如下代码可知，考勤打卡页面功能是通过调用 recog_views.mark_your_attendance 模块实现的。

```
path('mark_your_attendance',
recog_views.mark_your_attendance ,name='mark-your-attendance'),
```

在视图文件 views.py 中，函数 mark_your_attendance() 的功能是采集摄像头中的人脸，根据前面训练的模型识别出是哪一名员工，然后实现考勤打卡功能，并将打卡信息添加到数据库中。函数 mark_your_attendance() 的具体实现代码如下：

```
def mark_your_attendance(request):
    detector = dlib.get_frontal_face_detector()
    predictor = dlib.shape_predictor('face_recognition_data/shape_predictor_68_face_landmarks.dat')    #向形状预测器中添加路径#######稍后更改为相对路径
    svc_save_path="face_recognition_data/svc.sav"
    with open(svc_save_path, 'rb') as f:
            svc = pickle.load(f)
```

303

```python
fa = FaceAligner(predictor , desiredFaceWidth = 96)
encoder=LabelEncoder()
encoder.classes_ = np.load('face_recognition_data/classes.npy')

faces_encodings = np.zeros((1,128))
no_of_faces = len(svc.predict_proba(faces_encodings)[0])
count = dict()
present = dict()
log_time = dict()
start = dict()
for i in range(no_of_faces):
    count[encoder.inverse_transform([i])[0]] = 0
    present[encoder.inverse_transform([i])[0]] = False

vs = VideoStream(src=0).start()
sampleNum = 0
while(True):
    frame = vs.read()
    frame = imutils.resize(frame ,width = 800)
    gray_frame = cv2.cvtColor(frame, cv2.COLOR_BGR2GRAY)
    faces = detector(gray_frame,0)
    
    for face in faces:
        print("INFO : inside for loop")
        (x,y,w,h) = face_utils.rect_to_bb(face)
        face_aligned = fa.align(frame,gray_frame,face)
        cv2.rectangle(frame, (x,y), (x+w,y+h), (0,255,0),1)
        (pred,prob)=predict(face_aligned,svc)
        if(pred!=[-1]):
            
            person_name=encoder.inverse_transform(np.ravel([pred]))[0]
            pred=person_name
            if count[pred] == 0:
                start[pred] = time.time()
                count[pred] = count.get(pred,0) + 1
            
            if count[pred] == 4 and (time.time()-start[pred]) > 1.2:
                count[pred] = 0
            else:
            #if count[pred] == 4 and (time.time()-start) <= 1.5:
                present[pred] = True
                log_time[pred] = datetime.datetime.now()
                count[pred] = count.get(pred,0) + 1
                print(pred, present[pred], count[pred])
            cv2.putText(frame, str(person_name)+ str(prob), (x+6,y+h-6),
            cv2.FONT_HERSHEY_SIMPLEX,0.5,(0,255,0),1)
        else:
            person_name="unknown"
```

```
            cv2.putText(frame, str(person_name), (x+6,y+h-6),
            cv2.FONT_HERSHEY_SIMPLEX,0.5,(0,255,0),1)
        
        #cv2.putText()
        #在继续下一个循环之前，设置一个50毫秒的暂停等待键
        #cv2.waitKey(50)
    
    #在另一个窗口中显示图像
    #创建一个窗口，窗口名为Face，图像为img
    cv2.imshow("Mark Attendance - In - Press q to exit",frame)
    #在关闭它之前，需要给出一个wait命令，否则OpenCV将无法工作
    #下面的参数1表示延迟1毫秒
    #cv2.waitKey(1)
    #停止循环
    key=cv2.waitKey(50) & 0xFF
    if(key==ord("q")):
        break
#停止视频流
vs.stop()

#销毁所有窗体
cv2.destroyAllWindows()
update_attendance_in_db_in(present)
return redirect('home')
```

## 8.9.2 下班打卡签退

在系统主页中单击"打卡 - 下班打卡"上面的图标链接，来到下班打卡页面，根据文件 urls.py 中的如下代码可知，下班打卡页面功能是通过调用 recog_views.mark_your_attendance_out 模块实现的。

```
path('mark_your_attendance_out',
recog_views.mark_your_attendance_out ,name='mark-your-attendance-out'),
```

在视图文件 views.py 中，函数 mark_your_attendance_out()的功能是采集摄像头中的人脸，根据前面训练的模型识别出是哪一名员工，然后实现下班打卡功能，并将打卡信息添加到数据库中。函数 mark_your_attendance_out()的具体实现代码同前面的上班打卡。

## 8.10 查看员工考勤信息

管理员登录系统后，可以在考勤统计管理页面中查看员工的考勤信息。在本项目中，使用可视化工具绘制员工的考勤数据，使用折线图统计最近两周的员工考勤信息。

扫码看视频

## 8.10.1 统计最近两周的考勤信息

### 1. 视图函数

在后台主页中单击"查看员工考勤信息"上面的图表链接,在打开的网页中可以查看员工的考勤统计信息。根据文件 urls.py 中的如下代码可知,可视化考勤数据页面的功能是通过调用 recog_views.view_attendance_home 模块实现的。

```
path('view_attendance_home', recog_views.view_attendance_home ,
name='view-attendance-home')
```

在视图文件 views.py 中,函数 view_attendance_home()的功能是可视化展示员工的考勤信息,具体实现代码如下:

```
@login_required
def view_attendance_home(request):
    total_num_of_emp=total_number_employees()
    emp_present_today=employees_present_today()
    this_week_emp_count_vs_date()
    last_week_emp_count_vs_date()
    return render(request,"recognition/view_attendance_home.html",
{'total_num_of_emp' : total_num_of_emp, 'emp_present_today': emp_present_today})
```

在上述代码中用到了如下所示的 4 个函数。

(1) 函数 total_number_employees()的功能是统计当前系统中的考勤员工信息,具体实现代码如下:

```
def total_number_employees():
    qs=User.objects.all()
    return (len(qs) -1)
```

(2) 函数 employees_present_today()的功能是统计今日打卡的员工数量,具体实现代码如下:

```
def employees_present_today():
    today=datetime.date.today()
    qs=Present.objects.filter(date=today).filter(present=True)
    return len(qs)
```

(3) 函数 this_week_emp_count_vs_date()的功能是统计本周员工每天的打卡信息,并绘制可视化折线图,具体实现代码如下:

```
def this_week_emp_count_vs_date():
    today=datetime.date.today()
    some_day_last_week=today-datetime.timedelta(days=7)
```

```
    monday_of_last_week=some_day_last_week- datetime.timedelta(days=
(some_day_last_week.isocalendar()[2] - 1))
    monday_of_this_week = monday_of_last_week + datetime.timedelta(days=7)
    qs=Present.objects.filter(date__gte=monday_of_this_week).filter(date__lte=today)
    str_dates=[]
    emp_count=[]
    str_dates_all=[]
    emp_cnt_all=[]
    cnt=0

    for obj in qs:
        date=obj.date
        str_dates.append(str(date))
        qs=Present.objects.filter(date=date).filter(present=True)
        emp_count.append(len(qs))
    while(cnt<5):
        date=str(monday_of_this_week+datetime.timedelta(days=cnt))
        cnt+=1
        str_dates_all.append(date)
        if(str_dates.count(date))>0:
            idx=str_dates.index(date)
            emp_cnt_all.append(emp_count[idx])
        else:
            emp_cnt_all.append(0)

    df=pd.DataFrame()
    df["date"]=str_dates_all
    df["Number of employees"]=emp_cnt_all

    sns.lineplot(data=df,x='date',y='Number of employees')
    plt.savefig('./recognition/static/recognition/img/attendance_graphs/this_week/1.png')
    plt.close()
```

(4) 函数 last_week_emp_count_vs_date()的功能是统计上一周员工每天的打卡信息。

2. 模板文件

编写模板文件 view_attendance_home.html,该文件的功能是调用上面的视图函数,使用曲线图可视化展示最近两周的员工考勤数据。员工考勤数据可视化页面的执行效果如图 8-7 所示。

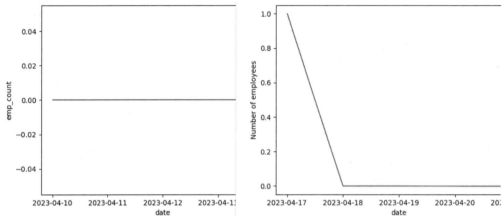

图 8-7　员工考勤数据可视化页面

## 8.10.2　查看某员工在指定时间范围内的考勤信息

**1. 视图函数**

管理员登录系统后，输入 URL 链接 http://127.0.0.1:8000/view_attendance_employee，打开如图 8-8 所示的页面。在此页面中可以输入员工的名字和时间段，单击"提交"按钮，可以查看这名员工在指定时间段内的考勤信息。

图 8-8　选择时间范围

根据文件 urls.py 中的如下代码可知，查看某员工在指定时间范围内的考勤统计图的功能是通过调用 recog_views.view_attendance_employee 模块实现的。

```
path('view_attendance_employee',
recog_views.view_attendance_employee ,name='view-attendance-employee'),
```

在视图文件 views.py 中，函数 view_attendance_employee()的功能是可视化展示某员工在指定时间段内的考勤信息，具体实现代码如下：

```
@login_required
def view_attendance_employee(request):
    if request.user.username!='admin':
        return redirect('not-authorised')
    time_qs=None
    present_qs=None
    qs=None
    
    if request.method=='POST':
        form=UsernameAndDateForm(request.POST)
        if form.is_valid():
            username=form.cleaned_data.get('username')
            if username_present(username):
                u=User.objects.get(username=username)
                time_qs=Time.objects.filter(user=u)
                present_qs=Present.objects.filter(user=u)
                date_from=form.cleaned_data.get('date_from')
                date_to=form.cleaned_data.get('date_to')
                if date_to < date_from:
                    messages.warning(request, f'Invalid date selection.')
                    return redirect('view-attendance-employee')
                else:
                    time_qs=time_qs.filter(date__gte=date_from).filter(date__lte=date_to).order_by('-date')
                    present_qs=present_qs.filter(date__gte=date_from).filter(date__lte=date_to).order_by('-date')
                    
                    if (len(time_qs)>0 or len(present_qs)>0):
                        qs=hours_vs_date_given_employee(present_qs,time_qs,admin=True)
                        return render(request,'recognition/view_attendance_employee.html', {'form' : form, 'qs' :qs})
                    else:
                        messages.warning(request, f'No records for selected duration.')
                        return redirect('view-attendance-employee')
            else:
                print("invalid username")
                messages.warning(request, f'No such username found.')
                return redirect('view-attendance-employee')
    else:
        form=UsernameAndDateForm()
```

```
        return render(request,'recognition/view_attendance_employee.html',
{'form' : form, 'qs' :qs})
```

在上述代码中，也可以使用视图函数 hours_vs_date_given_employee()绘制柱状考勤统计图。

### 2. 模板文件

编写模板文件 view_attendance_employee.html，其功能是创建设置员工用户名和选择时间段的表单页面。

## 8.10.3 查看指定日期的考勤信息

在管理员的考勤信息页面中单击 By Date 链接，来到查看指定日期的员工考勤信息页面，如图 8-9 所示。

在此页面中可以查看指定日期的所有员工的考勤信息，在视图文件 views.py 中，函数 view_attendance_date()的功能是可视化展示指定日期所有员工的考勤信息，具体实现代码如下：

图 8-9　查看指定日期的考勤信息

```
@login_required
def view_attendance_date(request):
    if request.user.username!='admin':
        return redirect('not-authorised')
    qs=None
    time_qs=None
    present_qs=None
    if request.method=='POST':
        form=DateForm(request.POST)
        if form.is_valid():
            date=form.cleaned_data.get('date')
            print("date:"+ str(date))
            time_qs=Time.objects.filter(date=date)
            present_qs=Present.objects.filter(date=date)
            if(len(time_qs)>0 or len(present_qs)>0):
                qs=hours_vs_employee_given_date(present_qs,time_qs)
                return render(request,'recognition/view_attendance_date.html',
{'form' : form,'qs' : qs })
            else:
                messages.warning(request, f'No records for selected date.')
                return redirect('view-attendance-date')
    else:
        form=DateForm()
        return render(request,'recognition/view_attendance_date.html', {'form' :
form, 'qs' : qs})
```

## 8.11 查看本人的考勤信息

普通员工登录系统后，单击"查看我的考勤"上面的图标链接，来到如图 8-10 所示的页面，在此页面中可以查看本人在指定时间段内的考勤统计信息。

扫码看视频

图 8-10 查看指定时间段内的考勤信息

### 8.11.1 视图函数

根据文件 urls.py 中的如下代码可知，查看本人指定时间范围内的考勤数据的功能是通过调用 recog_views.view_my_attendance_employee_login 模块来实现的。

```
path('view_my_attendance', recog_views.view_my_attendance_employee_login ,
name='view-my-attendance-employee-login')
```

在视图文件 views.py 中，函数 view_my_attendance_employee_login()的功能是可视化展示本人在指定时间段内的考勤信息，具体实现代码如下：

```
@login_required
def view_my_attendance_employee_login(request):
    if request.user.username=='admin':
        return redirect('not-authorised')
    qs=None
    time_qs=None
    present_qs=None
    if request.method=='POST':
```

```
            form=DateForm_2(request.POST)
            if form.is_valid():
                u=request.user
                time_qs=Time.objects.filter(user=u)
                present_qs=Present.objects.filter(user=u)
                date_from=form.cleaned_data.get('date_from')
                date_to=form.cleaned_data.get('date_to')
                if date_to < date_from:
                    messages.warning(request, f'Invalid date selection.')
                    return redirect('view-my-attendance-employee-login')
                else:
                    time_qs=time_qs.filter(date__gte=date_from).filter(date__lte=date_to).order_by('-date')
                    present_qs=present_qs.filter(date__gte=date_from).filter(date__lte=date_to).order_by('-date')

                    if (len(time_qs)>0 or len(present_qs)>0):
                        qs=hours_vs_date_given_employee(present_qs,time_qs,admin=False)
                        return render(request,'recognition/view_my_attendance_employee_login.html', {'form' : form, 'qs' :qs})
                    else:
                        messages.warning(request, f'No records for selected duration.')
                        return redirect('view-my-attendance-employee-login')
        else:
            form=DateForm_2()
            return render(request,'recognition/view_my_attendance_employee_login.html', {'form' : form, 'qs' :qs})
```

在上述代码中，调用函数 hours_vs_date_given_employee()，绘制在指定时间段内的考勤统计图，具体实现代码如下：

```
def hours_vs_date_given_employee(present_qs,time_qs,admin=True):
    register_matplotlib_converters()
    df_hours=[]
    df_break_hours=[]
    qs=present_qs
    for obj in qs:
        date=obj.date
        times_in=time_qs.filter(date=date).filter(out=False).order_by('time')
        times_out=time_qs.filter(date=date).filter(out=True).order_by('time')
        times_all=time_qs.filter(date=date).order_by('time')
        obj.time_in=None
        obj.time_out=None
        obj.hours=0
        obj.break_hours=0
        if (len(times_in)>0):
            obj.time_in=times_in.first().time
```

```python
            if (len(times_out)>0):
                obj.time_out=times_out.last().time
            if(obj.time_in is not None and obj.time_out is not None):
                ti=obj.time_in
                to=obj.time_out
                hours=((to-ti).total_seconds())/3600
                obj.hours=hours
            else:
                obj.hours=0
            (check,break_hourss)= check_validity_times(times_all)
            if check:
                obj.break_hours=break_hourss
                
            else:
                obj.break_hours=0
            df_hours.append(obj.hours)
            df_break_hours.append(obj.break_hours)
            obj.hours=convert_hours_to_hours_mins(obj.hours)
            obj.break_hours=convert_hours_to_hours_mins(obj.break_hours)
        df = read_frame(qs)
        df["hours"]=df_hours
        df["break_hours"]=df_break_hours
        
        print(df)
        sns.barplot(data=df,x='date',y='hours')
        plt.xticks(rotation='vertical')
        rcParams.update({'figure.autolayout': True})
        plt.tight_layout()
        if(admin):
            plt.savefig('./recognition/static/recognition/img/attendance_graphs/hours_vs_date/1.png')
            plt.close()
        else:
            plt.savefig('./recognition/static/recognition/img/attendance_graphs/employee_login/1.png')
            plt.close()
        return qs
```

## 8.11.2　模板文件

编写模板文件 view_my_attendance_employee_ login.html，功能是创建选择时间段的表单页面。主要实现代码如下：

```
<nav class="navbar navbar-expand-lg navbar-light bg-light">
  <a class="navbar-brand" href="{%url 'view-my-attendance-employee-login' %}">我的考勤</a>
```

```html
            <button class="navbar-toggler" type="button" data-toggle="collapse" 
data-target="#navbarNav" aria-controls="navbarNav" aria-expanded="false" 
aria-label="Toggle navigation">
    <span class="navbar-toggler-icon"></span>
  </button>
  <div class="collapse navbar-collapse" id="navbarNav">
    <ul class="navbar-nav">
      <li class="nav-item active" style="padding-left: 1740px">
        <a class="nav-link" href="{% url 'dashboard' %}">返回个人中心</a>
      </li>

    </ul>
  </div>
</nav>
      <div class="container">
 <div style="width:400px;">
    <form method="POST" >
      {% csrf_token %}
      <fieldset class="form-group">
          <legend class="border-bottom mb-4"> 选择时间范围 </legend>
          {{form| crispy}}
      </fieldset>
      <div class="form-group">
          <button class="btn btn-outline-info" type="submit">提交</button>
      </div>
    </form>
</div>

{%if qs%}
<table class="table" style="margin-top: 5em;">
    <thead class="thead-dark">
    <tr>
        <th scope="col">日期</th>

        <th scope="col" >员工</th>
        <th scope="col">迟到</th>
        <th scope="col">上班打卡</th>
        <th scope="col">下班打卡 </th>
        <th scope="col">上班时长 </th>
        <th scope="col"> 旷工时长 </th>
    </tr>
    </thead>
<tbody>
    {% for item in qs %}
    <tr>
        <td>{{ item.date }}</td>
        <td>{{ item.user.username}}</td>

        {% if item.present %}
```

```
            <td> P </td>
            {% else %}
            <td> A </td>
            {% endif %}
            {% if item.time_in %}
            <td>{{ item.time_in }}</td>
            {% else %}
            <td> - </td>
            {% endif %}
            {% if item.time_out %}
            <td>{{ item.time_out }}</td>
            {% else %}
            <td> - </td>
            {% endif %}
            <td> {{item.hours}}</td>
            <td> {{item.break_hours}}</td>
        </tr>
        {% endfor %}
</tbody>
</table>

<div class="card" style=" margin-top: 5em; margin-bottom: 10em;">
  <img class="card-img-top" src="{% static
'recognition/img/attendance_graphs/employee_login/1.png'%}" alt="Card image cap">
  <div class="card-body">
    <p class="card-text" style="text-align: center;">每天工作小时数</p>
  </div>
</div>

{% endif %}
{% if messages %}
    {% for message in messages%}
    <div class="alert alert-{{message.tags}}" > {{message}}
    </div>
    {%endfor %}
{%endif%}
</div>
```

## 8.12 调试运行

本项目系统主页面的执行效果如图 8-11 所示。

查看指定日期员工的考勤信息页面如图 8-12 所示。

某员工查看本人在指定时间段内的考勤信息页面，如图 8-13 所示。

扫码看视频

图 8-11 系统主页

图 8-12 查看指定日期员工的考勤信息页面

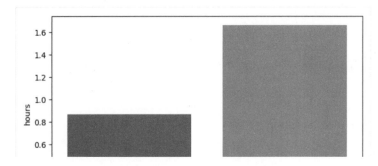

图 8-13 查看本人的考勤信息页面

# 第9章

## 网络舆情数据分析系统

网络舆情是以网络为载体,针对社会问题、现象以及事件等,广大网民情感、态度、意见、观点的表达、传播与互动,以及后续影响力的集合。在本章的内容中,将详细介绍使用 Python 语言开发一个网络舆情数据分析系统的过程,并使用可视化技术实现一个用户画像分析系统,具体流程由 Scikit-learn+Tornado+Celery+FastAPI+Pandas+Vue 实现。

## 9.1 系统介绍

互联网的飞速发展促进了很多新媒体的发展，不论是知名大 V、明星，还是围观群众都可以通过手机在微博、朋友圈或者点评网站上发表观点，分享自己的所见所想，使得"人人都有了麦克风"。不论是热点新闻还是娱乐八卦，传播速度已远超我们的想象。在短短数分钟内，就会有数万计转发、数百万的阅读量。如此海量的信息可以得到爆炸式的传播，如何能够实时地把握民情并做出相应的处理，对很多企事业单位来说是至关重要的，这时就需要把传统的舆情数据分析系统升级为大数据舆情采集和分析系统。

扫码看视频

### 9.1.1 舆情数据分析的方式和意义

舆情数据分析的常用方式有以下两种。

(1) 人工检索，借助商业搜索引擎等开放性工具、平台，进行实时监测，并筛选获取的数据。

(2) 使用专业网络舆情监测系统，实现跨屏、跨库、跨区域、跨媒介的全方位信息收集。如人民在线、方正智思、军犬、清博舆情系统、新浪舆情通、林克艾普、企鹅风讯、舆情雷达、鹰击舆情系统等。

网络舆情分析的意义主要有两个方面：一是还原舆情发展过程，找到舆情产生的根源；二是预测，分析网络舆情的未来走向，再根据预测结果提出应对方案。而针对这两方面的工作，网络舆情分析的重点在于舆情数据分析中的热度分析、倾向性分析和预测分析。

### 9.1.2 舆情热度分析

舆情热度分析，还原舆情发展过程，找到舆情产生的根源，它是网络舆情分析工作的重点。通过数据可以反映出信息的变化趋势，也能够监测出负面舆情扩散的严重程度。

1) 热度概况与全网声量分析

分析舆情热度，首先要看热度概况和全网声量，以便从总体上把握事件的热度情况。以新浪微热点的热度指数为例，该指数是从新闻媒体、微博、微信、客户端、网站、论坛等互联网平台采集海量信息的基础上，提取与指定事件、人物、品牌、地域等相关的信息，并对所提取的信息进行标准化计算后得出的热度指数。

2) 热度指数趋势分析

了解了总体热度指数后，再来分析热度指数的趋势。

3) 声量走势分析

声量走势分析是对某舆情事件的信息发布数量的趋势分析，它是对信息数据发布量的统计和展示。通过声量走势图，可以从网民和媒体生产和传递信息的角度观察事件热度。

4) 舆情信息来源、活跃媒体分析

舆情信息来源和活跃媒体分析是对舆情信息主要来源和传播时较活跃的媒体进行分析和统计。目前，网络舆情的产生和传播主要是在新闻网站、论坛、微博、移动客户端和微信等几类平台，来源于不同平台的热点舆情，在传播上也会呈现出不同的特征。

5) 地域热度分析

地域要素体现了舆情爆发的地域性特征。通过对舆情主要分布地域的分析，可以获知全国不同地区网民和媒体对事件的关注程度；同时，舆情的地域分布也可以反映出舆情的热度，一些地方性事件，由于其影响较大，讨论较多，其舆情分布可能遍及全国。

6) 舆情演化分析

舆情演化分析是从舆情内容和热度的双重方面对舆情进行分析。分析网络舆情热度，需要了解舆情爆发和演进的过程——潜伏期、爆发期、蔓延期、缓解期、反复期、消退期，从而梳理出舆情的起因、经过和结果。

## 9.2 架构设计

架构设计的目标是解决目前或者未来软件系统由于复杂度可能带来的问题。就目前而言，架构设计主要是为了识别、梳理用例模型交互、功能模块实现、接口设计和概念模型设计等涉及的复杂点，再针对这些复杂点制定处理方案，从而通过设计来增强效用，减少成本，降低复杂度。就未来而言，系统架构设计将随着业务发展不断演变、完善，以解决未来软件系统由于复杂度可能带来的问题。

扫码看视频

### 9.2.1 模块分析

1) 网络爬虫

使用网络爬虫技术获取微博中的各种类型数据信息，通过 Tornado 框架以协程运行方式完成数据分类。

- 推文搜索：快速获取指定关键字的信息。
- 推文展示：根据推文 id 搜索推文。
- 用户搜索：根据关键词搜索用户。
- 用户展示：根据用户 id 搜索用户。

❏ 用户朋友列表接口(朋友关注的人)：根据用户的 id 搜索用户朋友，同时也要返回他们的信息。

2) 系统后端

使用 FastAPI 框架搭建，结合分布式异步任务处理框架 Celery 实现任务发布和任务执行的解耦，使得异步、快速处理话题分析任务成为可能。

❏ 微博话题博文搜索：根据用户输入的关键字搜索出热度最高的 50 页近 1000 条博文。
❏ 微博话题分析：用户可以将相关话题加入分析任务队列中，后端会异步执行分析任务(基于 Celery)。
❏ 博文分析：对话题中热度前十的博文进行详细分析。

3) 系统前端

前端使用 Vue 框架，调用后端生成的 API 实现数据展示和可视化功能。

## 9.2.2 系统结构

本网络舆情数据分析系统的结构如图 9-1 所示。

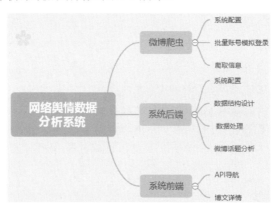

图 9-1 系统结构

# 9.3 微博爬虫

本项目将爬取某知名微博中的数据信息，通过 Tornado 框架以协程运行方式完成爬取微博信息，目的是为系统后台提供各种类型数据的 Web 服务。

扫码看视频

## 9.3.1 系统配置

在文件 settings.py 中设置程序运行的端口号和日志的记录位置，在 cookies 字段中填写由多个有效 cookie 组成的列表，在 proxies 字段中填写由多个代理组成的列表，每个代理的格式为[proxy_host, proxy_port]。代码如下：

```
import logging
PORT_NUM = 8000  # App 运行的端口号
# 发送一个 request 最多重新尝试的次数
RETRY_TIME = 3

# requests 的 headers
HEADERS = {
    "User-Agent": "MMozilla/5.0 (Windows NT 10.0; Win64; x64) AppleWebKit/537.36
    (KHTML, like Gecko) Chrome/84.0.4147.105 Safari/537.36 Edg/84.0.522.52",
}
HEADERS_WITH_COOKIR = HEADERS.copy()
HEADERS_WITH_COOKIR["Cookie"] = ""
# requests 的超时时长限制 das
REQUEST_TIME_OUT = 10
# 爬取结果正确时返回结果的格式
SUCCESS = {
    'error_code': 0,
    'data': None,
    'error_msg': None
}
# 日志
LOGGING = logging
```

## 9.3.2 批量账号模拟登录

为了提高爬取效率，本项目允许用多个微博账号实现模拟登录。编写实例文件 login.py，功能是对批量账号进行模拟登录并获取 cookie，代码如下：

```
class WeiboLogin():
    def __init__(self, username, password):
        self.url = 'https://passport.weibo.cn/signin/login?entry=mweibo&r=https://weibo.cn/'
        options = webdriver.ChromeOptions()
        options.add_argument('--no-sandbox')
        user_agent = "Mozilla/5.0 (Macintosh; Intel Mac OS X 10_14_2) AppleWebKit/537.36 (KHTML, like Gecko) Chrome/71.0.3578.98 Safari/537.36"
        mobile_emulation = {"deviceMetrics": {"width": 1050, "height": 840,
            "pixelRatio": 3.0}, "userAgent": user_agent}
```

```python
        options.add_experimental_option('mobileEmulation', mobile_emulation)
        options.add_argument('--headless')
        self.browser = webdriver.Chrome(chrome_options=options)
        self.username = username
        self.password = password

    def login(self):
        """
        open login page and login
        :return: None
        """
        self.browser.get(self.url)
        wait = WebDriverWait(self.browser, 5)
        username = wait.until(EC.presence_of_element_located((By.ID, 'loginName')))
        password = wait.until(EC.presence_of_element_located((By.ID, 'loginPassword')))
        submit = wait.until(EC.element_to_be_clickable((By.ID, 'loginAction')))
        username.send_keys(self.username)
        password.send_keys(self.password)
        submit.click()

    def run(self):
        try:
            self.login()
            WebDriverWait(self.browser, 20).until(
                EC.title_is('我的首页')
            )
            cookies = self.browser.get_cookies()
            cookie = [item["name"] + "=" + item["value"] for item in cookies]
            cookie_str = '; '.join(item for item in cookie)
            return cookie_str
        except Exception as e:
            print(e)
        finally:
            self.browser.quit()
        return None

if __name__ == '__main__':
    username = 'tnnmyvxj27431@sina.com'
    password = 'cxy6331i1'
    cookie_str = None
    try:
        cookie_str = WeiboLogin(username, password).run()
        print('Cookie:', cookie_str)
    except Exception as e:
        print(e)
```

## 9.3.3 爬取信息

(1) 编写文件 index_parser.py,获取指定 id 号的微博用户信息,代码如下:

```python
class IndexParser(BaseParser):
    def __init__(self, user_id, response):
        super().__init__(response)
        self.user_id = user_id

    def get_user_id(self):
        """获取用户id,使用者输入的user_id不一定是正确的,可能是个性域名等,需要获取真正的user_id"""
        user_id = self.user_id
        url_list = self.selector.xpath("//div[@class='u']//a")
        for url in url_list:
            if (url.xpath('string(.)')) == u'资料':
                if url.xpath('@href') and url.xpath('@href')[0].endswith(
                        '/info'):
                    link = url.xpath('@href')[0]
                    user_id = link[1:-5]
                    break
        self.user_id = user_id
        return user_id

    def get_user(self, user_info):
        """获取用户信息、微博数、关注数、粉丝数"""
        self.user = user_info
        try:
            user_info = self.selector.xpath("//div[@class='tip2']/*/text()")
            self.user['id'] = self.user_id
            self.user['weibo_num'] = int(user_info[0][3:-1])
            self.user['following'] = int(user_info[1][3:-1])
            self.user['followers'] = int(user_info[2][3:-1])
            return self.user
        except Exception as e:
            utils.report_log(e)
            raise HTMLParseException

    def get_page_num(self):
        """获取微博总页数"""
        try:
            if not self.selector.xpath("//input[@name='mp']"):
                page_num = 1
            else:
                page_num = int(self.selector.xpath("//input[@name='mp']")
                               [0].attrib['value'])
```

```python
            return page_num
        except Exception as e:
            utils.report_log(e)
            raise HTMLParseException

class InfoParser(BaseParser):
    def __init__(self, response):
        super().__init__(response)

    def extract_user_info(self):
        """提取用户信息"""
        user = USER_TEMPLATE.copy()
        nickname = self.selector.xpath('//title/text()')[0]
        nickname = nickname[:-3]
        # 检查cookie
        if nickname == u'登录 - 新' or nickname == u'新浪':
            LOGGING.warning(u'cookie 错误或已过期')
            raise CookieInvalidException()

        user['nickname'] = nickname
        # 获取头像
        try:
            user['head'] = self.selector.xpath('//div[@class="c"]/img[@alt="头像"]')[0].get('src')
        except:
            user['head'] = ''
        # 获取基本信息
        try:
            basic_info = self.selector.xpath("//div[@class='c'][3]/text()")
            zh_list = [u'性别', u'地区', u'生日', u'简介', u'认证', u'达人']
            en_list = [
                'gender', 'location', 'birthday', 'description',
                'verified_reason', 'talent'
            ]
            for i in basic_info:
                if i.split(':', 1)[0] in zh_list:
                    user[en_list[zh_list.index(i.split(':', 1)[0])]] = i.split(':', 1)[1].replace('\u3000', '')

            if self.selector.xpath(
                    "//div[@class='tip'][2]/text()")[0] == u'学习经历':
                user['education'] = self.selector.xpath(
                    "//div[@class='c'][4]/text()")[0][1:].replace(
                    u'\xa0', u' ')
                if self.selector.xpath(
                        "//div[@class='tip'][3]/text()")[0] == u'工作经历':
```

```
                    user['work'] = self.selector.xpath(
                        "//div[@class='c'][5]/text()")[0][1:].replace(
                        u'\xa0', u' ')
                elif self.selector.xpath(
                        "//div[@class='tip'][2]/text()")[0] == u'工作经历':
                    user['work'] = self.selector.xpath(
                        "//div[@class='c'][4]/text()")[0][1:].replace(
                        u'\xa0', u' ')
                return user
        except Exception as e:
            utils.report_log(e)
            raise HTMLParseException


# 封装一个用户信息的dict
USER_TEMPLATE = {
    'id': '',
    'nickname': '',
    'gender': '',
    'location': '',
    'birthday': '',
    'description': '',
    'verified_reason': '',
    'talent': '',
    'education': '',
    'work': '',
    'weibo_num': 0,
    'following': 0,
    'followers': 0
}
```

(2) 编写文件 fans_parser.py，爬取粉丝信息，代码如下：

```
class FansParser(BaseParser):
    def __init__(self, response):
        super().__init__(response)

    def get_fans(self):
        fans_list = list()
        fans_nodes = self.selector.xpath(r'//table')
        for node in fans_nodes:
            a_fans = self.get_one_fans(node)
            if a_fans is not None:
                fans_list.append(a_fans)
        return fans_list

    @staticmethod
    def get_one_fans(fans_node):
```

```python
        return utils.extract_from_one_table_node(fans_node)

    def get_max_page_num(self):
        """
        获取总页数
        """
        total_page_num = ''.join(self.selector.xpath(r'//div[@id="pagelist"]/form/div/text()'))
        total_page_num = total_page_num[total_page_num.rfind(r'/') + 1:total_page_num.rfind('页')]
        return int(total_page_num)
```

(3) 编写文件 search_users_parser.py，搜索指定关键字的用户信息，代码如下：

```python
class SearchUsersParser(BaseParser):
    """搜索用户页面的解析器"""

    USER_TEMPLATE = {
        'user_id': None,            # 用户的id
        'nickname': None,           # 昵称
        'head': None,
        'title': None,              # 所拥有的头衔
        'verified_reason': None,    # 认证原因
        'gender': None,             # 性别
        'location': None,           # 位置
        'description': None,        # 简介
        'tags': None,               # 标签
        'education': None,          # 教育信息
        'work': None,               # 工作信息
        'weibo_num': None,          # 微博数
        'following': None,          # 关注数
        'followers': None           # 粉丝数
    }

    @staticmethod
    def make_a_user():
        """生成一个用来存储user信息的dict"""
        return SearchUsersParser.USER_TEMPLATE.copy()

    def __init__(self, response):
        super().__init__(response)

    def parse_page(self):
        """解析网页"""
        try:
            user_list = self._get_all_user()
            return user_list
```

```python
        except Exception as e:
            utils.report_log(e)
            raise HTMLParseException

def _get_all_user(self):
    """获取全部用户信息"""
    user_list = list()
    user_nodes = self.selector.xpath('//div[@id="pl_user_feedList"]/div')
    for node in user_nodes:
        user = self._parse_one_user(node)
        # print(user)
        user_list.append(user)
    return user_list

def _parse_one_user(self, user_node):
    """解析单个用户的 selector 节点"""
    user = SearchUsersParser.make_a_user()
    # 获取用户头像
    try:
        user['head'] = user_node.xpath('.//div[@class="avator"]/a/img')[0].get('src')
    except:
        user['head'] = ''
    # 获取其他信息
    info_selector = user_node.xpath('./div[@class="info"]')[0]
    headers = info_selector.xpath('./div[1]/a')

    if len(headers) > 2:  # 拥有头衔的情况
        header_node = headers[1]
        title = header_node.get('title')
        if title is not None:
            user['title'] = title


    user_id = headers[-1].get('uid')
    if user_id is None:
        # 尝试另外一种方法获取 uid
        user_index_url = headers[0].get('href')
        pattern1 = re.compile(r'(?<=com/u/).+')
        # 正则匹配 '//weibo.com/u/61248565' 类型,提取出其中的'61248565'
        user_id = pattern1.search(user_index_url)
        if user_id is not None:
            user_id = user_id.group()
        else:
            # 尝试另外一种方法获取 uid
            pattern2 = re.compile(r'(?<=com/).+')
            # 正则匹配 'weibo.com/xiena' 类型,提取出其中的'xiena'
            user_id = pattern2.search(user_index_url)
```

```python
            if user_id is not None:
                user_id = user_id.group()
        user['user_id'] = user_id

        user['nickname'] = ''.join(headers[0].xpath(".//text()"))

        all_p_node = info_selector.xpath('./p')
        first_p = all_p_node[0]
        gender_info = first_p.xpath('./i')[0].get('class')
        user['gender'] = 0 if gender_info.rfind('female') != -1 else 1
        # 0 为女性，1 为男性
        user['location'] = ''.join(first_p.xpath('./text()')).strip()

        footer = None
        other_p_nodes = list()
        for p_node in all_p_node:
            if p_node is first_p:
                continue
            elif len(p_node.xpath('./span')) == 3:
                footer = p_node
            else:
                other_p_nodes.append(p_node)

        if footer is not None:
            spans = footer.xpath('./span')
            user['following'] = spans[0].xpath('./a/text()')[0]
            user['followers'] = spans[1].xpath('./a/text()')[0]
            user['weibo_num'] = spans[2].xpath('./a/text()')[0]

        for node in other_p_nodes:
            info = ''.join(node.xpath('.//text()'))
            info_type = info[0: 2]

            if info_type == '教育':
                user['education'] = info
            elif info_type == '职业':
                user['work'] = info
            elif info_type == '简介':
                user['description'] = info
            elif info_type == '标签':
                user['tags'] = node.xpath('./a/text()')
            else:
                user['verified_reason'] = info

        return user
```

(4) 编写文件 weibo_curl_api.py，具体实现流程如下。

① 创建 Tornado Web 程序，根据搜索关键字爬取指定的信息，并将信息保存到 MongoDB 数据库，代码如下：

```python
SEARCH_LIMIT_PAGES = 50  # 微博的搜索接口限制的最大页数
class BaseHandler(tornado.web.RequestHandler):
    def write(self, dict_data: dict):
        """在发送之前将编码方式转化成 Unicode"""
        data = json.dumps(dict_data, ensure_ascii=False)
        super().write(data)

    def args2dict(self):
        """
        将请求 url 中的请求查询字符串转换成 dict
        :return: 转化后的 dict
        """
        input_dict = dict()
        args = self.request.arguments
        for i in args:
            input_dict[i] = self.get_argument(i)
        return input_dict

    def get_json(self):
        """
        将获取 post 时的 json
        """
        json_str = self.request.body.decode('utf8')
        json_obj = json.loads(json_str)
        return json_obj

    def save_data_to_mongo(self, dict_data: list, table_name: str):
        """
        将获得的数据存到 Mongo 数据库中
        """
        mongo_client = pymongo.MongoClient('mongodb://localhost:27017')
        weibo_db = mongo_client['weibo']
        weibo_table = weibo_db[table_name]
        weibo_table.insert_many(dict_data)
        print('向 Mongo 写入数据成功')

class SearchTweetsHandler(BaseHandler):
    """
    微博搜索接口
        说明：根据关键词搜索微博
        路由：/weibo_curl/api/search_tweets
    """
    @gen.coroutine
    def get(self):
```

```python
# 获取参数
args_dict = self.args2dict()    # 查询参数 -> 参数字典
keyword, cursor, is_hot = args_dict.get('keyword'), args_dict.get(
    'cursor', '1'), args_dict.get('is_hot', False)
if keyword is None:
    self.write(WeiboCurlError.REQUEST_LACK_ARGS)    # 缺少参数
    return
try:
    cursor = 1 if not cursor else int(cursor)
except ValueError:
    self.write(WeiboCurlError.REQUEST_ARGS_ERROR)
    return

# 进行爬取
search_weibo_curl_result = yield weibo_web_curl(SpiderAim.search_weibo,
                        keyword=keyword, page_num=cursor, is_hot=is_hot)
if not search_weibo_curl_result['error_code']:
    self.response = search_weibo_curl_result['response']
else:
    error_res = curl_result_to_api_result(search_weibo_curl_result)
    self.write(error_res)
    return
# 构建解析器
searchWeiboParser = SearchWeiboParser(self.response)
# 获取微博信息
try:
    weibo_list = searchWeiboParser.parse_page()
    # print(weibo_list)
except HTMLParseException:
    self.write(WeiboCurlError.HTML_PARSE_ERROR)
    return

if weibo_list is None:
    self.write(WeiboCurlError.PAGE_NOT_FOUND)    # 页面找不到
    return
# 成功返回结果
success = settings.SUCCESS.copy()
success['data'] = {
    'result': weibo_list,
    'cursor': str(cursor + 1) if cursor < 50 else '0'
}
self.write(success)
print(success)
return
```

② 实现微博搜索接口类 SearchTweetsHandler，根据输入的关键词搜索对应的微博信息，代码如下：

```python
class SearchTweetsHandler(BaseHandler):
    """
    微博搜索接口
        说明：根据关键词搜索微博
        路由：/weibo_curl/api/search_tweets
    """
    @gen.coroutine
    def get(self):
        # 获取参数
        args_dict = self.args2dict()    # 查询参数 -> 参数字典
        keyword, cursor, is_hot = args_dict.get('keyword'), args_dict.get(
            'cursor', '1'), args_dict.get('is_hot', False)
        if keyword is None:
            self.write(WeiboCurlError.REQUEST_LACK_ARGS)    # 缺少参数
            return
        try:
            cursor = 1 if not cursor else int(cursor)
        except ValueError:
            self.write(WeiboCurlError.REQUEST_ARGS_ERROR)
            return

        # 进行爬取
        search_weibo_curl_result = yield weibo_web_curl(SpiderAim.search_weibo,
                            keyword=keyword, page_num=cursor, is_hot=is_hot)
        if not search_weibo_curl_result['error_code']:
            self.response = search_weibo_curl_result['response']
        else:
            error_res = curl_result_to_api_result(search_weibo_curl_result)
            self.write(error_res)
            return
        # 构建解析器
        searchWeiboParser = SearchWeiboParser(self.response)
        # 获取微博信息
        try:
            weibo_list = searchWeiboParser.parse_page()
            # print(weibo_list)
        except HTMLParseException:
            self.write(WeiboCurlError.HTML_PARSE_ERROR)
            return

        if weibo_list is None:
            self.write(WeiboCurlError.PAGE_NOT_FOUND)    # 页面找不到
            return
        # 成功返回结果
        success = settings.SUCCESS.copy()
        success['data'] = {
            'result': weibo_list,
```

```
            'cursor': str(cursor + 1) if cursor < 50 else '0'
        }
        self.write(success)
        #self.save_data_to_mongo(weibo_list, 'search_tweets')
        print(success)
        return
```

③ 创建微博推文信息展示接口类 StatusesShowHandler，根据推文的 id 搜索出对应的推文信息，代码如下：

```
class StatusesShowHandler(BaseHandler):
    """
    推文展示接口
    说明：根据推文id搜索推文
    路由：/weibo_curl/api/statuses_show
    """
    @gen.coroutine
    def get(self):
        # 获取参数
        args_dict = self.args2dict()
        weibo_id = args_dict.get('weibo_id')
        if weibo_id is None:
            self.write(WeiboCurlError.REQUEST_LACK_ARGS)
            return
        hot = args_dict.get('hot', False)   # 是否获取热评
        cursor = args_dict.get('cursor', '1')
        try:
            cursor = 1 if not cursor else int(cursor)
        except ValueError:
            self.write(WeiboCurlError.REQUEST_ARGS_ERROR)
            return
        if cursor > SEARCH_LIMIT_PAGES:
            results = settings.SUCCESS.copy()
            results['data'] = {
                'result': [],
                'cursor': '0'
            }
            self.write(results)
            return
        # 进行爬取
        comment_curl_result = yield weibo_web_curl(SpiderAim.weibo_comment,
weibo_id=weibo_id, page_num=cursor)
        if not comment_curl_result['error_code']:
            self.response = comment_curl_result['response']
        else:
            error_res = curl_result_to_api_result(comment_curl_result)
            self.write(error_res)
```

```python
        return
    # 构建解析器
    try:
        commonParser = CommentParser(weibo_id, response=self.response)
    except CookieInvalidException:
        self.write(WeiboCurlError.COOKIE_INVALID)
        return

    try:
        weibo_detail = yield commonParser.parse_one_weibo()
    except HTMLParseException as e:
        report_log(e)
        self.write(WeiboCurlError.HTML_PARSE_ERROR)
        return
    except Exception as e:
        report_log(e)
        self.write(WeiboCurlError.UNKNOWN_ERROR)
        return

    # 根据 hot 参数来确定获取 comment_list 的方式
    if not hot:
        comment_list = commonParser.get_all_comment()
    else:
        hot_comment_curl_result = yield weibo_web_curl(SpiderAim.hot_comment,
weibo_id=weibo_id, page_num=cursor)
        if not hot_comment_curl_result['error_code']:
            self.hot_comment_response = hot_comment_curl_result['response']
        else:
            error_res = curl_result_to_api_result(comment_curl_result)
            self.write(error_res)
            return

        try:
            comment_list = HotCommentParser(
                weibo_id, self.hot_comment_response).get_all_comment()
        except HTMLParseException:
            self.write(WeiboCurlError.HTML_PARSE_ERROR)
            return
        except Exception as e:
            report_log(
                (__class__.__name__, StatusesShowHandler.get.__name__), e)
            self.write(WeiboCurlError.UNKNOWN_ERROR)
            return
    # 成功时返回结果
    weibo_detail['weibo_id'] = weibo_id
    weibo_detail['comments'] = comment_list
    success = settings.SUCCESS.copy()
```

```
        success['data'] = {
            'result': weibo_detail,
            'cursor': str(cursor + 1) if cursor < weibo_detail['max_page'] else '0'
        }
        print(success)
        self.write(success)
        return
```

## 9.4 系统后端

本项目的系统后端使用 FastAPI 框架搭建实现，结合分布式异步任务处理框架 Celery 实现任务发布和任务执行的解耦，使得异步、快速处理话题分析任务成为可能。

扫码看视频

### 9.4.1 系统配置

编写文件 config_class.py，首先设置系统后端的主机地址和端口号，然后设置爬虫模块的 API 地址，代码如下：

```
from pydantic import BaseSettings
from typing import Union, List, Dict, Any

class AppConfig(BaseSettings):
    """
    app 启动的相关配置
    """
    HOST: str = '127.0.0.1'
    PORT: int = 81

class WeiBoConfig(BaseSettings):
    """weibo 爬虫 api 的相关配置"""
    BASEPATH: str = 'http://127.0.0.1:8000'
```

### 9.4.2 数据结构设计

本项目后端数据分析的数据是爬虫模块保存在数据库中的数据，在 models 目录中保存了实现数据库模型的程序文件。

1）话题任务

话题任务 tag_task 的设计结构如表 9-1 所示。

## 第 9 章　网络舆情数据分析系统

表 9-1　tag_task 的设计结构

| key | description | value type |
|---|---|---|
| tag_task_id | 话题任务 id：由时间戳+tag 的 MD5 构成 | str |
| tag_celery_task_id | celery_task id：任务初始化时生成 | str |
| tag_introduce_task_id | 话题基本信息 id：初始化任务时插入 tag_introduce 的返回值 | str |
| tag_hot_id | 话题热度信息 id：初始化任务时插入 tag_hot 的返回值 | str |
| tag_word_cloud_task_id | 话题词云 id：初始化任务时插入 tag_word_cloud 的返回值 | str |
| tag_character_task_id | 人物角色分类 id：初始化任务时插入 mongo 的返回值 | str |
| tag_relation_task_id | 话题传播关系 id：初始化任务时插入 mongo 的返回值 | str |
| tag_evolve_task_id | 话题演变 id：初始化任务时插入 mongo 的返回值 | str |
| tag_weibo_task_id | 博文任务 id：初始化任务时插入 mongo 的返回值 | str |
| tag_create_time | 创建时间：创建任务的时间 | datetime |
| status | 任务状态 | str |

编写文件 task_control_dto.py，实现话题任务 tag_task 的模型设计工作，代码如下：

```
class TaskCManage(BaseModel):
    tag_task_id: str = Field(None, title='话题任务 id',
                    description='由时间戳+tag 的 MD5 构成')
    tag: str = Field(None, title='话题内容',
                    description='搜索的话题关键字')
    tag_celery_task_id: str = Field(None, title='celery 任务 id',
                    description='任务执行时生成')
    tag_introduce_id: str = Field(None, title='话题基本信息 id',
                    description='由热度、用户数量等数据组成')
    tag_hot_id: str = Field(None, title='话题热度',
                    description='话题一天、一周、一个月内的热度数据')
    tag_word_cloud_task_id: str = Field(None, title='话题词云 id',
                    description='以分析后的词云为数据项')
    tag_character_task_id: str = Field(None, title='人物角色分类 id',
                    description='依据节点特征对人物进行分类')
    tag_relation_task_id: str = Field(None, title='话题传播关系 id',
                    description='以关系数据结构为数据项')
    tag_evolve_task_id: str = Field(None, title='tag_evolve_task_id',
                    description='以演变数据结构为数据项')
    tag_comment_task_id: str = Field(None, title='博文任务 id',
                    description='博文评论分析任务的 id')
    status: str = Field(None, title='任务状态')
    tag_create_time: datetime = Field(None, title='创建时间')
```

2) 人物分类

人物分类 character_category 的设计结构如表 9-2 所示。

表 9-2 character_category 的设计结构

| key | description | value type |
|---|---|---|
| tag_task_id | 话题任务 id：表示属于该话题任务 | str |
| Id | 人物角色分类 id：初始化任务时插入 mongo 的返回值 | str |
| user_id | 用户 id：微博用户 id | str |
| user_name | 用户昵称：微博用户昵称 | str |
| category | 人物类别：暂分为推手(0)、水军(1)、普通用户(2)、最具影响力用户(3) | int |

编写文件 character_category.py，实现人物分类 character_category 的模型设计工作，代码如下：

```python
from pydantic import Field, BaseModel
from enum import Enum

class Category(Enum):
    rumor = 0   # 推手
    faker = 1   # 水军
    user = 2    # 普通用户
    major = 3   # 最具影响力用户

class CharacterCategory(BaseModel):
    tag_task_id: str = Field(None, title='话题任务id', description='参照初始化任务时的任务id')
    tag_character_task_id: str = Field(None, title='人物角色分类id', description='依据节点特征对人物进行分类')
    user_id: str = Field(None, title='微博用户id')
    category: Category = Field(2, title='人物类别', description='暂分为推手(0)、水军(1)、普通用户(2)、最具影响力用户(3)')
```

3) 话题基本信息

话题基本信息 tag_introduce 的设计结构如表 9-3 所示。

表 9-3 tag_introduce 的设计结构

| key | description | value type |
|---|---|---|
| id | 话题基本信息 id：初始化任务时插入 mongo 的返回值 | str |
| tag | 话题名称 | str |
| tag_task_id | 话题任务 id：表示属于该话题任务 | str |

续表

| key | description | value type |
|---|---|---|
| user_count | 涉及用户数量：话题涉及用户数量 | int |
| weibo_count | 涉及博文数量：话题涉及博文数量 | int |
| vital_user | 重要用户 | dict |

在表 9-3 中，vital_user 字段的设计结构如表 9-4 所示。

表 9-4　vital_user 字段的设计结构

| key | description | value type | example |
|---|---|---|---|
| user_id | 用户真实 id | str | '1669879400' |
| head | 用户头像 | str | "..." |
| nickname | 昵称 | str | 'Dear-迪丽热巴' |
| gender | 性别 | str | '女' |
| location | 用户所在地 | str | '上海' |
| birthday | 生日 | str | '0001-00-00' |
| description | 用户简介 | str | '一只喜欢默默表演的小透明。工作联系...' |
| verified_reason | 认证信息 | str | '嘉行传媒签约演员' |
| education | 学习经历 | str | '上海戏剧学院' |
| work | 工作经历 | str | '嘉行传媒' |
| weibo_num | 微博数 | int | 1178 |
| following | 关注数 | int | 257 |
| followers | 粉丝数 | int | 72325060 |
| max_page | 个人微博的最大页数 | int | 200 |

编写文件 introduce_dto.py，实现话题基本信息 tag_introduce 的模型设计工作，代码如下：

```
class User(BaseModel):
    """
    重要用户信息
    """
    user_id: str = Field(None, title='用户的id')
    head: str = Field(None, title='头像url')
    nickname: str = Field(None, title='用户名')
    birthday: str = Field(None, title='生日')
    verified_reason: str = Field(None, title='认证信息')
    gender: str = Field(None, title='性别')
    location: str = Field(None, title='位置')
```

```
    description: str = Field(None, title='简介')
    education: str = Field(None, title='受教育信息')
    work: str = Field(None, title='工作信息')
    weibo_num: str = Field(None, title='微博数')
    following: str = Field(None, title='关注数')
    followers: str = Field(None, title='粉丝数')
    max_page: int = Field(None, title='个人微博的最大页数')

class ProgressTask(BaseModel):
    """
    执行中的任务
    """
    tag_task_id: str = Field(None, title='话题任务id')
    tag: str = Field(None, title='话题名称', description='用户输入的查询话题')
    status: str = Field(None, title='任务状态', description='后端实时任务状态')

class TagBase(BaseModel):
    """
    话题的基本信息
    """
    tag_task_id: str = Field(None, title='话题任务id')
    tag: str = Field(None, title='话题名称', description='用户输入的查询话题')
    user_count: int = Field(None, title='涉及用户数量', description='话题涉及用户数量')
    weibo_count: int = Field(None, title='涉及博文数量', description='话题涉及博文数量')
    vital_user: User = Field(None, title='最具影响力用户基本信息',
                            description='dict(用户名、头像url、昵称、认证、标签、粉丝数)')
```

4) 话题热度

话题热度 tag_hot 的设计结构如表 9-5 所示。

表 9-5 话题热度 tag_hot 的设计结构

| key | description | value type |
| --- | --- | --- |
| id | 话题基本信息 id：初始化任务时插入 mongo 的返回值 | str |
| tag | 话题名称 | str |
| tag_task_id | 话题任务 id：表示属于该话题任务 | str |
| one_day | 一天内热度 | dict |
| one_month | 一个月内热度 | dict |
| three_month | 三个月内热度 | dict |

编写文件 tag_hot.py，实现话题热度 tag_hot 的模型设计工作，代码如下：

```
class TagHot(BaseModel):
    tag_task_id: str = Field(None, title='任务id',
```

```
                    description='话题任务id')
    tag: str = Field(None, title='话题',
                    description='该话题内容')
    one_day: dict = Field(None, title='一天热度',
                    description='话题一天的热度数据')
    one_month: dict = Field(None, title='一月热度',
                    description='话题一个月内的热度数据')
    three_month: dict = Field(None, title='三月热度',
                    description='话题三个月内的热度数据')
```

5) 词云信息

词云信息 tag_word_cloud 的设计结构如表 9-6 所示。

表 9-6 词云信息 tag_word_cloud 的设计结构

| key | description | value type |
|---|---|---|
| tag_task_id | 话题任务 id：表示属于该话题任务 | str |
| id | 词云 id：初始化任务时插入 mongo 的返回值 | str |
| data | list 类型，每一个元素都由 key, value 组成，key 表示聚类后的关键字，value 表示出现过的次数，如 [ {'key':奥运会, 'count': 7}, {'key':中国, 'count':5},……] | list |

编写文件 word_cloud.py，实现词云信息 tag_word_cloud 的模型设计工作，代码如下：

```
class WordCloud(BaseModel):
    tag_task_id: str = Field(None, title='话题任务id')
    tag_word_cloud_task_id: str = Field(None, title='词云的id')
    data: List[dict] = Field(None, title='数据内容',
                    description='每一个元素都由key,value组成,key表示聚类后的关键
字,value表示出现过的次数')
```

## 9.4.3 数据处理

将爬取到的信息保存到数据库后，需要对数据进行进一步的处理，删除敏感数据和非法数据。

(1) 编写文件 task.py 清洗数据，功能是根据预先设置的停用词，将数据库中的非法信息删除，代码如下：

```
# 去除中文、英文和标点符号以外的字符
def cleantxt(raw):
    fil = re.compile(u'[^0-9a-zA-Z\u4e00-\u9fa5.，,。""]+', re.UNICODE)
    return fil.sub(' ', raw)
```

```python
def filter_emoji(desstr, restr=''):
    # 过滤表情
    try:
        co = re.compile(u'[\U00010000-\U0010ffff]')
    except re.error:
        co = re.compile(u'[\uD800-\uDBFF][\uDC00-\uDFFF]')
    return co.sub(restr, desstr)


def Traditional2Simplified(sentence):
    '''
    将 sentence 中的繁体字转换为简体字
    :param sentence: 待转换的句子
    :return: 将句子中繁体字转换为简体字之后的句子
    '''
    sentence = Converter('zh-hans').convert(sentence)
    return sentence


# 加载停用词
stopword_list = set()
with open(os.path.join(os.path.dirname(os.path.abspath(__file__)),
'stopwords.txt'), encoding="utf8") as f:
    for line in f:
        item = line.strip()
        stopword_list.add(item)
stopword_list.add("转发")

# 加载色情词
pron_list = set()
with open(os.path.join(os.path.dirname(os.path.abspath(__file__)),
'pronography.txt'), encoding="utf8") as f:
    for line in f:
        item = line.strip()
        pron_list.add(item)


def tokenize_text(text):
    tokens = jieba.lcut(text)
    tokens = [token.strip() for token in tokens if len(token.strip()) > 0]
    return tokens


# 去停用词
def remove_stopwords(text):
    tokens = tokenize_text(text)
```

```python
    filtered_tokens = [token for token in tokens if token not in stopword_list]
    return filtered_tokens


def normalize_corpus(twitter_data):
    # 清洗数据
    fulltext = []
    pron_pttn = r"|".join(pron_list)            # 色情词
    url_pattern = re.compile(r'https://[a-zA-Z0-9.?/&=:]*', re.S)  # 过滤网址
    name_pattern = re.compile(r'RT @[a-z,A-Z,0-9,_]+:|@[a-z,A-Z,0-9,_]+')
    # 过滤 @微博名
    for item in twitter_data['fulltext']:
        dd = Traditional2Simplified(item)       # 将繁体字转化为中文
        dd = url_pattern.sub("", dd)            # 去除推文后面的链接
        dd = filter_emoji(dd)                   # 去除表情
        dd = name_pattern.sub("", dd)
        dd = cleantxt(dd)                       # 去除外文
        dd = dd.replace('&amp', '')
        dd = dd.replace('RT', '')
        dd = dd.strip()                         # 去除空格
        fulltext.append(dd)

    # 去停用词
    normalized_corpus = []
    for text in fulltext:
        noStopWords = remove_stopwords(text)
        text = " ".join(noStopWords)
        normalized_corpus.append(text)
    return normalized_corpus, fulltext

    # 去停用词
    normalized_corpus = []

    for text in fulltext:
        noStopWords = remove_stopwords(text)

        text = " ".join(noStopWords)
        normalized_corpus.append(text)

    words = []
    for i in normalized_corpus:
        words.append(i.split(' '))
    return words
```

(2) 编写文件 tfidf.py，使用 TF-IDF 算法处理数据，获取文本信息的权重数据，代码如下：

```python
"""
    TF-IDF 权重：
        1、CountVectorizer 构建词频矩阵
        2、TfidfTransformer 构建 TF-IDF 权值计算
        3、文本的关键字
        4、对应的 TF-IDF 矩阵
"""
# 数据预处理操作：分词，去停用词，词性筛选
def dataPrepos(text, stopkey):
    l = []
    pos = ['n', 'nz', 'v', 'vd', 'vn', 'l', 'a', 'd']  # 定义选取的词性
    seg = jieba.posseg.cut(text)  # 分词
    for i in seg:
        if i.word not in stopkey and i.flag in pos:  # 去停用词 + 词性筛选
            l.append(i.word)
    return l

# TF-IDF 获取文本 top10 关键词
def getKeywords_tfidf(data,stopkey,topK):
#     idList, titleList, abstractList = data['id'], data['title'], data['abstract']
    idList,textList,categoryList = data['_id'],data['Cleaned_fulltext'],data['cluster_num']
    corpus = []  # 将所有文档输出到一个 list 中，一行就是一个文档
    for index in range(len(categoryList)):
#         text = '%s。%s' % (titleList[index], abstractList[index]) # 拼接标题和摘要
        fulltext=''.join(textList[index]).replace(' ','')
        text = dataPrepos(fulltext,stopkey) # 文本预处理
        text = " ".join(text) # 连接成字符串，空格分隔
        corpus.append(text)

    # 1.构建词频矩阵，将文本中的词语转换成词频矩阵
    vectorizer = CountVectorizer()
    X = vectorizer.fit_transform(corpus) # 词频矩阵，a[i][j]:表示 j 词在第 i 个文本中的词频
    # 2.统计每个词的 tf-idf 权值
    transformer = TfidfTransformer()
    tfidf = transformer.fit_transform(X)
    # 3.获取词袋模型中的关键词
    word = vectorizer.get_feature_names()
    # 4.获取 tf-idf 矩阵，a[i][j]表示 j 词在 i 篇文本中的 tf-idf 权重
    weight = tfidf.toarray()
    # 5.打印词语权重
    ids, keys,texts = [], [], []
    for i in range(len(weight)):
#         # print("-------这里输出第", i+1 , "篇文本的词语 tf-idf------")
        ids.append(idList[i])
        texts.append(textList[i])
#         titles.append(titleList[i])
```

```
        df_word,df_weight = [],[] # 当前文章的所有词汇列表、词汇对应权重列表
        for j in range(len(word)):
            # 从tiidf中排除"转发"这个词作为关键词
            if word[j] == "转发":
                continue
            df_word.append(word[j])
            df_weight.append(weight[i][j])
        df_word = pd.DataFrame(df_word,columns=['word'])
        df_weight = pd.DataFrame(df_weight,columns=['weight'])
        word_weight = pd.concat([df_word, df_weight], axis=1) # 拼接词汇列表和权重列表
        word_weight = word_weight.sort_values(by="weight",ascending = False)
        # 按照权重值降序排列
        keyword = np.array(word_weight['word']) # 选择词汇列并转成数组格式
        word_split = [keyword[x] for x in range(0,topK)] # 抽取前topK个词汇作为关键词
        word_split = " ".join(word_split)
        keys.append(word_split)

    result = pd.DataFrame({"id": ids, "key": keys,
"content":texts},columns=['id','key',"content"])
    return result
```

(3) 编写文件 rule_cluster.ipynb，具体实现流程如下。

① 读取数据库中的数据，代码如下：

```
client = pymongo.MongoClient('10.245.142.249', 27017)
# 连接所需数据库、集合，twitter_search 为数据库名，mingyan 为表名
mongodata = client['twitter_search']['user_timeline'].find()
twitter_data=[]
for item in mongodata:
    t1=item['user_post']['full_text']
    data={}
    data['_id']=str(item["_id"])
    data['fulltext']=t1
    twitter_data.append(data)
#将twitter 数据转换为Dataframe 数据格式
twitter_data=pd.DataFrame(twitter_data)
```

② 实现分词去停用词等文本处理功能，代码如下：

```
documents,cleaned_fulltext=normalize_corpus(twitter_data)
print(documents[:10])
print(cleaned_fulltext[:10])
```

③ 迭代采样聚类处理，代码如下：

```
def compute_V(texts):
    V = set()
    for text in texts:
```

```
        for word in text:
            V.add(word)
    return len(V)

texts = [text.split() for text in documents]
V = compute_V(texts)
mgp=MovieGroupProcess(K=6,alpha=0.01,beta=0.02,n_iters=40)
y=mgp.fit(texts,V)
```

④ 获取簇数据，分别获取簇中心、每个簇的关键特征和每个簇的 id，代码如下：

```
def get_cluster_data(twitter_data,num_clusters):
    cluster_details={}
    #获取簇的中心
    
    #获取每个簇的关键特征
    #获取每个簇的id
    for cluster_num in range(num_clusters):
        cluster_details[cluster_num]={}
        cluster_details[cluster_num]['cluster_num']=cluster_num#簇序号
        
        id_s=twitter_data[twitter_data['cluster']==cluster_num]['_id'].values.tolist()
        cluster_details[cluster_num]['_id']=id_s
        
        fulltext=twitter_data[twitter_data['cluster']==cluster_num]['cleaned_fulltext'].values.tolist()
        cluster_details[cluster_num]['Cleaned_fulltext']=fulltext
    
    return cluster_details
```

⑤ 打印输出簇数据的详细信息，代码如下：

```
def print_cluster_data(cluster_data):
    
    for cluster_num,cluster_details in cluster_data.items():
        print('Cluster{} details'.format(cluster_num))
        print('-'*20)
#        print('Key features:',cluster_details['key_features'])
        print('sentence in this clusters:')
        print('》'.join(cluster_details['Cleaned_fulltext']).replace('\n',''))
        print('='*40)
```

⑥ 打印输出聚类结果文本，代码如下：

```
num_clusters=len(set(y))

cluster_data=get_cluster_data(twitter_data,num_clusters)
print_cluster_data(cluster_data)
```

⑦ 提取每一类的关键词，代码如下：

```
#获取聚类后的数据
cluster_data=get_cluster_data(twitter_data,num_clusters)

#把聚好类的数据转换成dataframe数据格式
cluster_dfdata=pd.DataFrame(cluster_data)
cluster_dfdata_T=cluster_dfdata.T #获得矩阵的转置

#使用TF-IDF法提取关键词

stopkey = [w.strip() for w in codecs.open('./stopwords.txt', 'r').readlines()]
# TF-IDF关键词抽取
result = getKeywords_tfidf(cluster_dfdata_T,stopkey,10)
result
```

## 9.4.4 微博话题分析

本项目将实现微博话题数据分析功能，用户可以将相关话题加入分析任务队列中，后端会异步执行分析任务(基于Celery)。

(1) 编写文件task.py，创建异步任务处理，实现异步处理的初始化操作功能，代码如下：

```
async def init_task(tag: str, mongo_db: AsyncIOMotorDatabase) -> dict:
    """
    初始化并执行任务
    :param mongo_db: mongo数据库
    :param tag: 话题
    :return: 初始化后生成的任务各部分id
    """
    time_str = str(time.time())    # 当前时间戳
    tag_task_id = md5((time_str + tag).encode('utf-8')).hexdigest()
    task = task_schedule.delay(tag_task_id, tag=tag)
    tag_introduce_id = await mongo_db['tag_introduce'].insert_one({"tag_task_id": tag_task_id})
    tag_hot_id = await mongo_db['tag_hot'].insert_one({"tag_task_id": tag_task_id})
    tag_word_cloud_task = await mongo_db['tag_word_cloud'].insert_one({"tag_task_id": tag_task_id})
    tag_character_task = await mongo_db['character_category'].insert_one({"tag_task_id": tag_task_id})
    tag_relation_task = await mongo_db['tag_relation_graph'].insert_one({"tag_task_id": tag_task_id})
    tag_evolve_task = await mongo_db['tag_evolve'].insert_one({"tag_task_id": tag_task_id})
    tag_weibo_task = await mongo_db['tag_weibo_task'].insert_one({"tag_task_id": tag_task_id})
```

```python
    tag_user_task = await mongo_db['tag_user'].insert_one({'tag_task_id':
tag_task_id})
    tag_create_time = datetime.now()

    task_init = {'tag_task_id': tag_task_id,
                 'tag': tag,
                 'tag_celery_task_id': task.id,
                 'tag_word_cloud_task_id': str(tag_word_cloud_task.inserted_id),
                 'tag_hot_task_id': str(tag_hot_id.inserted_id),
                 'tag_introduce_task_id': str(tag_introduce_id.inserted_id),
                 'tag_character_task_id': str(tag_character_task.inserted_id),
                 'tag_relation_task_id': str(tag_relation_task.inserted_id),
                 'tag_evolve_task_id': str(tag_evolve_task.inserted_id),
                 'tag_weibo_task_id': str(tag_weibo_task.inserted_id),
                 'tag_user_id': str(tag_user_task.inserted_id),
                 'status': 'PENDING',
                 'tag_create_time': str(tag_create_time)}
    await mongo_db['tag_task'].insert_one(task_init)

    print(task_init)
    return task_init
```

(2) 编写文件 tag_introduce_task.py，构建基本话题任务，代码如下：

```python
def introduce(tag_data: dict, tag_task_id: str):
    """
    :param tag_task_id: 话题任务 id
    :param tag_data: 话题下的微博数据
    :return:
    """
    data_list = tag_data['data']
    weibo_count = len(data_list)        # 微博数
    weibo_userid = set()                # 用户集合
    vital_user_id = str()               # 用户 id
    hot = 0
    for weibo in data_list:
        weibo_userid.add(weibo['weibo_id'])
        if int(weibo['hot_count']) > hot:
            vital_user_id = weibo['user_id']
            hot = int(weibo['hot_count'])
    vital_user = get_user_data(vital_user_id)
    tag_introduce_dict = {'tag_task_id': tag_task_id,
                          'tag': tag_data['tag'],
                          'user_count': len(weibo_userid),
                          'weibo_count': weibo_count,
                          'vital_user': vital_user}
    query_by_task_id = {'tag_task_id': tag_task_id}
    update_data = {"$set": tag_introduce_dict}
```

```python
    mongo_client.db[mongo_conf.INTRODUCE].update_one(query_by_task_id, update_data)
    # with Mongo('tag_introduce', 'test') as mongo_db:
    #     mongo_db.collect.update_one(query_by_task_id, update_data)

def get_user_data(user_id) -> json:
    """
    获取 user 详细信息
    :param user_id:微博 user_id
    :return:
    """
    url = f'http://127.0.0.1:8000/weibo_curl/api/users_show?user_id={user_id}'
    response = requests.get(url)
    response_dict = json.loads(response.text)
    if response_dict.get('data'):
        user_dict = response_dict.get('data').get('result')
        return user_dict
    else:
        return user(user_id)
```

(3) 编写文件 tag_relaton_task.py，创建话题关系任务，代码如下：

```python
def tag_relation(weibo_data: dict, tag_task_id: str, user_mark_data: dict):
    """
    处理话题人物关系网的函数
    :param user_mark_data：用户分类数据
    :param weibo_data:微博数据
    :param tag_task_id:话题任务
    :return:
    """
    node_list = list()
    link_list = list()
    weibo_list = reduce(lambda x, y: x if y in x else x + [y], [[], ] + weibo_data['data'])
    screen_name_set = set(i['screen_name'] for i in weibo_list)
    relation_data = list()
    for screen_name in screen_name_set:
        at_users_list = list()
        user_id = 0
        hot_count = 0
        for weibo in weibo_list:
            if weibo.get('screen_name') == screen_name:
                at_users_list.extend(weibo.get('at_users'))
                hot_count += int(weibo.get('hot_count'))
                user_id = weibo['user_id']
        relation_data.append({'screen_name': screen_name,
                              'user_id': user_id,
                              'at_users': at_users_list,
                              'hot_count': hot_count
                              }
```

```
        )
    for data in relation_data:
        category = -1
        for user_mark in user_mark_data.get('data'):
            if user_mark.get('user_id') == data['user_id']:
                category = user_mark.get('category')
                break
        node = {'category': category, 'name': data['screen_name'], 'userId':
data['user_id'], 'value': int(data['hot_count'])}
        node_list.append(node)
        for i in data['at_users']:
            link = {'source': data['screen_name'], 'target': i, 'weight':
data['at_users'].count(i)}
            if link not in link_list:
                link_list.append(link)
            if i not in screen_name_set:
                node = {'category': -1, 'name': i, 'userId': None,
                    'value': int(data['hot_count'])}
                node_list.append(node)
                screen_name_set.add(i)
            else:
                for node_item in node_list:
                    if node_item['name'] == i:
                        node_item['value'] += int(data['hot_count'])
                        break
    query_by_task_id = {'tag_task_id': tag_task_id}
    update = {"$set": {'nodes_list': node_list, 'links_list': link_list,
'categories': user_mark_data.get('categories')}}
    mongo_client.db[mongo_conf.RELATION].update_one(query_by_task_id, update)
    mongo_client.db[mongo_conf.CHARACTER].update_one(query_by_task_id, {'$set':
{'detail': node_list}})
```

(4) 编写文件 tag_hot_task.py 创建话题热度任务，展示该话题近一天、一个月、三个月的关注热度，代码如下：

```
def hot_task(tag: str, tag_task_id: str):
    """
    获取话题发展趋势信息
    :param tag:
    :param tag_task_id:
    :return:
    """
    try:
        hot_one_day = tendency(tag, '1day')
        dict_one_day = hot_one_day.to_dict()
        hot_one_month = tendency(tag, '1month')
        dict_one_month = hot_one_month.to_dict()
```

```python
            hot_three_month = tendency(tag, '3month')
            dict_three_month = hot_three_month.to_dict()
            final_dict = {'tag': tag, 'one_day': {str(time): value for time, value in dict_one_day[tag].items()},
                          'one_month': {str(time): value for time, value in dict_one_month[tag].items()},
                          'three_month': {str(time): value for time, value in dict_three_month[tag].items()}}
            query_by_id = {'tag_task_id': tag_task_id}
            update_data = {"$set": final_dict}
            mongo_client.db[mongo_conf.HOT].update_one(query_by_id, update_data)
        except AttributeError as exc:
            query_by_id = {'tag_task_id': tag_task_id}
            mongo_client.db[mongo_conf.HOT].update_one(query_by_id, {"$set": {"tag": tag}})
            # TODO log
            print(exc)
            # logging.log(exc)
```

(5) 编写文件 tag_word_cloud_task.py，构建话题词云任务，根据话题中爬取到的所有博文内容分析出与话题关联的词语，代码如下：

```python
def word_cloud(weibo: dict, tag_task_id: str):
    """
    词云构建函数
    :param weibo: 微博信息
    :param tag_task_id:话题任务id
    :return:
    """
    # 读取博文数据
    weibo_list = list()
    for weibo_item in weibo['data']:
        weibo_list.append(weibo_item['text'])

    # 词云构建
    my_cloud = MyCloud(weibo_list)
    word_cloud_list = my_cloud.GetWordCloud()
    if len(word_cloud_list) > 200:
        word_cloud_list = word_cloud_list[0: 200]

    # 存储词云
    query_by_task_id = {'tag_task_id': tag_task_id}
    update_data = {"$set": {'data': word_cloud_list}}
    mongo_client.db[mongo_conf.CLOUD].update_one(query_by_task_id, update_data)
```

在上述文件 tag_word_cloud_task.py 中调用了文件 my_cloud.py 中的功能函数，代码如下：

```python
def Traditional2Simplified(sentence):
    '''
    将 sentence 中的繁体字转换为简体字
    :param sentence: 待转换的句子
    :return: 将句子中繁体字转换为简体字之后的句子
    '''
    sentence = Converter('zh-hans').convert(sentence)
    return sentence

def Sent2Word(sentence):
    """将一个句子变成标记化单词列表,并删除使用jieba标记化的中文"""
    global stop_words
    words = jieba.cut(sentence)
    words = [w for w in words if w not in stop_words]
    return words

def Match(content):
    content_comment = []
    advertisement = ["王者荣耀", "券后", "售价", '¥', "￥", '下单', '转发微博', '转发', '微博']
    words = []
    for k in range(0, len(content)):
        judge = []
        print('Processing train ', k)
        content[k] = Traditional2Simplified(content[k])
        for adv in advertisement:
            if adv in content[k]:
                judge.append("True")
                break
        if re.search(r"买.*赠.*", content[k]):
            judge.append("True")
            continue
        if content[k] == "":
            judge.append("True")
            continue
        # 通过上面的两种模式判断是不是广告
        if "True" not in judge:
            # 数据清洗
            a2 = re.compile(r'#.*?#')
            content[k] = a2.sub('', content[k])
            a3 = re.compile(r'\[组图共.*张\]')
            content[k] = a3.sub('', content[k])
            a4 = re.compile(r'http:.*')
            content[k] = a4.sub('', content[k])
            a5 = re.compile(r'@.*? ')
            content[k] = a5.sub('', content[k])
            a6 = re.compile(r'\[.*?\]')
```

```python
            content[k] = a6.sub('', content[k])
            words.append(Sent2Word(content[k]))
    return words

def getRepostSent(tag_comment_task_id):
    sent = []
    my_query = {"tag_comment_task_id": tag_comment_task_id}
    my_doc = mongo_client.db[mongo_conf.COMMENT_REPOSTS].find(my_query)
    for item in my_doc:
        sent.append(item['content'])
    return sent

def countWords(sent_words):
    words_dict = dict()
    for words in sent_words:
        for word in words:
            if len(word) == 0 or word == '':
                continue
            elif word not in words_dict.keys():
                words_dict[word] = 1
            else:
                words_dict[word] += 1
    return words_dict

# 改变形式 + 排序
def reshapeDict(words_dict):
    words_list = []
    for key in words_dict:
        item = {}
        item['name'] = key
        item['value'] = words_dict[key]
        words_list.append(item)
    words_list.sort(key=lambda i: i['value'], reverse=True)
    return words_list

def preContent(tag_comment_task_id=None, doc_id=None):
    print("回复读取")
    content = getRepostSent(tag_comment_task_id)

    print("分词")
    # sent_words_a = Match(content)
    sent_words_b = normalize_corpus_part(content)

    print("统计")
    # words_dict_a = countWords(sent_words_a)
    words_dict_b = countWords(sent_words_b)

    print("排序")
```

```
    # words_dict_a = reshapeDict(words_dict_a)
    words_dict_b = reshapeDict(words_dict_b)

    if len(words_dict_b) >= 200:
        # words_dict_a = words_dict_a[0:200]
        words_dict_b = words_dict_b[0:200]

    print(words_dict_b)
    mongo_client.db[mongo_conf.COMMENT_CLOUD].update_one({"tag_comment_task_id":
tag_comment_task_id}, {"$set": {"data": words_dict_b}})
```

## 9.5 系统前端

本项目的前端使用 Vue 框架实现，调用后端生成的 API 实现数据展示和可视化功能。

### 9.5.1 API 导航

扫码看视频

在文件 index.js 中设置了后端 API 导航和前端 URL 的映射，代码如下：

```
import VueRouter from 'vue-router'

import wbAnalyze from '../pages/wbAnalyze';
import home from '../pages/home';
import login from '../pages/login';
import blog_detail from '../pages/blog_detail';
import person_list from '../pages/person_list';

export default new VueRouter({
    routes: [{
        path: "/",
        component: home,
        meta: {
            title: "舆情系统"
        }
    },
    {
        path: "/home",
        component: home,
        meta: {
            title: "舆情系统"
        }
    },
    {
```

```
            path: "/wb",
            component: wbAnalyze,
            meta: {
                title: "微博舆情分析"
            }
        },
        {
          path: "/login",
          name: 'login',
          component: login,
          meta: {
              title: "登录注册"
          }
        },
        {
          path: "/blog_detail",
          component: blog_detail,
          meta: {
              title: "博文详情"
          }
        },
        {
           path: "/person_list",
           component: person_list,
           meta: {
               title: "用户列表"
           }
        }
    ],
    mode: "history"
})
```

## 9.5.2 博文详情

文件 blog_info.vue 实现博文详情页面，展示博文的文本内容和发布微博用户的粗略信息，文件 blog_info.vue 的具体实现代码如下：

```
<template>
  <div class="hot_point">
    <div class="hot_point_title">热点转发</div>
    <div class="hot_point_contents">
      <div
        class="hot_point_content"
        v-for="(comment, index) in comments"
        :key="index">
        <div class="hot_point_number red" v-if="index == 0">
```

```
          {{ index + 1 }}
        </div>
        <div class="hot_point_number orange" v-if="index == 1">
          {{ index + 1 }}
        </div>
        <div class="hot_point_number blue" v-if="index == 2">
          {{ index + 1 }}
        </div>
        <div class="hot_point_number" v-if="index >= 3">{{ index + 1 }}</div>
        <span>{{ comment.content | snippet }}</span>
      </div>
    </div>
  </div>
</template>

<script>
export default {
  name: "hot_point",
  data() {
    return {
      comments: [],
    };
  },
  filters: {
    snippet(value) {
      if (value.length > 25) value = value.slice(0, 25) + "...";
      return value;
    },
  },
  methods: {
    getHotPoint() {
      let query = this.$route.query;
      this.$axios
        .get("comment/key_node?tag_task_id="+query.tag_task_id+"&weibo_id="+query.weibo_id)
        .then((res) => {
          console.log(res)
          this.comments = res.data.comments;
        });
    },
  },
  mounted() {
    this.getHotPoint();
  },
};
</script>
```

## 第 9 章 网络舆情数据分析系统

为了节省本书篇幅，系统前端页面的具体实现过程不再讲解，其余内容请读者参阅本书配套资料中的源代码和视频。执行后的主页界面效果如图 9-2 所示。

图 9-2　系统主页

本项目可以根据用户在不同话题下发表的微博文本的聚类主题为用户打标签，逐步完善用户画像。用户画像效果如图 9-3 所示。

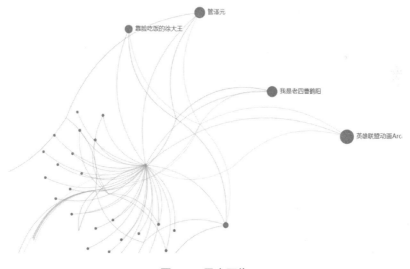

图 9-3　用户画像